SR-71A FLIGHT MANUAL (U)

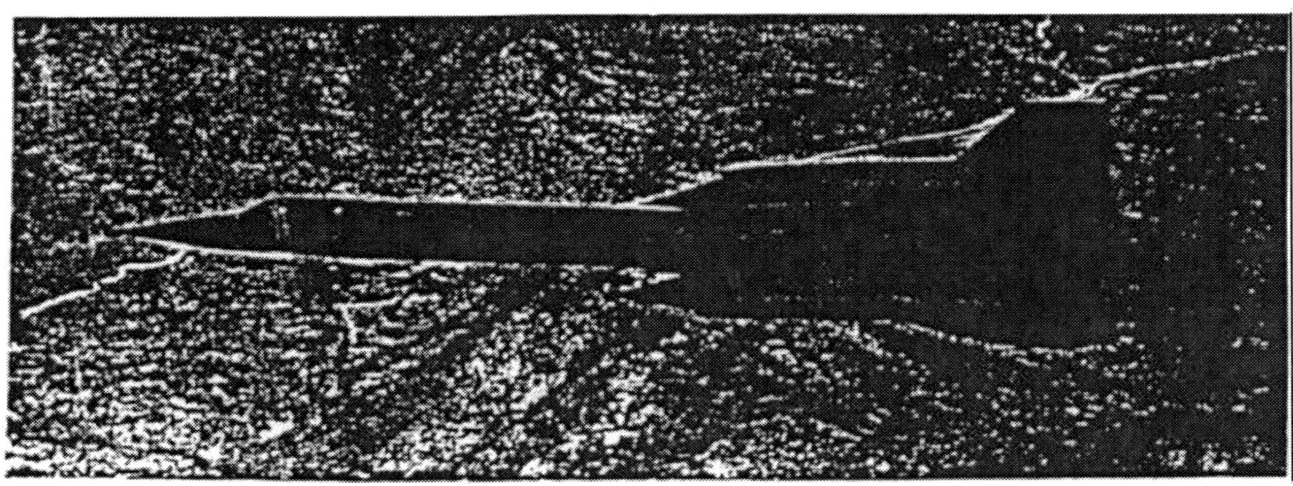

CLASSIFICATION STATEMENT: This document contains security classifcation markings in accordance with DOD 5200.1R/AFR 205-1 or DOD 5220.22M. Classification is based on the compilation of information in Sections 1 through 7, and appendices. NOTICE - information extracted from this document must be classified in accordance with the Senior Crown Program Security Classification Guide.

FURTHER DISSEMINATION OF THESE DOCUMENTS SHALL ONLY BE MADE BY DET 6, 2762 LS (Sp), NORTON AFB CA, 92409-6437

CHANGE NOTICE

LATEST CHANGED PAGES SUPERSEDE THE SAME PAGES OF PREVIOUS DATE

Insert changed pages into basic publication. Destroy superseded pages.

PUBLISHED UNDER AUTHORITY OF T

Retention Granted Through Sept 1992 per Final DD254 Dtd. 15 Aug 1990

THIS PAGE IS UNCLASSIFIED

ISSUE E: 31 OCTOBER 1986
Change 2: 31 July 1989

UNCLASSIFIED

THIS MATERIAL HAS BEEN DECLASSIFIED

LIST OF EFFECTIVE PAGES

TOTAL NUMBER OF PAGES IN THIS PUBLICATION IS 1052 CONSISTING OF THE FOLLOWING

Page No.	Change No.
*Title	2
*A	2
*B	2
*C	2
i	1
ii	1
iii	0
iv	0

SECTION I

Page No.	Change No.
1-1	0
1-2	1
1-3, 1-4	0
1-5	1
1-6 thru 1-22	0
*1-23	2
1-24	1
1-25	0
1-26 thru 1-28	1
1-28A	1
1-28B	1
1-29, 1-30	0
*1-31	2
1-32 thru 1-40	0
1-41	1
*1-42	2
1-42A	1
1-42B Blank	1
1-43 thru 1-49	0
*1-50	2
1-51	0
*1-52 thru 1-55	2
1-56	0
1-57	0
*1-58, 1-59	2
1-60	0
1-61, 1-62	1
1-62A	1
1-62B Blank	1
1-63	0
1-64	1
1-65	0
1-66	1
1-67, 1-68	0
1-69, 1-70	1
1-71 thru 1-73	0
*1-74	2
1-75, 1-76	1
1-77 thru 1-83	0
1-84, 1-85	1
*1-86	2
1-87, 1-88	1
*1-89	2
1-90, 1-91	0
1-92 thru 1-94	1
1-94A	1
1-94B Blank	1
1-95	1
1-96	0
1-97	1
1-98 thru 1-102	0
1-103, 1-104	1
1-104A	1
1-104B Blank	1
1-105 thru 1-110	0
*1-111	2
1-112 thru 1-114	0
1-115 thru 1-118	1
*1-118A	2
*1-118B Blank	2
*1-119, 1-120	2
1-121	1
*1-122	2
1-123 thru 1-127	0
1-128	1
1-129	0
1-130	1
1-130A Dltd	1
1-130B Blank Dltd	1
1-131, 1-132	0
1-133	1
1-134 thru 1-138	0
1-139 thru 1-141	1
1-142 thru 1-144	0
1-145	1
1-146	0
1-147	1
1-148 thru 1-165	0
1-166	1
*1-167	2
1-168	0
*1-169 thru 1-171	2
1-172 thru 1-181	0
1-182 thru 1-188	1
1-189, 1-190	0
1-191	1
1-192 thru 1-196	0
1-197, 1-198	1
1-199, 1-200	0
1-201	1
1-202 thru 1-208	0
1-209, 1-210	1
1-211	0
1-212 Blank	0

SECTION IA

Page No.	Change No.
1A-1 thru 1A-12	0
1A-13	1
1A-14	0
1A-15, 1A-16	1
1A-17 thru 1A-24	0

SECTION II

Page No.	Change No.
2-1 thru 2-6	0
2-7	1
2-8 thru 2-10	0
2-11, 2-12	1
2-13	0
2-14	1
2-15 thru 2-21	0
*2-22	2
2-23, 2-24	1
2-24A Dltd	1
2-24B Blank Dltd	1
2-25	1
2-26 thru 2-32	[illegible]
2-33, 2-34	1
2-35, 2-36	0
2-37, 2-38	1
2-39, 2-40	0
2-41, 2-42	1
2-43	0
2-44	1
2-45 thru 2-48	0
*2-49	2
2-50	0
2-51	1
*2-52	2
2-53, 2-54	1
2-55 thru 2-60	0
*2-61	2
2-62 thru 2-64	0
*2-65	2
2-66, 2-67	0
2-68	1
2-69 thru 2-74	0
2-75, 2-76	1
*2-77	2
2-78	0
2-79	1
2-80, 2-81	0
2-82, 2-83	1
2-84 Blank	1

SECTION III

Page No.	Change No.
3-1	0
3-2	1
3-3	0
3-4, 3-5	1
3-6 thru 3-8	0
3-9, 3-10	1
3-11	0
3-12	1
3-13, 3-14	0
3-15	1
3-16	0

CURRENT FLIGHT CREW ABBREVIATED CHECKLIST

Pilot
SR-71A-1CL-1 - 16-01-89 Change 1 31-07-89

RSO
SR-71A-1CL-2 - 16-01-89 Change 1 21-07-89

Dates of issue for original and changed pages are:

Original .... 31 October 1986
Change 1 .... 16 January 1989
Change 2 .... 31 July 1989

The asterisk indicates pages changed, added or deleted by the current change. Insert changed and/or added pages; destroy superseded pages.

NOTE: The portion of text affected by the change is indicated by a vertical line in the outer margins of the page. ⟶ indicates deletion of text.

LIST OF EFFECTIVE PAGES Cont.

Page No.	Change No.
3-17 thru 3-20	1
3-21 thru 3-29	0
3-30, 3-31	1
3-32, 3-33	0
3-34	1
3-34A	1
3-34B Blank	1
3-35 thru 3-40	0
*3-41	2
3-42	0
*3-43	2
3-44	0
*3-44A/(3-44B Blank) Added	2
3-45 thru 3-47	0
3-48	1
3-48A	1
3-48B Blank	1
3-49	0
3-50, 3-51	1
3-52 thru 3-54	0
3-55	1
3-56 thru 3-58	0
*3-59	2
3-60, 3-61	0
*3-62	2
3-63	1
3-64	0
3-65, 3-66	1
3-66A	1
3-66B Blank	1
3-67, 3-68	1
*3-69	2
3-70 thru 3-74	0
3-75, 3-76	1
3-77, 3-78	0
3-79 thru 3-81	1
3-82	0
*3-83, 3-84	2
*3-84A/(3-84B Blank) Added	2
3-85 thru 3-87	0
3-88 thru 3-94	1
3-95, 3-96	0
*3-97	2
3-98 thru 3-100	1
3-100A	1
3-100B Blank	1
3-101 thru 3-106	1
3-106A	1

Page No.	Change No.
3-106B Blank	1
3-107, 3-108	0
3-109	1
3-110	0
3-111	1
3-112 thru 3-114	0
3-115	1
3-116	0
3-117, 3-118	1
3-118A	1
3-118B Blank	1
3-119 thru 3-121	1
3-122	0
3-123 thru 3-125	1
3-126	0
3-127 thru 3-131	1
3-132 thru 3-134	0
3-135	1
3-136	0
*3-137	2
3-138	1
3-139	0
3-140 Blank	0
SECTION IV	
4-1, 4-2	1
4-2A	1
4-2B Blank	1
4-3, 4-4	0
*4-5	2
4-6 thru 4-8	0
4-9 thru 4-11	1
4-12 thru 4-24	0
4-25	1
4-26 thru 4-28	0
4-29	1
4-30, 4-31	0
*4-32	2
4-32A Dltd	1
4-32B Blank Dltd	1
4-33 thru 4-35	0
4-36	1
4-37 thru 4-45	0
4-46	1
4-47 thru 4-56	0
4-57	1
4-58	0
4-59, 4-60	1
4-60A Dltd	1

Page No.	Change No.
4-60B Blank Dltd	1
4-61 thru 4-64	0
4-65	1
4-66 thru 4-71	0
4-72	1
4-73 thru 4-76	0
4-77	1
4-78	0
4-79 thru 4-82	1
4-83 thru 4-85	0
4-86 thru 4-110	1
4-110A thru 4-110G/(4-110H Blank) Dltd	1
4-111 thru 4-159	1
*4-160	2
4-161, 4-162	1
*4-163	2
4-164 thru 4-190	1
*4-191, 4-192	2
4-193 thru 4-227/(4-228 Blank)	1
SECTION V	
5-1	0
5-2	1
5-3, 5-4	0
5-5	1
5-6, 5-7	0
*5-8	2
5-9	0
*5-10	2
5-11 thru 5-15	0
5-16	1
5-17 thru 5-23	0
5-24 Blank	0
SECTION VI	
6-1 thru 6-12	0
6-13	1
6-14 thru 6-25	0
6-26 Blank	0
SECTION VII	
7-1 thru 7-4	0
7-5	1

Page No.	Change No.
7-6 thru 7-9	0
7-10 Blank	0
APPENDIX I	
A-1 thru A-3	0
A-4 Blank	0
Appendix 1 Part 1	
A1-1 thru A1-19	0
A1-20 Blank	0
Appendix 1 Part 2	
A2-1 thru A2-38	0
Appendix 1 Part 3	
A3-1 thru A3-23	0
A3-24 Blank	0
A3-25	0
A3-26 Blank	0
A3-27	0
A3-28 Blank	0
A3-29	0
A3-30 Blank	0
A3-31	0
A3-32 Blank	0
A3-33	0
A3-34 Blank	0
A3-35	0
A3-36 Blank	0
Appendix 1 Part 4	
A4-1 thru A4-27	0
A4-28 Blank	0
A4-29 thru A4-31	0
A4-32 Blank	0
Appendix 1 Part 5	
A5-1 thru A5-60	0
Appendix 1 Part 6	
A6-1 thru A6-5	0
A6-6 Blank	0
A6-7	0

* The asterisk indicates pages changed, added or deleted by the current change. Insert changed and/or added pages; destroy superseded pages.

NOTE: The portion of text affected by the change [illegible] indicated by a vertical [illegible] page. + indicates deletion of text.

LIST OF EFFECTIVE PAGES Cont.

Page No.	Change No.
A6-8	0
A6-9	0
A6-10 Blank	0
A6-11	0
A6-12 Blank	0
A6-13	0
A6-14 Blank	0
A6-15	0
A6-16 Blank	0
A6-17	0
A6-18 Blank	0
A6-19	0
A6-20 Blank	0
A6-21	0
A6-22 Blank	0
A6-23	0
A6-24 Blank	0
A6-25	0
A6-26 Blank	0
A6-27	0
A6-28 Blank	0
A6-29	0
A6-30 Blank	0
A6-31	0
A6-32 Blank	0
A6-33	0
A6-34 Blank	0
A6-35	0
A6-36 Blank	0
A6-37	0
A6-38 Blank	0
A6-39	0
A6-40 Blank	0
A6-41	0
A6-42 Blank	0
A6-43	0
A6-44 Blank	0
A6-45	0
A6-46 Blank	0
A6-47	0
A6-48 Blank	0
A6-49	0
A6-50 Blank	0
A6-51	0
A6-52 Blank	0
A6-53	0
A6-54 Blank	0
A6-55	0
A6-56 Blank	0
A6-57	0
A6-58 Blank	0
A6-59 thru A6-64	0
INDEX	
Index 1 thru Index 12	1

* The asterisk indicates pages changed, added or deleted by the current change. Insert changed and/or added pages; destroy superseded pages.

NOTE: The portion of text affected by the change is indicated by a vertical line in the outer margins of the page. + indicates deletion of text.

SR-71A FLIGHT MANUAL

SECURITY CLASSIFICATION

The Department of Defense security classification of this manual is:

SECRET

Special Access Required

Classified by:	SENIOR CROWN Program Security Classification Guide, 25 May 1987.
Declassify on:	OADR.

"CLASSIFICATION STATEMENT - This document contains security classification markings in accordance with DOD 5200.1R/AFR 205-1 or DOD 5220.22M. Classification is based on the compilation of information in Sections 1 through 7, and appendices. NOTICE - Information extracted from this document must be classified in accordance with the Senior Crown Security Classification Guide."

"WARNING - This document contains technical data whose export is restricted by the Arms Export Control Act (Title 22, U.S.C. Sec. 2751, et seq) or Executive Order 12470. Violators of these export laws are subject to severe criminal penalties."

"DISTRIBUTION NOTICE - Distribution authorized to U.S. Government agencies and their contractors for administrative or operational use (22 Nov 88). Other requests for this document shall be referred to Det 6, 2762 LS (Special)/ME, Norton AFB CA., 92409-6437."

"DESTRUCTION NOTICE - For classified documents, follow the procedures in DOD 5220.22M, Industrial Security Manual, Section II-19 or DOD 5200.1R/AFR 205-1, Information Security Program Regulation, Chapter IX. For Unclassified, limited documents, destroy by any methods that will prevent disclosure of contents or reconstruction of the document."

The front and back exterior covers for this manual will bear markings at the top and bottom as follows:

SECRET ** SPECIAL ACCESS REQUIRED ** SENIOR CROWN PROGRAM

Strict accountability will be maintained of all contents of this manual.

FOREWORD

SCOPE. This manual contains the information necessary for safe and efficient operation of the SR-71A and the SR-71B aircraft. These instructions provide you with a general knowledge of the aircraft, its characteristics and specific normal and emergency operation procedures. Your prerequisite flying experience is recognized, and it is not considered necessary to discuss basic flight principles.

SOUND JUDGMENT. Instructions in this manual provide the best operating instructions under most circumstances, but they are not a substitute for sound judgment. Multiple emergencies, adverse weather, terrain, etc. may require modification of the procedures.

ARRANGEMENT. The manual is divided into eight nearly independent sections to simplify using it as a reference manual. Sections II, III and V must be thoroughly understood before attempting to fly the aircraft. Section I is subdivided into two parts. The first section provides information which is applicable to the basic SR-71A aircraft. Section IA provides special descriptive information for the SR-71B trainer aircraft. The remaining sections provide important information for safe and efficient mission accomplishment.

CHANGES. This manual will be revised by numbered changes to reflect information gained from tests and operational experience and to augment the data provided. Operational or safety of flight supplements indicated as -1S or -1SS respectively, will be issued when necessary and incorporated in the succeeding numbered change.

CHECKLISTS. The Flight Manual contains only amplified checklists. Abbreviated checklists are issued separately. Line items in the Flight Manual and checklists are identical with respect to arrangement and item number.

WARNINGS, CAUTIONS, AND NOTES. The following definitions apply to "Warnings", "Cautions", and "Notes" found throughout the manual.

WARNING — **Operating procedures, practices, etc., which may result in personal injury or loss of life if not correctly followed.**

CAUTION — **Operating procedures, practices, etc., which if not strictly observed may result in damage to or destruction of equipment.**

NOTE — **An operating procedure, condition, etc., which it is essential to highlight.**

YOUR RESPONSIBILITY - TO LET US KNOW. Comments, corrections, and questions regarding this manual are welcome. Air Force change recommendations will be submitted to 2762 LS Det 6/FT IAW SR-71 Logistics Support Manual - 400 for approval with simultaneous coordination copy furnished to Lockheed and other Air Force agencies using established channels and contact points. Contractor recommendations may be forwarded through the manufacturer's senior representative present or directly to manuals engineers in the flight test organization.

Table Of Contents

SR-71

Section I

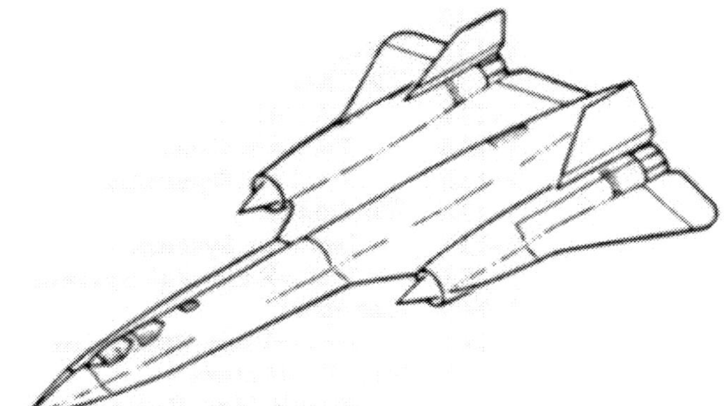

Description And Operation

TABLE OF CONTENTS

INTRODUCTION

This section describes SR-71A reconnaissance aircraft. Subsection 1A describes the SR-71B.

THE AIRCRAFT

The SR-71 is a delta-wing, two-place aircraft powered by two axial-flow turbojet engines. The aircraft, built by the Lockheed California Company, features titanium construction and is designed to operate at high altitudes and high supersonic speeds. The aircraft has very thin wings, twin canted rudders mounted on top of the engine nacelles, and a pronounced fuselage "chine" extending from the nose to the leading edge of the wing. The propulsion system uses movable spikes to vary air inlet geometry. Surface controls are elevons and rudders, operated by irreversible hydraulic actuators with artificial pilot control feel. The aircraft can be refueled either in-flight or on the ground through separate receptacles that feed into a common refueling line. A drag chute is provided to augment the six-mainwheel brakes. The aircraft is painted black to reduce internal temperatures when at high speed.

Dimensions

Length (overall)	107.4	ft.
Height (to top of rudders)	18.5	ft.
Wing span	55.6	ft.
Wing area (reference)	1605 sq.	ft.
Tread (MLG middle wheel centerlines)	16.67	ft.

Gross Weight

The loaded gross weight of the aircraft varies from approximately 135,000 to over 140,000 pounds. Zero fuel weight varies from 56,500 to more than 60,000 pounds.

NOTE

Use SR-71 manual of weight and balance data applicable to specific aircraft to compute aircraft performance.

FUEL

The operating envelope of the JT11D-20 engine requires special fuel. The fuel is not only the source of energy but is also used in the engine hydraulic system. During high Mach flight, the fuel is also a heat sink for the various aircraft and engine accessories which would otherwise overheat at the high temperatures encountered. This requires a fuel having high thermal stability so that it will not break down and deposit coke and varnishes in the fuel system passages. A high luminometer number (brightness of flame index) is required to minimize transfer of heat to the burner parts. Other items are also significant, such as the amount of sulfur impurities tolerated. Advanced fuels, JP-7 (PWA 535) and PWA 523E, were developed to meet the above requirements.

JP-7 and PWA 523E contain one gallon of PWA 536 lubricity additive per 5200 gallons of fuel to insure adequate lubrication of fuel hydraulic pumps.

ENGINE AND AFTERBURNER

Thrust is supplied by two Pratt & Whitney JT11D-20 bleed bypass turbojet engines with afterburners. (See Figure 1-2.) The engines are designed for continuous operation at compressor inlet temperatures above 400°C, which are associated with high Mach flight. The engine has a single-rotor, nine-stage, 8.8:1 pressure ratio compressor utilizing a compressor bleed bypass at high Mach. When opened, bypass valves bleed air from the fourth stage of the compressor, and six ducts route it around the rear stages of

GENERAL ARRANGEMENT AND BAY LOCATOR DIAGRAM

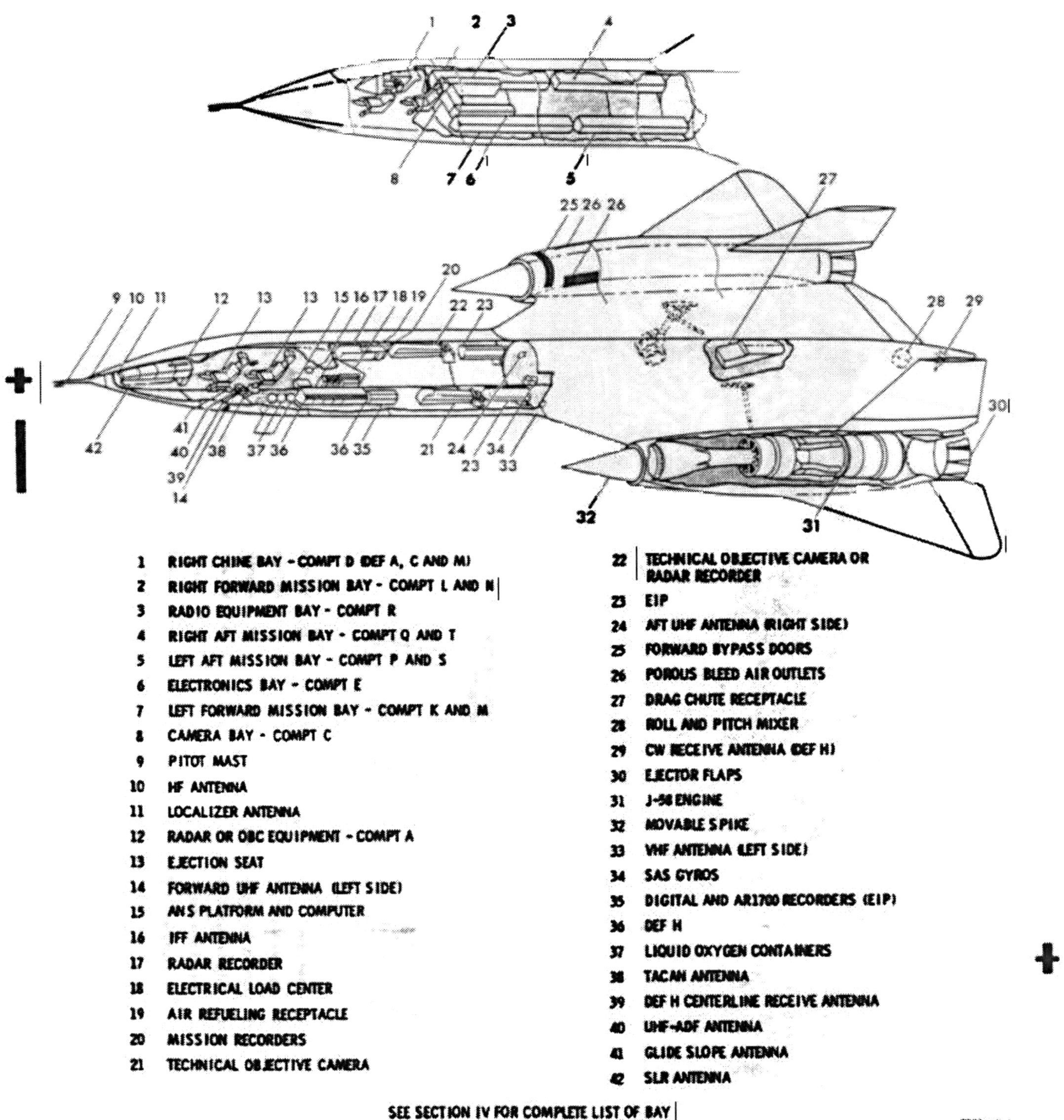
1 RIGHT CHINE BAY - COMPT D (DEF A, C AND M)
2 RIGHT FORWARD MISSION BAY - COMPT L AND N
3 RADIO EQUIPMENT BAY - COMPT R
4 RIGHT AFT MISSION BAY - COMPT Q AND T
5 LEFT AFT MISSION BAY - COMPT P AND S
6 ELECTRONICS BAY - COMPT E
7 LEFT FORWARD MISSION BAY - COMPT K AND M
8 CAMERA BAY - COMPT C
9 PITOT MAST
10 HF ANTENNA
11 LOCALIZER ANTENNA
12 RADAR OR OBC EQUIPMENT - COMPT A
13 EJECTION SEAT
14 FORWARD UHF ANTENNA (LEFT SIDE)
15 ANS PLATFORM AND COMPUTER
16 IFF ANTENNA
17 RADAR RECORDER
18 ELECTRICAL LOAD CENTER
19 AIR REFUELING RECEPTACLE
20 MISSION RECORDERS
21 TECHNICAL OBJECTIVE CAMERA
22 TECHNICAL OBJECTIVE CAMERA OR RADAR RECORDER
23 EIP
24 AFT UHF ANTENNA (RIGHT SIDE)
25 FORWARD BYPASS DOORS
26 POROUS BLEED AIR OUTLETS
27 DRAG CHUTE RECEPTACLE
28 ROLL AND PITCH MIXER
29 CW RECEIVE ANTENNA (DEF H)
30 EJECTOR FLAPS
31 J-58 ENGINE
32 MOVABLE SPIKE
33 VHF ANTENNA (LEFT SIDE)
34 SAS GYROS
35 DIGITAL AND AR1700 RECORDERS (EIP)
36 DEF H
37 LIQUID OXYGEN CONTAINERS
38 TACAN ANTENNA
39 DEF H CENTERLINE RECEIVE ANTENNA
40 UHF-ADF ANTENNA
41 GLIDE SLOPE ANTENNA
42 SLR ANTENNA
SEE SECTION IV FOR COMPLETE LIST OF BAY DESIGNATIONS AND EQUIPMENT LOCATIONS
F203-4(m)

JT11D -20 ENGINE

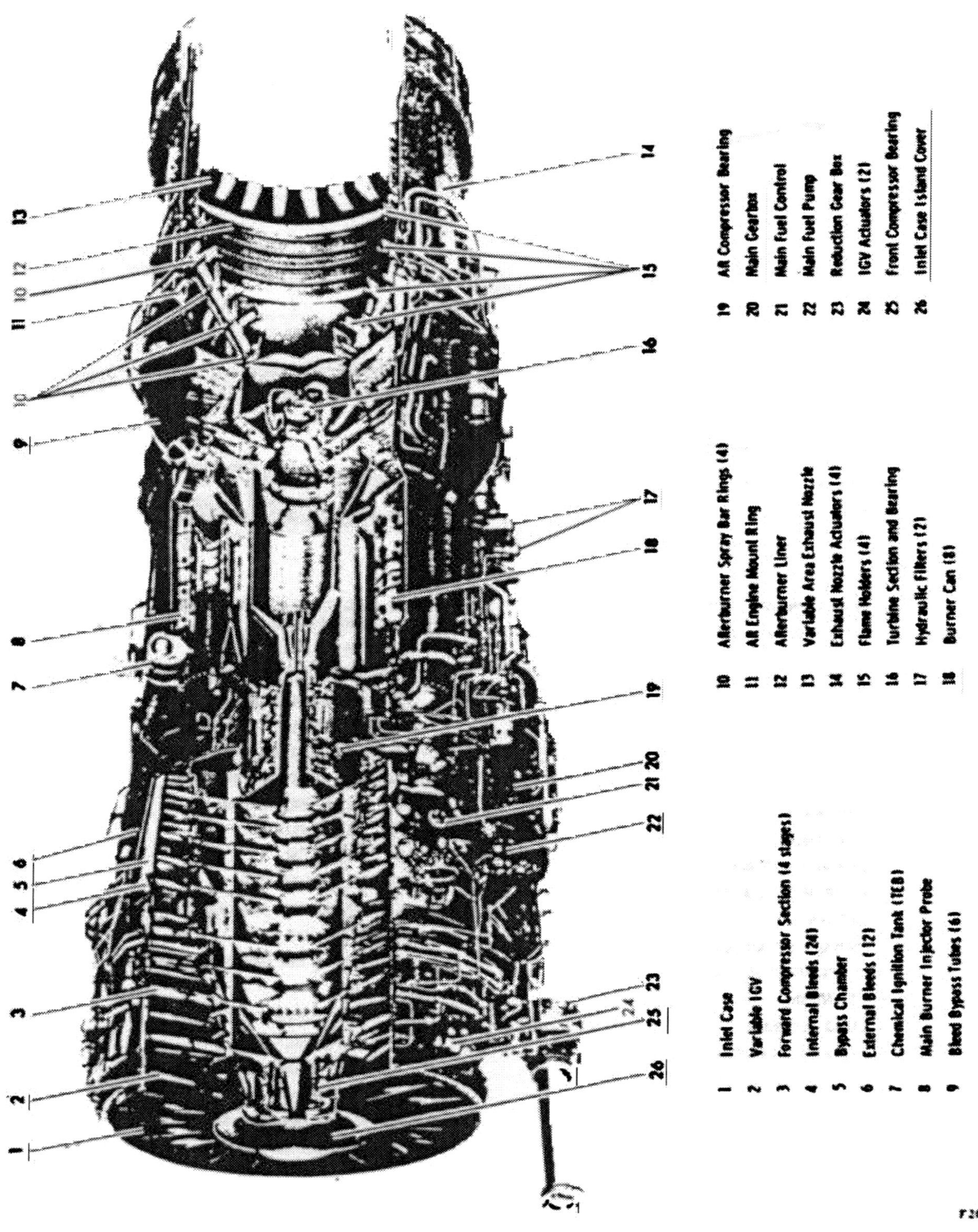
1 Inlet Case
2 Variable IGV
3 Forward Compressor Section (4 stages)
4 Internal Bleeds (24)
5 Bypass Chamber
6 External Bleeds (12)
7 Chemical Ignition Tank (TEB)
8 Main Burner Injector Probe
9 Bleed Bypass Tubes (6)
10 Afterburner Spray Bar Rings (4)
11 AB Engine Mount Ring
12 Afterburner Liner
13 Variable Area Exhaust Nozzle
14 Exhaust Nozzle Actuators (4)
15 Flame Holders (4)
16 Turbine Section and Bearing
17 Hydraulic Filters (2)
18 Burner Can (8)
19 AB Compressor Bearing
20 Main Gearbox
21 Main Fuel Control
22 Main Fuel Pump
23 Reduction Gear Box
24 IGV Actuators (2)
25 Front Compressor Bearing
26 Inlet Case Island Cover

Figure 1-2

the compressor, the combustion section, and the turbine. The bleed air re-enters the turbine exhaust around the front of the afterburner where it is used for increased thrust and cooling. The transition to bypass operation is scheduled by the main fuel control as a function of compressor inlet temperature and engine speed. The transition normally occurs in a CIT range of 85° to 115°C, corresponding to a Mach range of 1.8 to 2.0.

When on the ground, or at low Mach numbers, engine speed varies with throttle movement when the throttle is between IDLE and slightly below the Military stop. At higher settings, up to maximum afterburner, the main fuel control schedules engine speed as a function of CIT and modulates the variable area exhaust nozzle to maintain approximately constant rpm. Throttle movement in the afterburning range only changes the afterburner fuel flow, nozzle position, and thrust. At high Mach number and constant inlet conditions, engine speed is essentially constant for all throttle positions down to and including IDLE. At a fixed throttle position, engine speed will vary according to main fuel control schedule when CIT (Mach) changes.

The engine contains a two-stage turbine. Turbine discharge temperatures are monitored by exhaust gas temperature indication. A chemical ignition system is used to ignite the low vapor pressure fuel. A separate engine-driven hydraulic system, using fuel as hydraulic fluid, operates the exhaust nozzle, chemical ignition system dump, compressor bypass, starting bleed systems, and Inlet Guide Vanes (IGV). The main fuel pump, engine hydraulic pump and tachometer are driven by the main engine gearbox. The afterburner fuel pump is powered by an air turbine, driven by compressor discharge air.

Maximum Rated Thrust

Maximum rated thrust is obtained in afterburning by placing the throttle against the quadrant forward stop. The maximum afterburning uninstalled thrust of each engine at sea level, static condition, and standard day is 34,000 pounds. Takeoff thrust in maximum afterburner is illustrated in Figure 1-3 at sea level pressure altitude. It shows the variation in thrust with ambient temperature and the effect of airspeed during the takeoff acceleration.

Partial Afterburning Thrust

Afterburning fuel flow and thrust are modulated by moving the throttle between the Military detent and the quadrant forward stop. Minimum afterburning thrust is obtained with the throttle just forward of Military and is approximately 85% of maximum afterburning thrust at sea level and

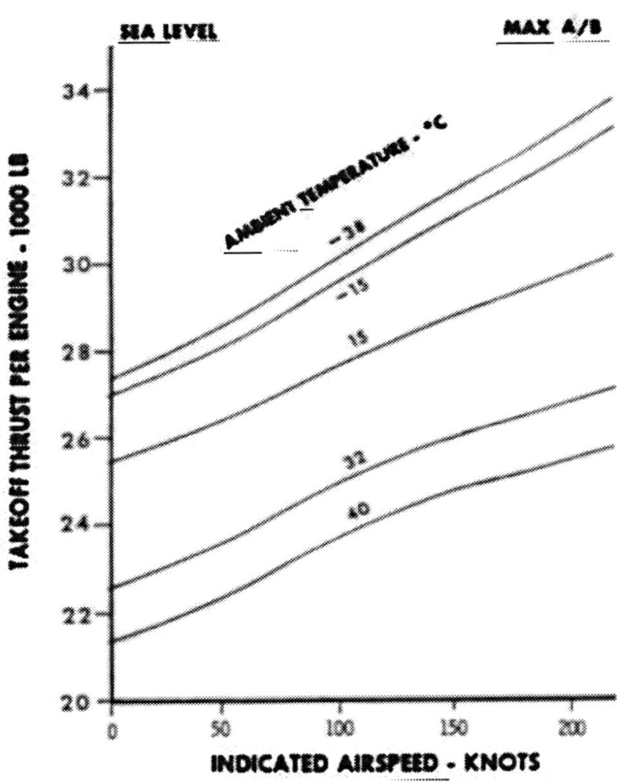

Figure 1-3

approximately 55% at high altitude. Afterburner ignition is automatically actuated when the throttle is advanced past the detent. The time required to obtain afterburner ignition after moving the throttle past the detent is a function of afterburner fuel manifold fill time. The fill time can be up to three seconds at sea level and up to seven seconds at altitude. Afterburner fuel flow is terminated when the throttle is retarded below the detent. The basic engine operates at Military rated thrust during all afterburning operation.

MILITARY THRUST

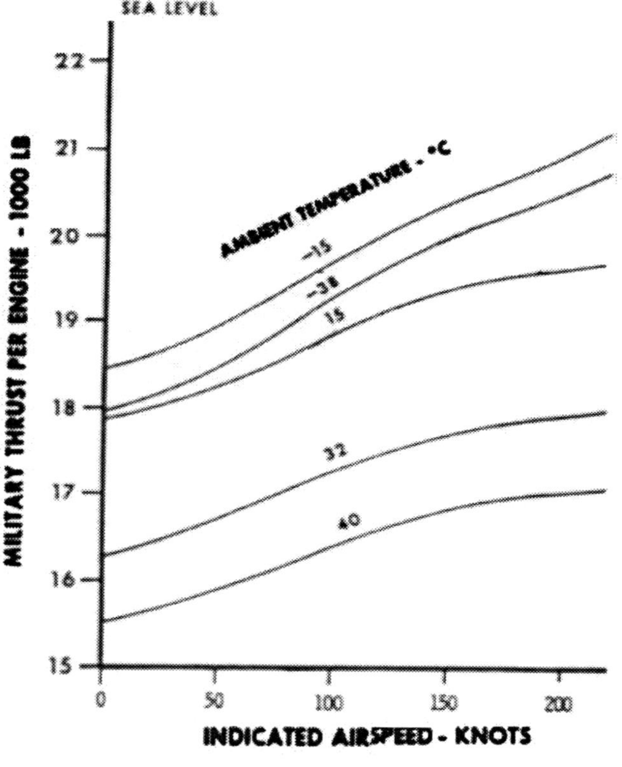

Figure 1-4

Military Thrust

Military thrust is the maximum non-afterburning thrust and is obtained by placing the throttle against the aft side of the Military detent. At sea level static conditions, military thrust is approximately 70% of maximum thrust. At high altitude, military thrust is approximately 28% of the maximum thrust available. Figure 1-4 illustrates the variation in military thrust with ambient temperature and airspeed at sea level pressure altitude.

Idle Thrust

Idle thrust is the minimum non-afterburning thrust level. With the throttle in IDLE, the engine operates at approximately 3975 rpm up to 60°C (140°F). At higher ambient temperatures, rpm increases approximately 50 rpm per 1°C. Idle thrust is illustrated in Figure 1-5, at sea level pressure altitude, for airspeeds typical of a landing and deceleration.

IDLE THRUST

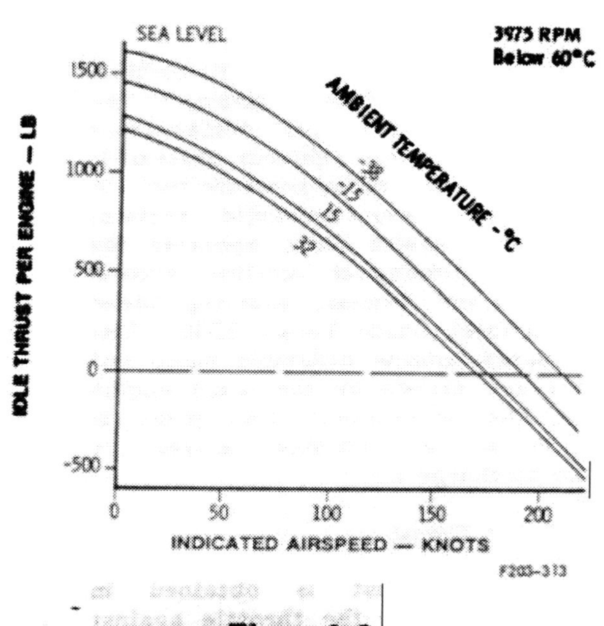

Figure 1-5

THROTTLES AND THROTTLE SETTINGS

Two throttle levers are located in a quadrant on the pilot's left forward console. The right throttle is mechanically linked to the right engine main fuel control. The left throttle is linked to the left engine afterburner fuel control. The afterburner and the main fuel controls are interconnected by a closed loop cable. The throttle quadrant has three labeled positions, OFF, IDLE, and AFTERBURNER, and an unlabeled Military power stop. See Figure 1-6. The non-afterburning operating range of the engine is between IDLE and Military.

A spring-loaded pushbutton switch is located on the right throttle knob. When depressed, the pilot's microphone is connected to the selected radio transmitter.

A throttle air inlet restart switch is located on the inboard side of the right throttle. The switch is used for restarting both air inlets simultaneously.

Off

In the OFF position, the windmill bypass valve cuts off fuel from the burner cans and routes it back to the aircraft system. This provides cooling for engine oil, fuel pump, and fuel hydraulic pump when an engine is windmilling.

Idle

When a throttle is moved forward from OFF to IDLE, a roller drops over a hidden ledge at the IDLE position. This ledge prevents the engine from being inadvertently cut off when the throttles are retarded to IDLE. The throttles must be lifted to be moved from IDLE to OFF.

Afterburner

The throttle must be slightly raised and pushed forward to clear the Military stop before additional forward movement of the throttle can initiate afterburner ignition. The AFTERBURNER range extends from the Military stop to the quadrant forward stop.

Start

There is no distinct throttle position for starting. Starting is accomplished by moving the throttle from OFF to IDLE as the engine is accelerated by the starter. As the proper engine speed is reached, fuel is directed to the engine burners by actuation of the windmill bypass valve and the chemical ignition system is actuated by fuel pressure.

Throttle Friction Lever

Throttle friction is controlled by a lever located on the inboard side of the throttle quadrant. Moving the lever forward, as the INCREASE FRICTION label indicates, progressively increases friction.

TEB Remaining Counters

Mechanical digital counters, aft of each throttle,indicate the number of TEB shots remaining for each engine. The counters are spring wound and set to 16 prior to engine start. Each time a throttle is moved forward from OFF to IDLE, or from Military to AFTERBURNER, the corresponding counter indication decreases by 1.

Tachometers

Tachometers for each engine are mounted on the right side of the pilot's instrument panel. They indicate engine speed in revolutions per minute by means of a main pointer and dial calibrated to 10,000 rpm, and a smaller dial and subpointer which make one complete revolution for each 1000 rpm. The tachometers are self-energized and operate independently of the aircraft electrical system.

ENGINE FUEL SYSTEM

Engine fuel system components include the engine-driven fuel pump, main fuel control,

THROTTLE QUADRANT

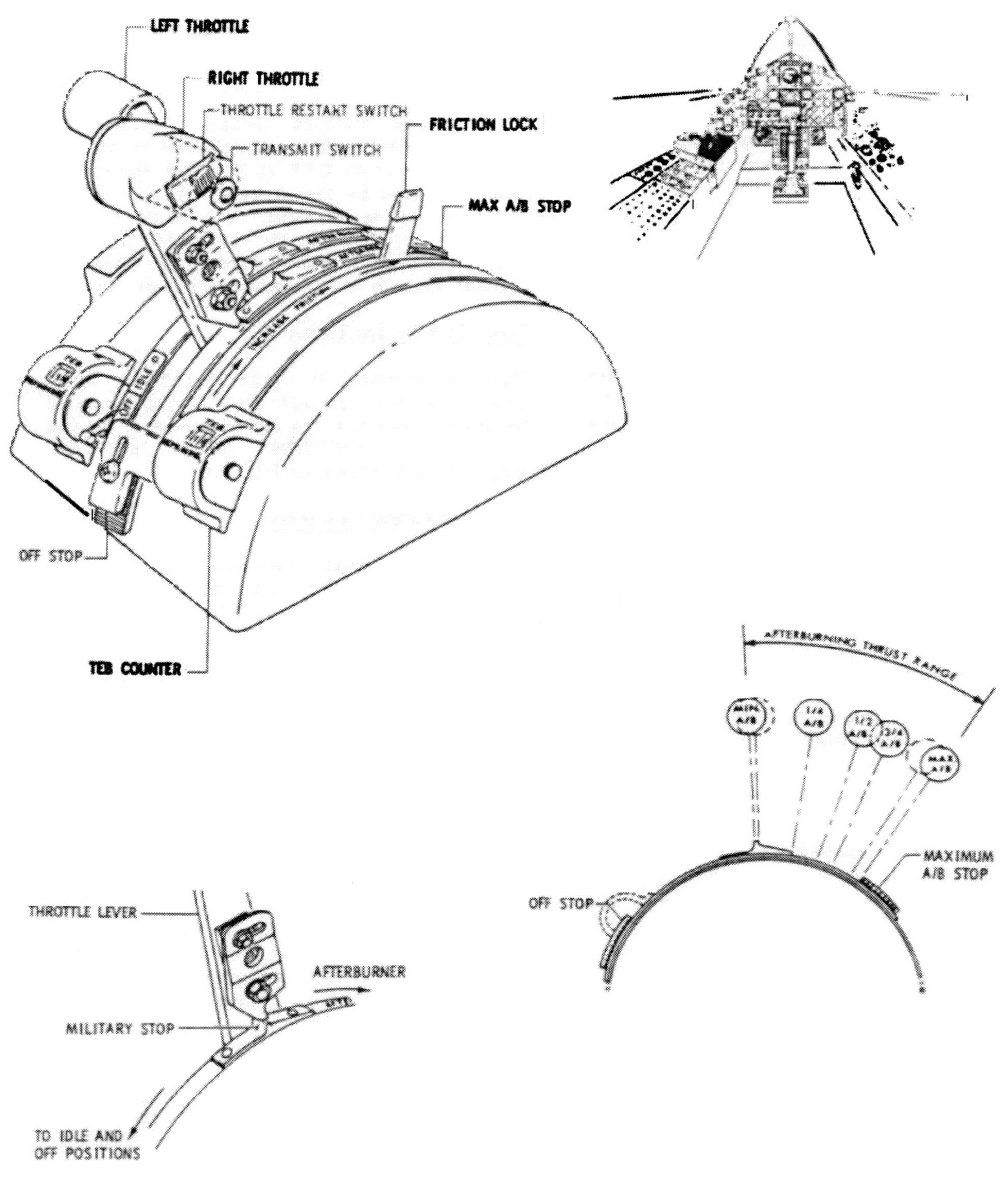

Figure 1-6

windmill bypass valve and variable area fuel nozzles in the main burner section. (See Figure 1-7.)

Main Fuel Pump

The engine-driven main fuel pump is a two-stage unit. The first stage, a single centrifugal pump, acts as a boost stage. The second stage consists of two parallel gear-type pumps with discharge check valves. The parallel pump and check valve arrangement permits continued operation if either pump fails.

Main Fuel Control

The main fuel control meters main burner fuel flow, controls the bleed bypass, start bleed valves, IGV, and exhaust nozzle modulation. It regulates main engine thrust as a function of throttle position, compressor inlet air temperature, main burner pressure, and engine speed. The bypass, start-bleed valve positions and IGV are controlled as a function of engine speed, biased by CIT. Afterburner operation is always at Military-rated engine speed and EGT. The control has a remote trimmer for EGT regulation. There is no emergency fuel control system.

Windmill Bypass and Dump Valve

The windmill bypass and dump valve directs fuel to the engine burners for normal operation or bypasses fuel to the recirculation system for accessory, engine component, and engine oil cooling during windmilling operation. The valve responds to signals from the main fuel control. The valve opens to drain the engine fuel manifold when the engine is shut down.

Fuel Nozzles

The engine has an eight-unit can-annular combustion section with 48 variable-area, dual-orifice fuel nozzles. The nozzles are arranged in clusters of six nozzles per burner. Each nozzle has a fixed area primary metering orifice and a variable area secondary metering orifice, discharging through a common opening. The secondary orifice opens as a function of primary pressure drop.

Combustion Chamber Drain Valve

The main engine ignition system plumbing is equipped with a fuel purge or "Dribble Tee". This allows fuel from the main fuel pump interstage to flush residual ignition fluid (TEB) from the ignition probe. It prevents "coking" from occurring which would restrict the ignition probe and prevent engine ignition. Hence, fuel in small quantity should drain from the main burner case overboard drain fitting anytime there is fuel pressure to the engine pump inlet, due to fuel boost pump operation or tank pressure developed by the LN_2 fuel tank pressurization system. If fuel does not drain normally, either the chemical ignition system probe is plugged or the burner drain has malfunctioned. The normal leakage from the main burner case overboard drain should be confirmed before start. If the overboard drain is restricted, it will increase the "wetted" fuel area in the burner and could result in severe torching during engine start.

Fuel Flow Indicators

Fuel flow indicators for each engine, mounted on the right side of the pilot's instrument panel, display total fuel flow (engine and afterburner) plus tank return flow, if any. Dial calibrations are provided in 5000 pound per hour increments to 95,000 pph. Five center digit windows show fuel flow to the nearest 100 pph. Power for the indicators is supplied from the essential ac bus through the L and R FLOW circuit breakers on the pilot's right console.

ENGINE AND AFTERBURNER FUEL SYSTEM

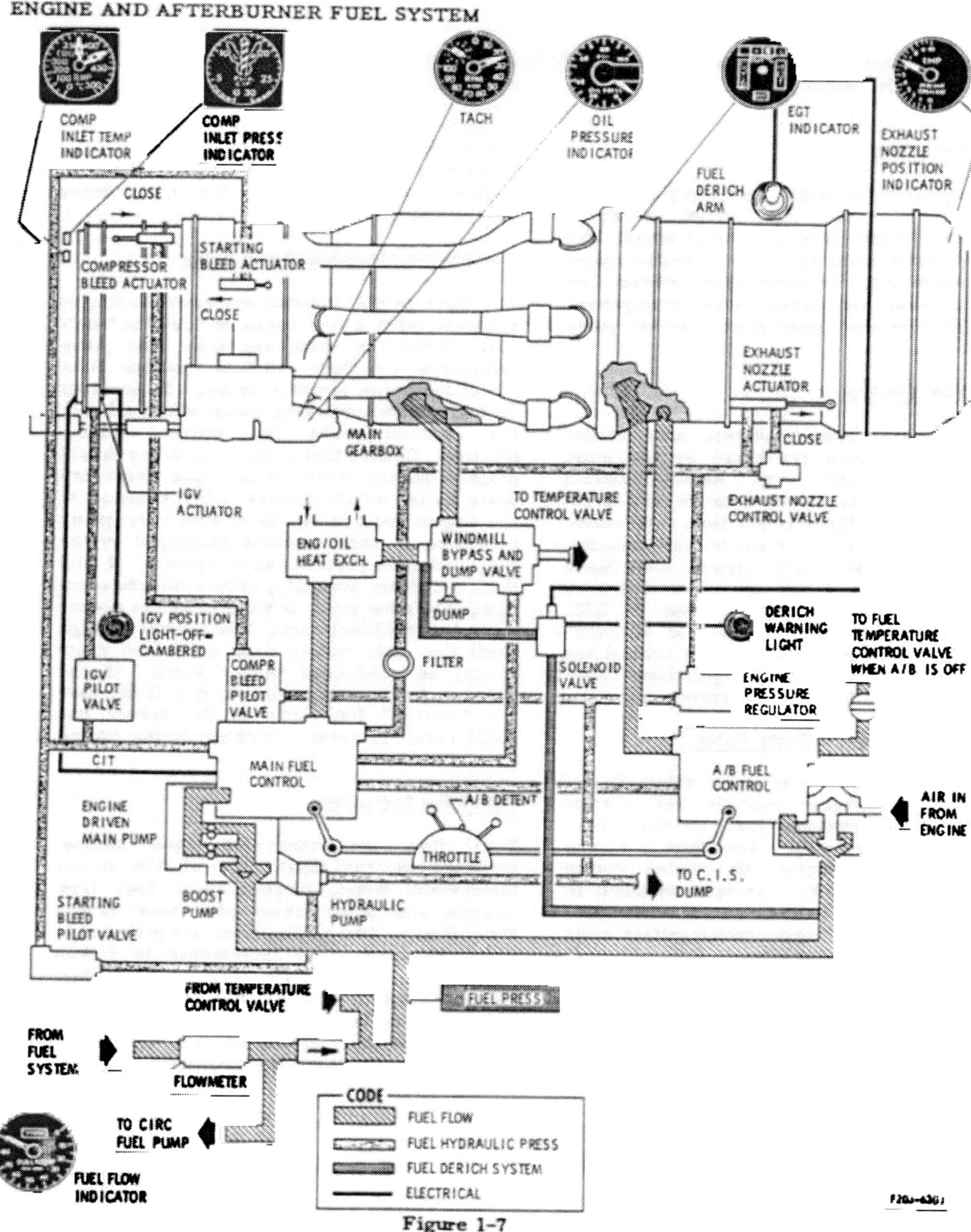

Figure 1-7

NOTE

Tank return flow forms an appreciable portion of the indication when at or below Military.

AFTERBURNER FUEL SYSTEM

Afterburner Fuel Pump

The afterburner fuel pump is a high speed, single stage centrifugal pump. The pump is driven by an air turbine, operated by engine compressor discharge air. The compressor discharge air supply is regulated by a butterfly valve in response to the demand of the afterburner fuel control.

Afterburner Fuel Control

The hydromechanical afterburner fuel control schedules fuel flow as a function of throttle position, main burner pressure, and compressor inlet temperature. Fuel flow is metered to discharge fuel from the four concentric afterburner spraybar rings.

FUEL DERICH SYSTEM

A derichment system on each engine protects against severe turbine overtemperature. If the respective EGT indicator reaches 860°C while the system is armed, the fuel/air ratio in the engine burner cans is automatically reduced (deriched). A signal from the EGT gage actuates a solenoid-operated valve which bypasses metered engine fuel from the fuel/oil cooler to the afterburner fuel pump inlet. See Figures 1-7 and 1-8. Once actuated, the solenoid valve is held open until the system is turned off or rearmed. A red, flashing FUEL DERICH warning light illuminates when the valve for the respective engine is open.

Derichment at sea level decreases thrust approximately 5% in maximum afterburner; 7% at Military. Derichment during supersonic cruise decreases engine thrust approximately 45% in maximum afterburner (overall engine/inlet thrust loss is about 10% for the deriched engine since derichment has no effect on CIP) and may cause the afterburner to blow out. Continuous operation with the engine deriched does not harm the engine (provided derichment can reduce EGT to normal limits).

After an unstart, do not move the fuel derich switch from ARM unless inlet roughness has cleared and the inlet has restarted (CIP recovered); otherwise, severe overtemperature can result. Do not attempt to relight the afterburner while deriched. Lighting the afterburner while deriched can result in engine speed suppression of up to 750 rpm.

Fuel Derich Arming Switch

A three-position FUEL DERICH arming toggle switch is located on the pilot's left instrument side panel. In the ARM (center) position, the derich circuit is armed and the derich solenoid valve will open and remain open if the respective EGT indication reaches 860°C. In the OFF (down) position, the derich solenoid valve is closed and the system cannot provide derichment. The REARM (up) position is spring-loaded and allows the pilot to rearm the fuel derich system without moving the switch to OFF. Power for the switch is furnished by the essential dc bus through the L and R FUEL DERICH circuit breakers on the pilot's left console.

Fuel Derich System Test Switch

A three-position toggle switch, labeled FUEL DERICH SYSTEM, is located on the pilot's left console. The switch is labeled L (left), R (right) and is spring-loaded to the center (off) position. When the switch is moved to the L or R position, the digital indication on the respective EGT gage slews toward 1198°C. When the EGT gage indication exceeds 860°C, the red jewel light in the gage illuminates; and, if the FUEL DERICH switch is in ARM, the respective fuel derich warning light flashes and the derich solenoid valve opens.

Fuel Derich Warning Light

Two red, flashing fuel derich warning lights are located on the right side of the pilot's instrument panel. A light flashes when the fuel derich system for the respective engine is activated (derich solenoid valve open) and continues flashing until the fuel derich system is rearmed or off.

EXHAUST GAS TEMPERATURE (EGT) TRIM SYSTEM

EGT Gages

Two EGT gages, one for each engine, are located on the right side of the pilot's instrument panel. Each gage has a digital indicator that shows turbine discharge temperature from 0° to 1198°C, a HOT (red) and COLD (yellow) condition flag, a red "jewel" overtemperature warning light that illuminates when the EGT digital indication reaches 860°C, and a power OFF warning flag. Each indicator receives power from the essential ac bus through its respective L or R EGT IND circuit breaker on the pilot's right console.

EGT Trim Switches

Two four-position EGT trim switches, one for each engine, are located on the pilot's left console. The positions are labeled AUTO (left), INCR (up), DECR (down) and HOLD (center). When a switch is held in INCR, a small electric motor on the engine fuel control increases the ratio of main burner fuel flow to main burner combustion pressure and thus increases the turbine discharge temperature (EGT). Holding a switch in DECR runs the motor in the opposite direction and decreases EGT. The switches have no effect on rpm as long as the nozzles are modulating to provide the scheduled engine speed. However, engine speed will increase (or decrease) with increasing (or decreasing) EGT when nozzle position is limited full closed (or open).

An EGT "permission" circuit prevents automatic trimming in either direction and manual uptrim when the respective throttle is positioned below Military or the engine is deriched. Manual EGT downtrim remains available even when the permission circuit is on (throttle below Military or engine deriched). Power for the trim motors is furnished by the essential ac bus through the L and R EGT TRIM circuit breakers on the pilot's right console. Power for the permission circuit is furnished by the essential dc bus through the L and R EGT circuit breakers on the pilot's left console.

Automatic EGT Trim

When an EGT trim switch is in AUTO, the respective throttle is positioned at or above Military, and that engine is not deriched, the permission circuit is opened (off) to allow automatic EGT trim. The electric trim motor for the respective engine regulates its main fuel control and automatically provides EGT within a 10°C nominal deadband. See Figure 1-9.

If the EGT for an engine is either above or below the deadband, the system trims EGT toward its deadband. The rate of trimming depends on temperature deviation from the deadband. If EGT is more than 10°C above the deadband, the system downtrims at its maximum rate (approximately 8°C per second), and the HOT flag is displayed on the EGT gage. When the EGT is more than 10°C below the deadband, the system uptrims at 1°C per second and the COLD flag is displayed. When EGT is 10°C or less from its deadband, the system trims at only 1/3°C per second and the condition flags retract.

CAUTION

If EGT tends to hunt or has an abnormal tendency to uptrim or downtrim while AUTO is selected, the corresponding engine should be operated in manual trim and the condition reported following flight. An EGT overtemperature may occur if continued operation in AUTO is attempted.

EGT INDICATION AND CONTROL SYSTEM

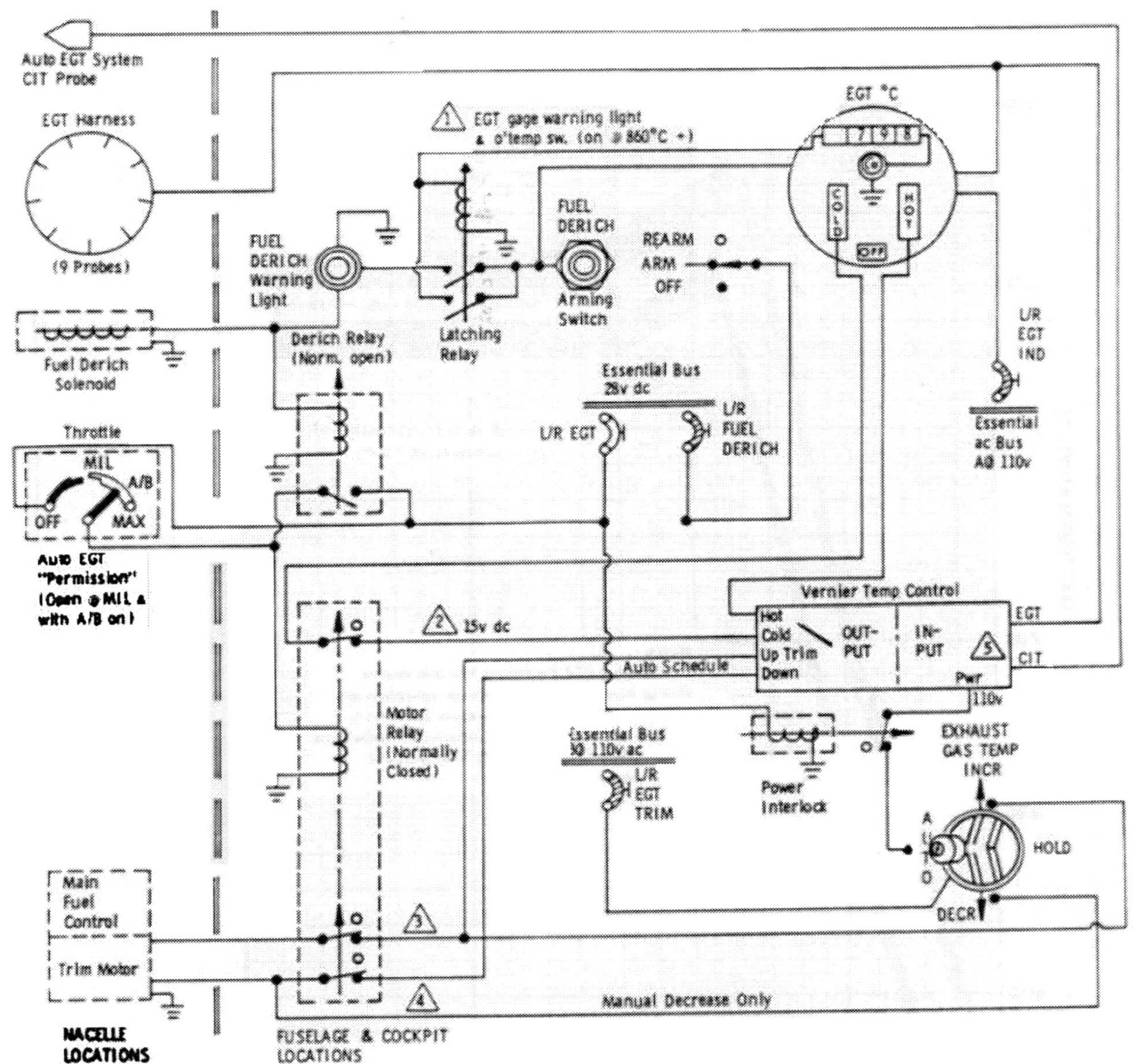

NOTE

1. Switch closes and light illuminates as 860°C reached. Switch stays closed and light remains on above 860°C. When EGT then decreases below 860°C, jewel light is extinguished but "latching" relay maintains power to derich solenoid until derich arming switch is cycled.
2. COLD flag does not operate when below Military power or when deriched. HOT and COLD flag operating power (15v dc) is produced within the vernier temperature control.
3. Direction of trim is controlled by ac power phase-sequencing.
4. Auto trim and manual uptrim are inoperative when below Military power or when deriched. Manual downtrim is always available if three-phase power is supplied.
5. Auto EGT vernier temperature control is powered only when AUTO is selected.

F203-1396(a)

Figure 1-8

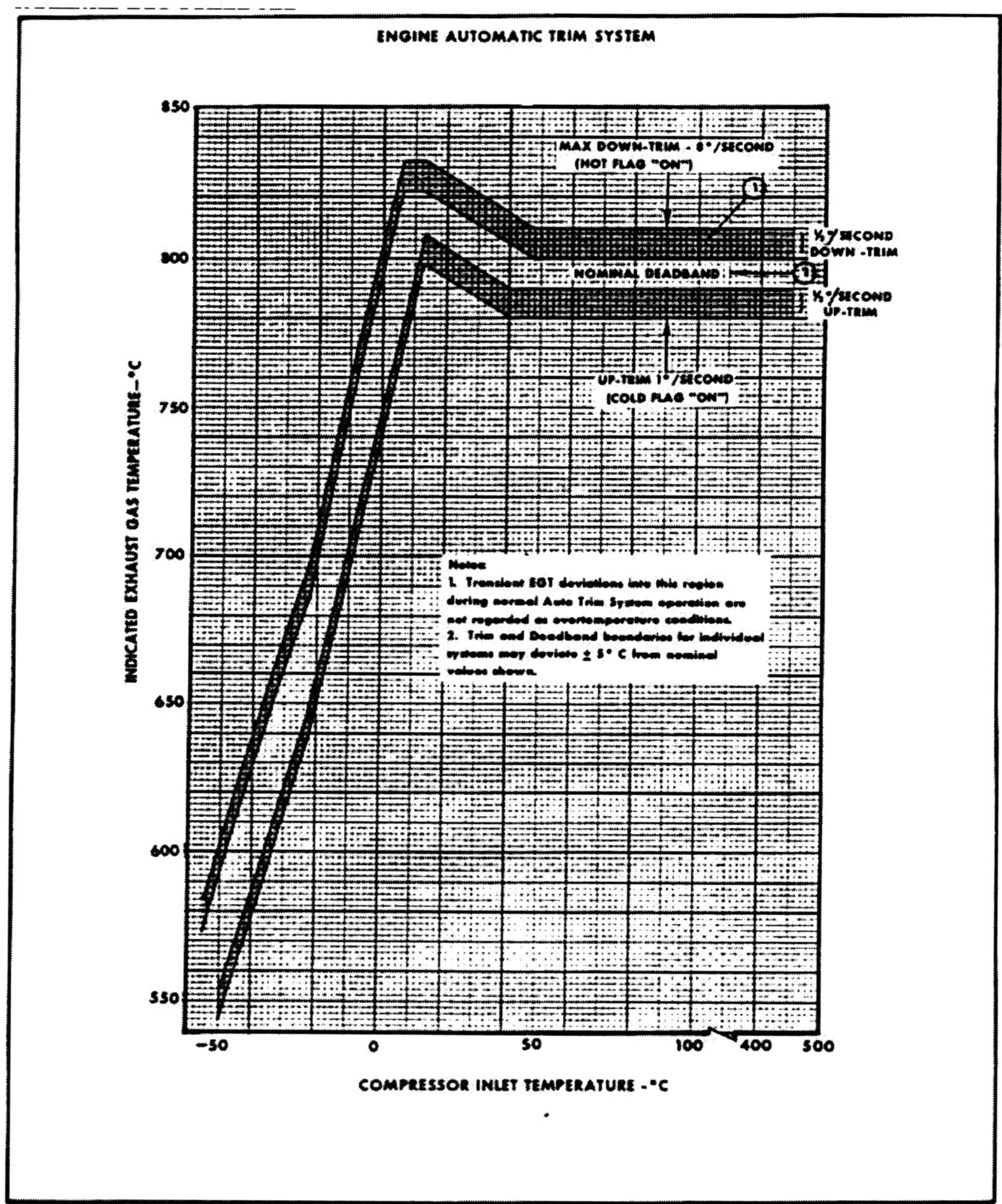
ENGINE AUTOMATIC TRIM SYSTEM
MAX DOWN-TRIM - 8°/SECOND
(HOT FLAG "ON")
½°/SECOND
DOWN-TRIM
NOMINAL DEADBAND
½°/SECOND
UP-TRIM
UP-TRIM 1°/SECOND
(COLD FLAG "ON")
Notes:
1. Transient EGT deviations into this region during normal Auto Trim System operation are not regarded as overtemperature conditions.
2. Trim and Deadband boundaries for individual systems may deviate ± 5° C from nominal values shown.
INDICATED EXHAUST GAS TEMPERATURE - °C
850
800
750
700
650
600
550
-50
0
50
100
400
500
COMPRESSOR INLET TEMPERATURE - °C

Figure 1-9

NOTE

- The EGT condition flags do not operate when manual EGT trim is used.
- With AUTO EGT selected, the COLD condition flag will not operate when the permission circuit is on (throttle below Military or deriched).
- With AUTO EGT selected, the HOT condition flag will operate regardless of the condition of the permission circuit.

Auto EGT is normally used during the engine trim check; however, if the system is obviously uptrimmed above normal limits, reduce power and manually downtrim prior to resetting military power.

The hot flag overtemperature warning circuit is continually energized when in AUTO EGT. A hot flag may appear at throttle positions below the permission switch and should be corrected by manually downtrimming or retarding the throttle.

Repeated rapid throttle movements forward and back through the permission switch operating range can cause military EGT to increase sufficiently to produce derichment. This EGT ratcheting is caused by a lag in thermocouple response and can be avoided by selecting manual trim when repeated rapid throttle movements to and below Military are expected.

Selecting the spring-loaded HOLD (center) position deactivates automatic trimming and maintains the existing EGT trimmer setting.

NOTE

When AUTO EGT is selected, a brief spurious flag indication may occur in response to the EGT and CIT conditions that existed when AUTO EGT trim was last used. Valid flag indications and normal AUTO EGT trim operation should occur within five seconds.

INLET PARAMETER INDICATIONS

Compressor Inlet Pressure (CIP) Gage

The CIP gage, on the left of the pilot's instrument panel, has an L and R needle which indicate inlet static pressure at the face of each compressor, and a striped third needle that indicates expected normal CIP. The position of the striped pointer is governed by DAFICS using pressures sensed by the pitot-static system. The indication varies with Mach and KEAS so that the striped needle shows "normal" CIP for the flight condition. A substantial difference between the striped needle and actual CIP indicates improper inlet operation. Higher actual pressure at normal speeds and altitudes may produce unstarts. Lower pressure indicates poor pressure recovery due to improper spike and/or bypass door settings except at abnormal angles of attack or yaw conditions, where inlet operation is automatically biased to less than normal recovery. The spread between inlet CIP indications (L and R needles) should not exceed 1 psi. The needle can be used as a guide for bypass door settings during manual operation of one or both inlets; however, it is preferable to keep the L and R needles slightly below the "normal" indication to maintain a margin below unstart pressures. Automatic or manual inlet operation at pressures below the "normal" indication reduces aircraft range.

NOMINAL CIP RANGE

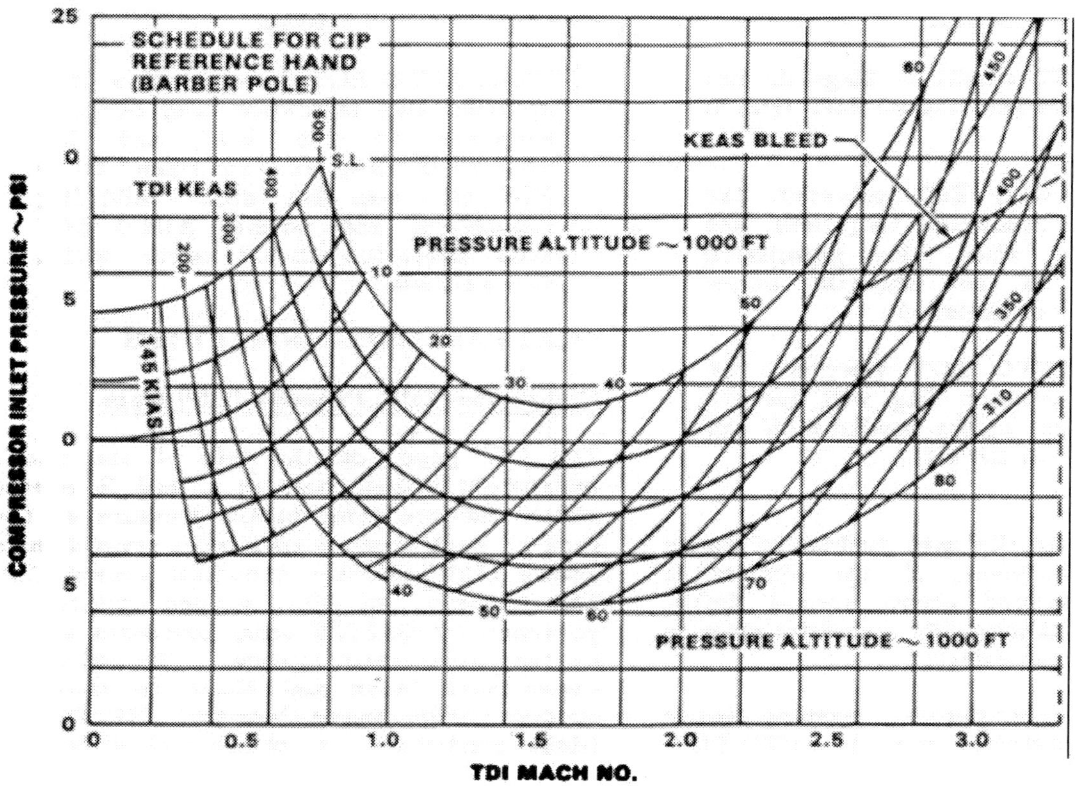

Figure 1-10

Power is supplied by the essential ac bus through the CIP circuit breaker on the pilot's right console.

Compressor Inlet Temperature (CIT) Gage

A dual indicating CIT gage is mounted on the left side of the pilot's instrument panel. L (left) and R (right) needles indicate the total (ram) air temperature forward of the first compressor stage in the corresponding engine inlet. Major calibrations are marked from 0°C to 500°C. The dial has 50° incremental markings below 300°C. Above 300°C, the gage has increased sensitivity and incremental marks are supplied for each 10°C. Slight differences between left and right CIT indications can be expected; however, differences of more than 15°C while at supersonic cruise speeds should be reported as a discrepancy. Power is furnished by the essential ac bus through the L and R CIT circuit breakers on the pilot's right console.

EXHAUST NOZZLE AND EJECTOR SYSTEM

The variable-area, iris-type, engine afterburner nozzle is comprised of segments operated by a cam and roller mechanism and four hydraulic actuators. The actuators are operated by fuel hydraulic system pressure. The engine afterburner nozzle is enclosed by a fixed-contour, convergent-divergent ejector nozzle to which free floating trailing edge flaps are attached. In flight, the inlet shock trap bleed and aft bypass doors (when open) supply secondary airflow between the engine and nacelle for cooling. During ground operation, suck-in doors in the aft nacelle area provide cooling air. Intake doors around the nacelle, just forward of the ejector, normally supply tertiary air to the ejector nozzle when subsonic. The tertiary doors and trailing edge flaps are free to open and close with varying internal nozzle pressure (a function of Mach and engine thrust).

Nozzle Actuation

The exhaust nozzle control and actuation system is composed of four actuators to position the exhaust nozzle, and an exhaust nozzle control which modulates pressure at the actuators in response to engine speed signals from the main fuel control. The exhaust nozzle control is mounted on the aft portion of the engine.

Exhaust Nozzle Position (ENP) Indicators

Two ENP indicators, one for each engine, are located on the right side of the pilot's instrument panel. They are marked from 0 to 10 as an index of nozzle position from closed to open. Each indicator responds to an electrical transducer located near its exhaust nozzle. The transducer is cooled by fuel and is operated by the afterburner nozzle feedback link. Power for the indicators is supplied from the essential ac bus through the L and R ENP circuit breakers on the pilot's right console.

ENGINE EXTERNAL AND INTERNAL BLEEDS

The internal bypass bleed control and actuation system consists of four two-position actuators to move the bleed valves and a pilot valve, within the main fuel control, to establish the pressure to the actuators. The pilot valve controls the bleed valve position in response to a mechanical signal from the main fuel control. Bleed valve position is scheduled within the main fuel control as a function of engine speed and CIT. The external bleed control and actuation system is similar to the internal bleed system, except that three actuators are used.

ENGINE INLET GUIDE VANES (IGV)

The engine compressor inlet case houses a two-position inlet guide vane (IGV) system. The guide vanes can be either in the cambered position, which is normal for cruise, or in the axial position which is normal for takeoff and acceleration to intermediate supersonic speed. The IGV axial position (parallel to the airflow) results in more thrust. Actuation to the cambered position occurs at a CIT of 85° to 115°C (about Mach 1.9) during acceleration. The cambered position is mandatory when operating continuously at CIT above 125°C (approximately Mach 2.0). Shifting is normally controlled by the main engine fuel control; however, the shift to the axial position from cambered is prevented if the IGV Lockout Switch is positioned to LOCKOUT. Refer to Figure 1-11 for IGV shift scheduling information.

IGV Lockout Switches

The shift schedules for the internal bypass bleeds and the inlet guide vanes are identical. There is a positive locking feature which prevents unscheduled IGV shift to axial after the cambered position has been reached. In addition, two-position IGV lockout switches (one for each engine) on the pilot's right console, can lock out IGV shift from cambered to axial. With a lift-loc switch in LOCKOUT, the respective IGV is maintained in the cambered position regardless of internal bleed position. The LOCKOUT position is ineffective until the guide vanes are in the cambered position. The switches cannot cause or prevent IGV shift from axial to cambered. With IGV NORM selected, IGV shift occurs with internal bleed shift. Power for the IGV lockout solenoid circuits is supplied from the essential dc bus through the L and R IGV circuit breakers on the pilot's left console.

Inlet Guide Vane Position Lights

Two rotate-to-dim, amber inlet guide vane (IGV) position lights are installed on the right side of the pilot's instrument panel. An indicator is provided for each engine, identified by L or R adjacent to the appropriate light. An IGV light illuminates when the inlet guide vanes of the respective engine shift to the axial position as scheduled by the main fuel control. The light extinguishes when the IGV reaches the cambered position. Inlet guide vane position is sensed by a switch on the engine compressor case which operates when the

COMPRESSOR BLEED AND IGV SHIFT SCHEDULE
IGV/Bypass Bleed Functions

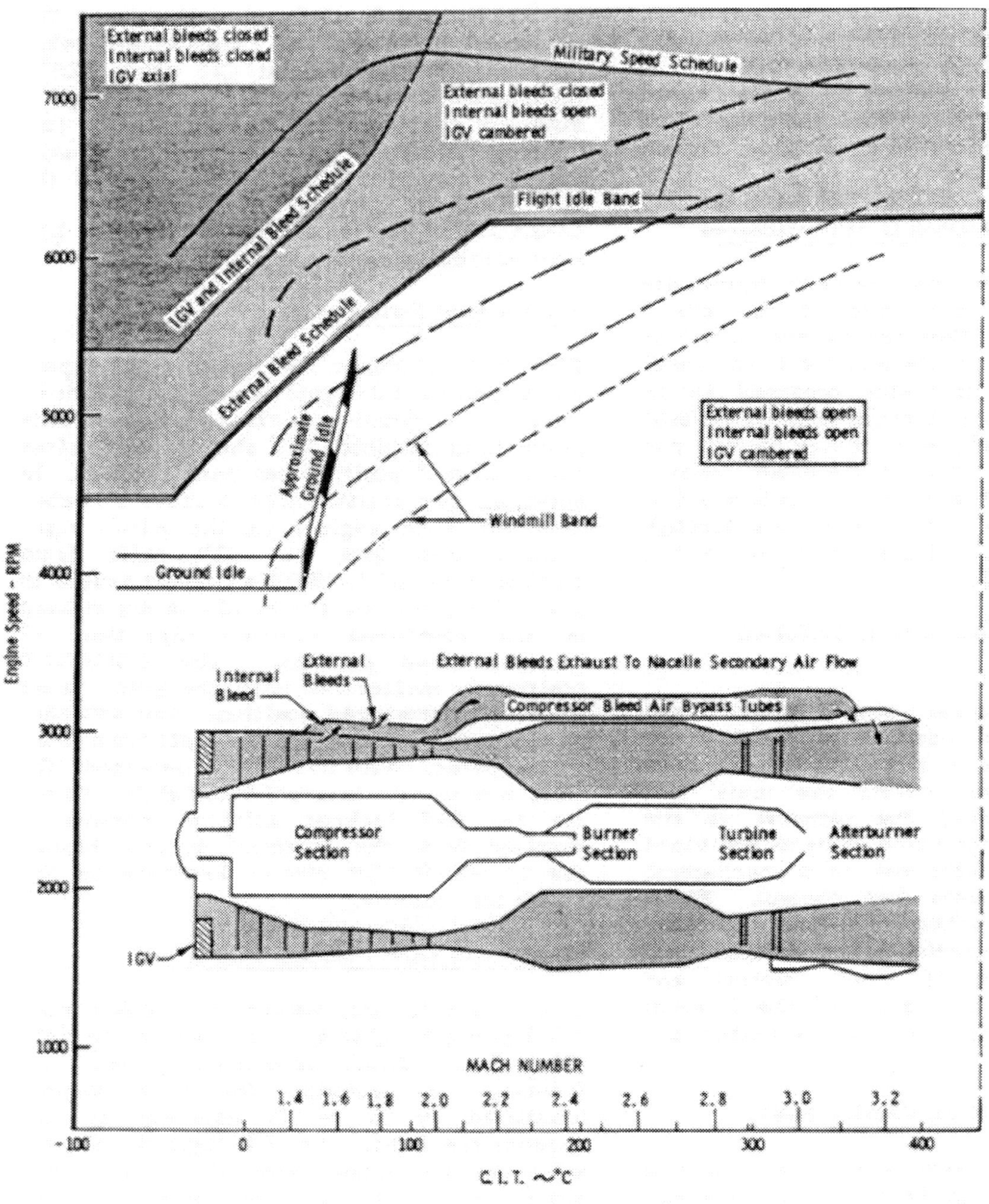

F303-10460

Figure 1-11

guide vanes reach or leave the cambered position. Power for the lights is furnished by the essential dc bus through the WARN 2 light circuit breaker on the pilot's left console.

a. The IGV lights must be off (IGV cambered) during start and at idle.

b. The IGV lights must be on (IGV axial) for takeoff.

c. The IGV lights must be off (IGV cambered) above 150°C (approximately Mach 2.2).

OIL SUPPLY SYSTEM

The engine and speed-reduction gearbox are lubricated by an engine-contained, "hot tank", closed system. The oil is cooled by circulation through an engine fuel-oil cooler. The oil tank is mounted on the lower right side of the engine compressor case. Tank volume is 6.7 US gallons. The oil tank is serviced to 5.15 US gallons. The oil is gravity-fed to the main oil pump which forces the oil through a filter and the fuel-oil cooler. The filter is equipped with a bypass in case of clogging. The oil is distributed to the engine bearings and gears from the fuel-oil cooler. Oil screens are installed at the lubricating jets for additional protection. Scavenge pumps return the oil to the tank where it is de-aerated. A pressure regulating valve keeps flow and pressure relatively constant during all flight conditions. Oil quantity warning lights are provided.

The approved oil is MIL-L-87100 (PWA 524). At low ambient temperatures, oil may be diluted with trichlorethylene, Federal Specification O-T-634, Type 1.

Main Fuel-Oil Cooler

Engine oil temperature is controlled by engine fuel which passes through the main fuel-oil heat exchanger. A bypass valve in the cooler passes additional fuel around the cooler when engine requirements are greater than the flow capacity of the cooler (approximately 12,000 pounds per hour).

Engine Oil Pressure Gages

An oil pressure gage for each engine is provided on the right side of the pilot's instrument panel. Each gage indicates output pressure of the respective engine oil pump in pounds per square inch, using electrical signals from a fuel-cooled transmitter. The dials are calibrated from 0 to 100 psi in 5-psi increments. Power for the gages is furnished by the essential ac bus 26 volt instrument transformer through the L and R OIL PRESS circuit breakers on the pilot's annunciator panel.

Engine Oil Temperature Lights

The L and R OIL TEMP annunciator lights are not functional. The OIL TEMP lights only illuminate when the IND & LT TEST push-button switch is depressed.

Low Oil Quantity Lights

L and R OIL QTY warning lights, on the pilot's annunciator panel, illuminate when oil quantity in the respective engine oil tank is less than 2-1/4 gallons.

ENGINE FUEL HYDRAULIC SYSTEM

Each engine is provided with a fuel hydraulic system for actuation of the afterburner exhaust nozzle, IGVs, and the start and bypass bleed valves. Fuel hydraulic pressure is also required to dump the chemical ignition system. An engine-driven pump maintains system pressures up to 1800 psi with a maximum flow of 50 gpm (approximately 19,700 pph) for transient requirements. Engine fuel is supplied to the pump from the main fuel pump boost stage. Some high-pressure fuel is diverted from the hydraulic system to cool the nonafterburning recirculation line and the windmill bypass valve discharge line. This fuel is returned to the aircraft fuel system. Low-pressure fuel from the hydraulic pump case is returned to the main fuel pump boost stage. Hydraulic system loop cooling is provided by the compensating fuel supplied by the main fuel pump.

ACCESSORY DRIVE SYSTEM (ADS)

An ADS is mounted forward of the engine in each nacelle. Its three major components include a constant speed drive, an accessory gearbox, and an all-attitude oil reservoir. Input power from the engine is transmitted to the ADS through a reduction gearbox on the engine and a flexible drive shaft. At the ADS, a constant speed drive unit converts the variable shaft speed to a constant rotational speed to power the ac generator. Two hydraulic pumps and a fuel circulating pump are also mounted on the ADS gearbox. The two hydraulic pumps supply power for the (A and L) or (B and R) hydraulic systems. The fuel circulating pump supplies fuel to the aircraft heat sink system. The speeds of these pumps vary directly with engine rpm.

The ADS is lubricated by an independent dry sump system with its own pump, using oil from an all-attitude reservoir. The reservoir is pressurized with nitrogen gas from the aircraft LN_2 system and supplies oil to the accessory gearbox, the constant speed drive, and the ac generator regardless of flight attitude. (Loss of the LN_2 supply to the ADS does not affect ADS operation.) The oil is cooled by circulation through a fuel-oil heat exchanger which is part of the heat sink system. Reservoir capacity is approximately 8 quarts.

EXTERNAL STARTER SYSTEM

An external starting unit is required for ground starts. This may be a compressed air supply, a self-contained gas engine cart, or a multiple air-turbine cart. The output drive gear of either cart connects to a starter gear on the main gearbox at the bottom of the engine. There are no aircraft controls for this system. It is turned on and off by the ground crew in response to instructions from the pilot.

CHEMICAL IGNITION (TEB) SYSTEM

Triethylborane (TEB) is used for starting ignition of main burner and afterburner fuel. Catalytic igniters attached to the afterburner flame holders tend to maintain afterburner operation after initial ignition.

A 600 cc (1-1/4 pint) TEB storage tank is installed on each engine. The tanks are pressurized with nitrogen gas prior to flight to provide inerting and operating pressure. Special handling procedures are required for TEB as it will burn spontaneously with exposure to air above -5°C. The TEB tank is cooled by main burner fuel flow. A rupture disk is provided for each tank which will allow vaporized TEB and nitrogen gas to be discharged through the afterburner section if tank pressure exceeds a safe level. No indication of TEB tank discharge is provided to the flight crew.

At least 16 metered TEB injections can be made with one full tank of TEB. The system is controlled by engine fuel pressure signals while the engine is rotating. Throttle advancement from OFF to IDLE provides main burner ignition, and from Military into the AFTERBURNER range provides afterburner ignition. Main burner ignition occurs almost immediately during an airstart with the engine windmilling. The time required to obtain afterburner ignition is a function of the afterburner fuel manifold fill time, (up to three seconds at sea level and seven seconds at altitude).

Igniter Purge Switch

An IGNITER PURGE toggle switch is located on the pilot's right instrument side panel. When the switch is held in the up position, a solenoid-operated valve supplies fuel-hydraulic system pressure to the chemical ignition system dump valve if the engine is rotating. This allows the TEB to be dumped into the afterburner section. While dumping, engine speed should be above 5000 rpm to avoid afterburner liner damage. The purge switch must be actuated for at least 40 seconds to dump a full load of TEB. At the end of the dump period, the switch should be released and re-cycled to clean out the lines. Electrical power for purging is furnished from the essential dc bus through the IGN PURGE circuit breaker on the pilot's left console.

CENTER INSTRUMENT PANEL - Forward Cockpit

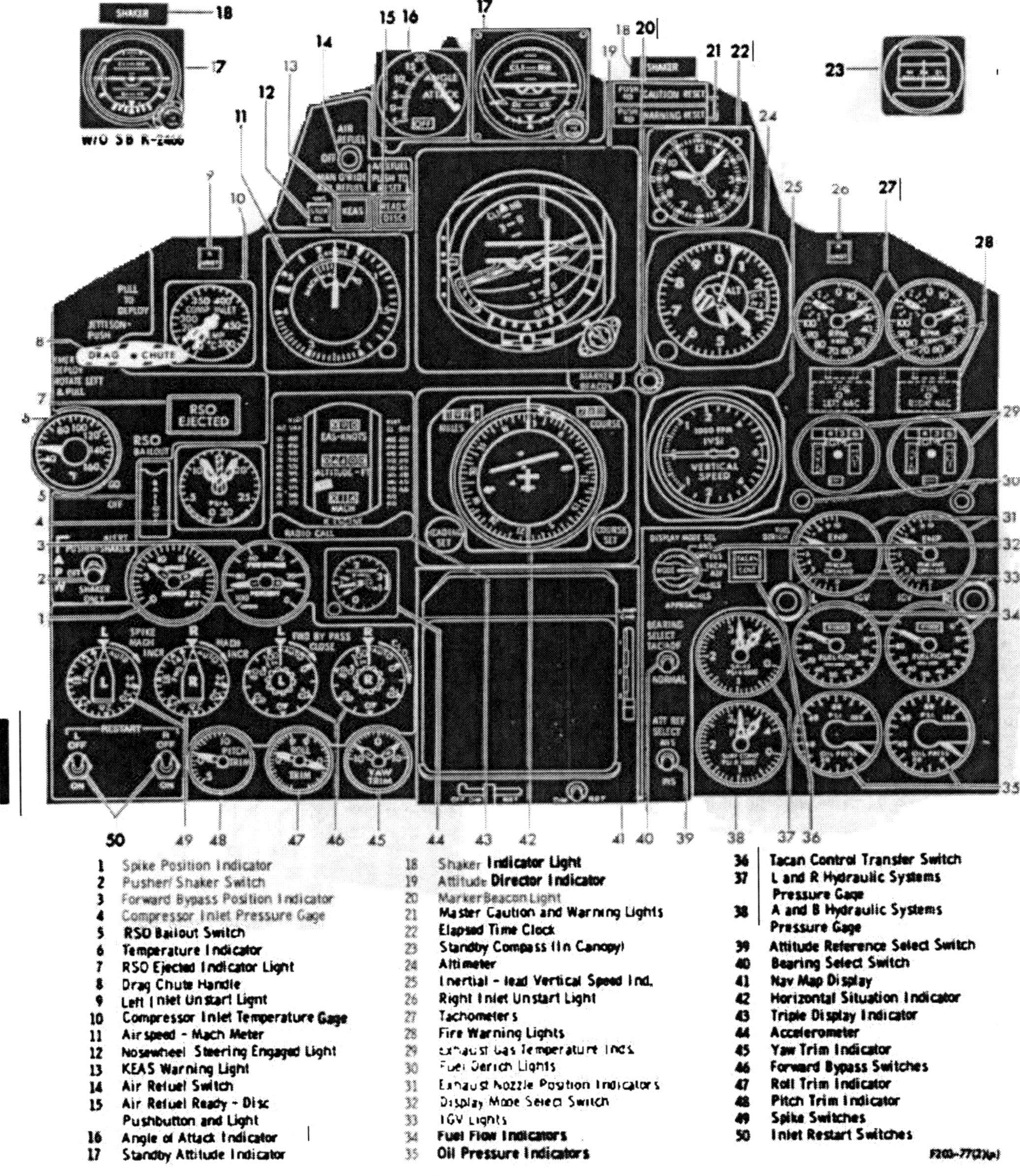

1 Spike Position Indicator
2 Pusher/Shaker Switch
3 Forward Bypass Position Indicator
4 Compressor Inlet Pressure Gage
5 RSO Bailout Switch
6 Temperature Indicator
7 RSO Ejected Indicator Light
8 Drag Chute Handle
9 Left Inlet Unstart Light
10 Compressor Inlet Temperature Gage
11 Airspeed - Mach Meter
12 Nosewheel Steering Engaged Light
13 KEAS Warning Light
14 Air Refuel Switch
15 Air Refuel Ready - Disc Pushbutton and Light
16 Angle of Attack Indicator
17 Standby Attitude Indicator
18 Shaker Indicator Light
19 Attitude Director Indicator
20 Marker Beacon Light
21 Master Caution and Warning Lights
22 Elapsed Time Clock
23 Standby Compass (In Canopy)
24 Altimeter
25 Inertial - lead Vertical Speed Ind.
26 Right Inlet Unstart Light
27 Tachometers
28 Fire Warning Lights
29 Exhaust Gas Temperature Inds.
30 Fuel Derich Lights
31 Exhaust Nozzle Position Indicators
32 Display Mode Select Switch
33 IGV Lights
34 Fuel Flow Indicators
35 Oil Pressure Indicators
36 Tacan Control Transfer Switch
37 L and R Hydraulic Systems Pressure Gage
38 A and B Hydraulic Systems Pressure Gage
39 Attitude Reference Select Switch
40 Bearing Select Switch
41 Nav Map Display
42 Horizontal Situation Indicator
43 Triple Display Indicator
44 Accelerometer
45 Yaw Trim Indicator
46 Forward Bypass Switches
47 Roll Trim Indicator
48 Pitch Trim Indicator
49 Spike Switches
50 Inlet Restart Switches

F203-77(3)(a)

Figure 1-12

INSTRUMENT SIDE PANELS - Forward Cockpit

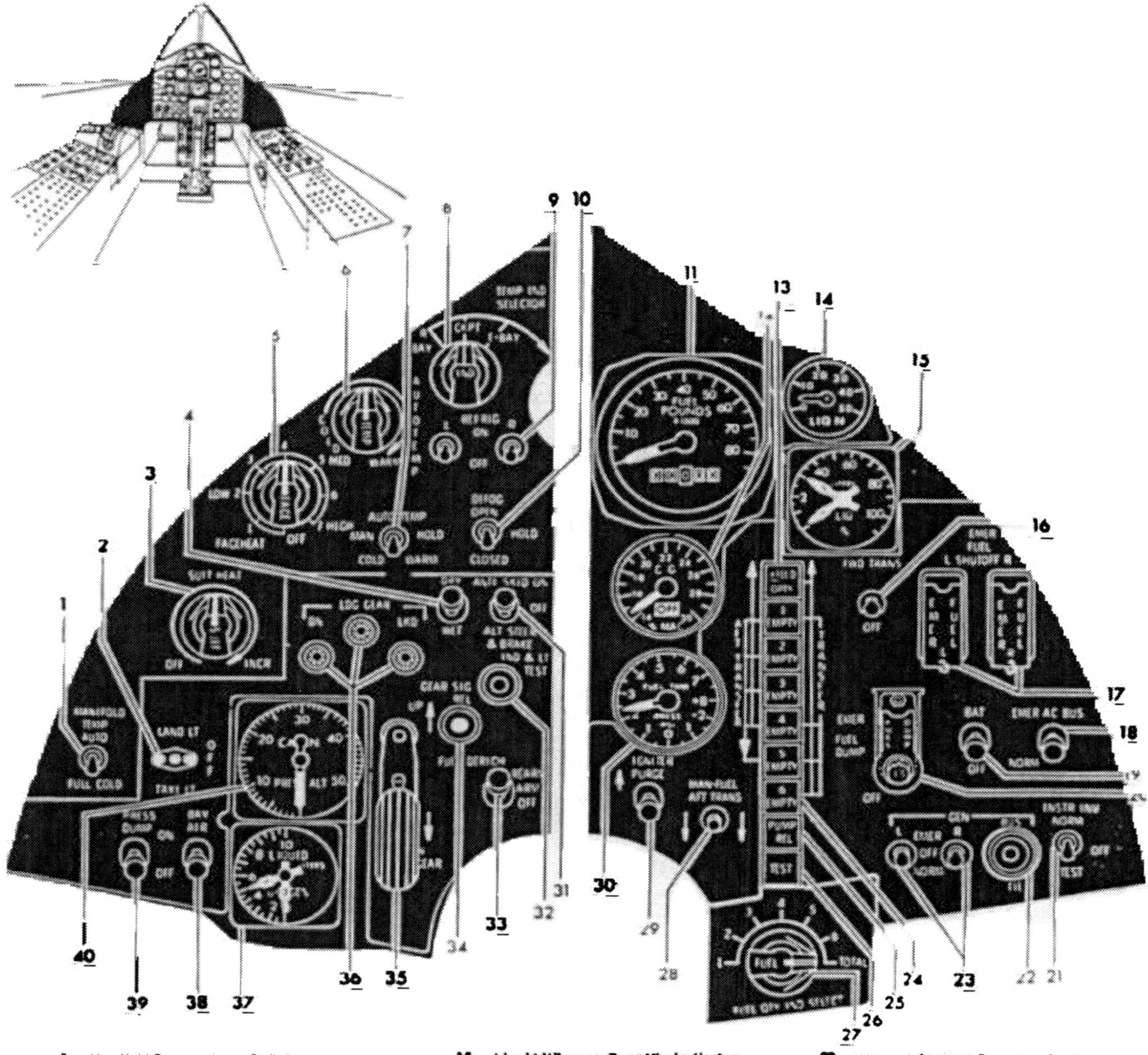

1 Manifold Temperature Switch
2 Landing and Taxi Light Switch
3 Suit Heat Rheostat
4 Wet-Dry Switch
5 Face Heat Rheostat
6 Cockpit Temperature Control
7 Cockpit Temperature Control and O-Ride
8 Temperature Indicator Selector Switch
9 L and R Refrigeration Switches
10 Defog Switch
11 Fuel Quantity Indicator
12 Center Of Gravity Indicator
13 Fuel Crossfeed Switch
14 System 3 Nitrogen Quantity Indicator
15 Liquid Nitrogen Quantity Indicator
16 Fuel Forward Transfer Switch
17 Emergency Fuel Shutoff Switches
18 Emergency AC Bus Switch
19 Battery Switch
20 Fuel Dump Switch
21 Instrument Inverter Switch
22 Generator Bus Tie Switch
23 L and R Generator Switches
24 Fuel Boost Pump Switches
25 Pump Release Switch
26 Fuel Boost Pump Light Test Switch
27 Fuel Quantity Indicator Selector Switch
28 Manual Fuel Aft Transfer Switch
29 Igniter Purge Switch
30 Fuel Tank Pressure Indicator
31 Brake Switch
32 Indicators and Light Test Switch
33 Fuel Derichment Switch
34 Gear Signal Release Switch
35 Landing Gear Lever
36 Landing Gear Indicator Lights
37 Liquid Oxygen Quantity Indicator
38 Bay Air Switch
39 Cockpit Pressure Dump Switch
40 Cabin Altitude Indicator

F203-314(a)

Figure 1-13

LEFT CONSOLE – Forward Cockpit

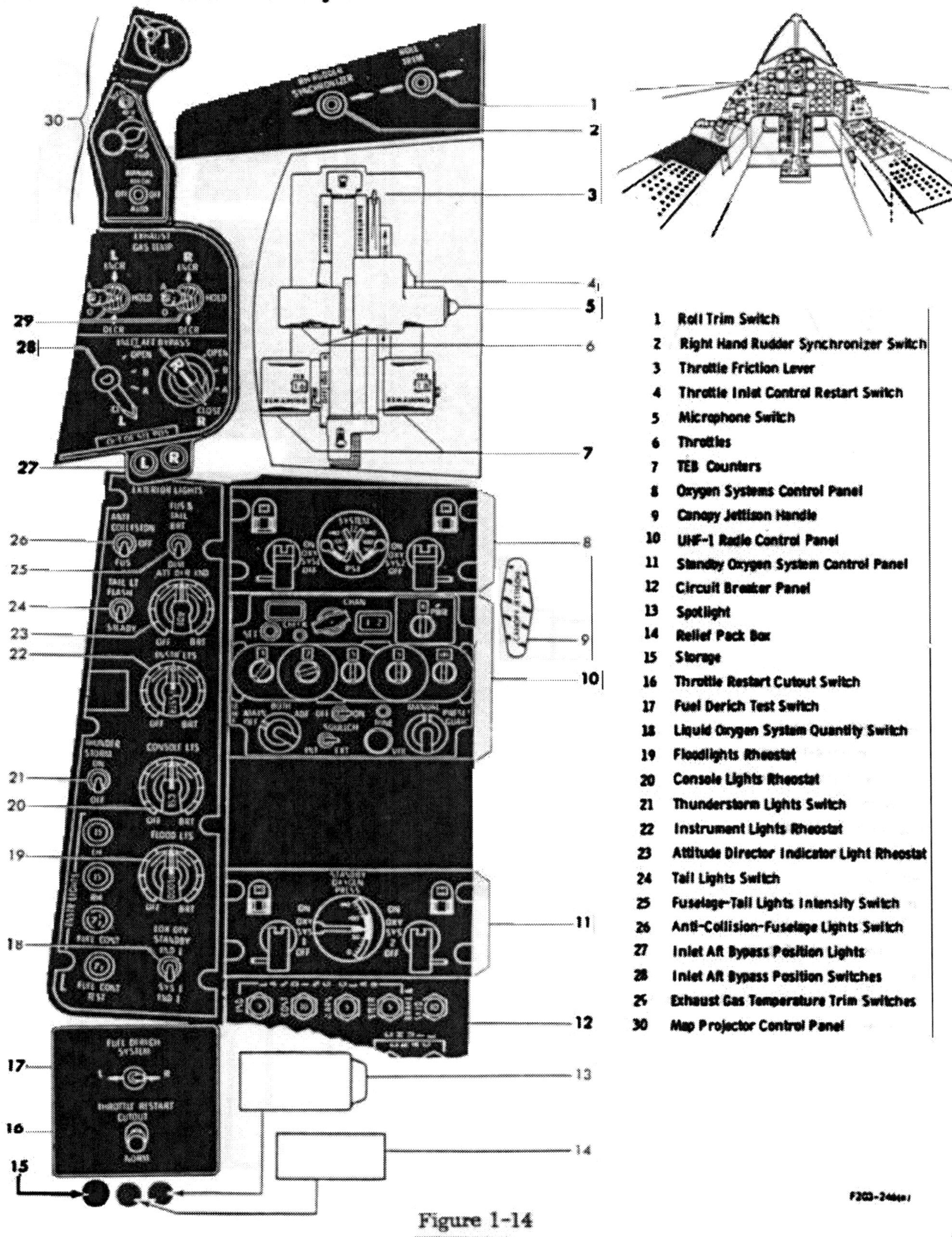
1 Roll Trim Switch
2 Right Hand Rudder Synchronizer Switch
3 Throttle Friction Lever
4 Throttle Inlet Control Restart Switch
5 Microphone Switch
6 Throttles
7 TEB Counters
8 Oxygen Systems Control Panel
9 Canopy Jettison Handle
10 UHF-1 Radio Control Panel
11 Standby Oxygen System Control Panel
12 Circuit Breaker Panel
13 Spotlight
14 Relief Pack Box
15 Storage
16 Throttle Restart Cutout Switch
17 Fuel Derich Test Switch
18 Liquid Oxygen System Quantity Switch
19 Floodlights Rheostat
20 Console Lights Rheostat
21 Thunderstorm Lights Switch
22 Instrument Lights Rheostat
23 Attitude Director Indicator Light Rheostat
24 Tail Lights Switch
25 Fuselage-Tail Lights Intensity Switch
26 Anti-Collision-Fuselage Lights Switch
27 Inlet Aft Bypass Position Lights
28 Inlet Aft Bypass Position Switches
29 Exhaust Gas Temperature Trim Switches
30 Map Projector Control Panel
F203-24(a)

Figure 1-14

RIGHT CONSOLE - Forward Cockpit

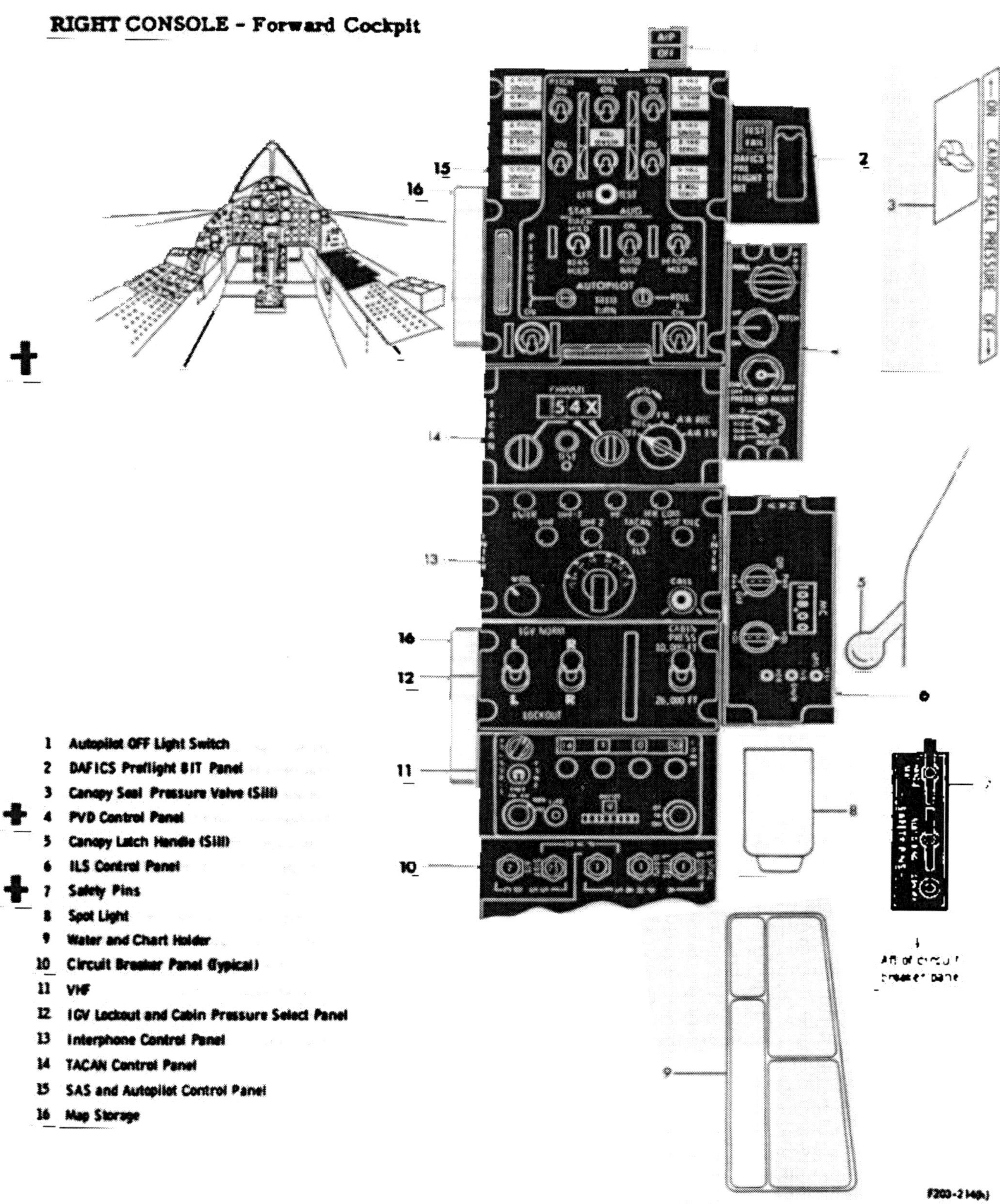

1 Autopilot OFF Light Switch
2 DAFICS Preflight BIT Panel
3 Canopy Seal Pressure Valve (Sill)
4 PVD Control Panel
5 Canopy Latch Handle (Sill)
6 ILS Control Panel
7 Safety Pins
8 Spot Light
9 Water and Chart Holder
10 Circuit Breaker Panel (typical)
11 VHF
12 IGV Lockout and Cabin Pressure Select Panel
13 Interphone Control Panel
14 TACAN Control Panel
15 SAS and Autopilot Control Panel
16 Map Storage

F200-2 14(A)

Figure 1-15

ANNUNCIATOR PANEL - Forward Cockpit

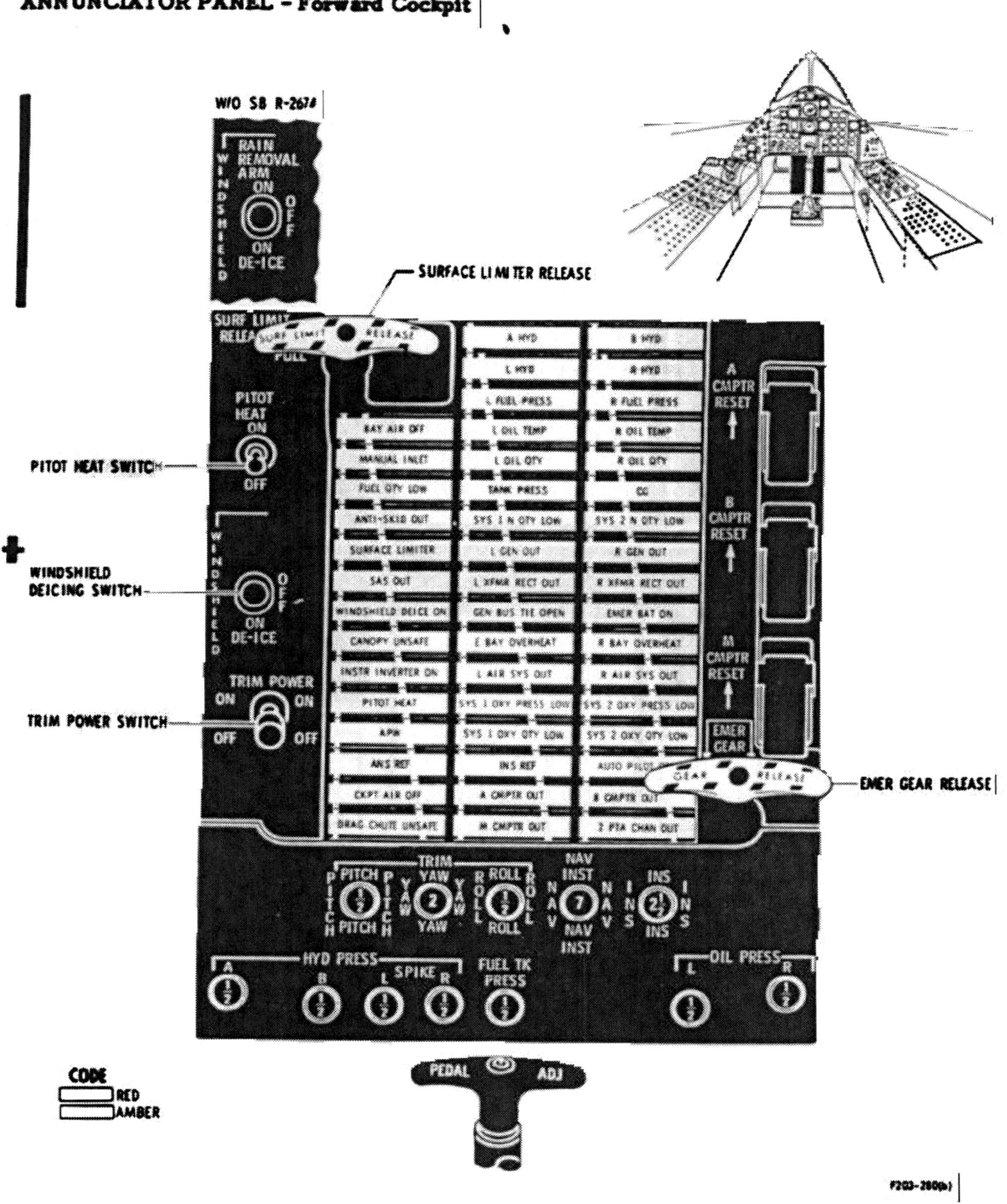

Figure 1-16

INSTRUMENT PANEL - Aft Cockpit

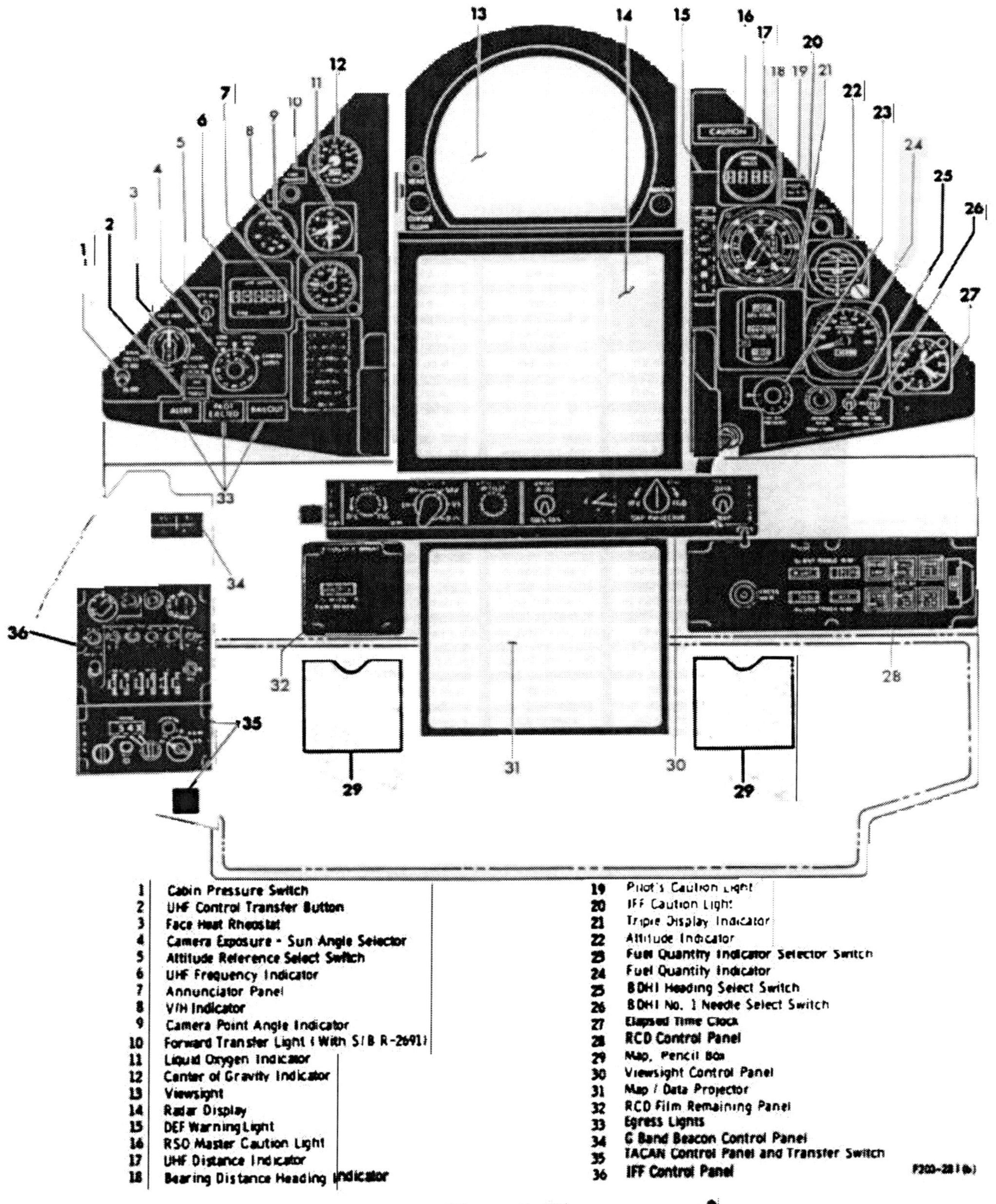

1. Cabin Pressure Switch
2. UHF Control Transfer Button
3. Face Heat Rheostat
4. Camera Exposure - Sun Angle Selector
5. Attitude Reference Select Switch
6. UHF Frequency Indicator
7. Annunciator Panel
8. V/H Indicator
9. Camera Point Angle Indicator
10. Forward Transfer Light (With S/B R-2691)
11. Liquid Oxygen Indicator
12. Center of Gravity Indicator
13. Viewsight
14. Radar Display
15. DEF Warning Light
16. RSO Master Caution Light
17. UHF Distance Indicator
18. Bearing Distance Heading Indicator
19. Pilot's Caution Light
20. IFF Caution Light
21. Triple Display Indicator
22. Attitude Indicator
23. Fuel Quantity Indicator Selector Switch
24. Fuel Quantity Indicator
25. BDHI Heading Select Switch
26. BDHI No. 1 Needle Select Switch
27. Elapsed Time Clock
28. RCD Control Panel
29. Map, Pencil Box
30. Viewsight Control Panel
31. Map / Data Projector
32. RCD Film Remaining Panel
33. Egress Lights
34. C Band Beacon Control Panel
35. TACAN Control Panel and Transfer Switch
36. IFF Control Panel

F300-28 I (b)

LEFT CONSOLE - Aft Cockpit

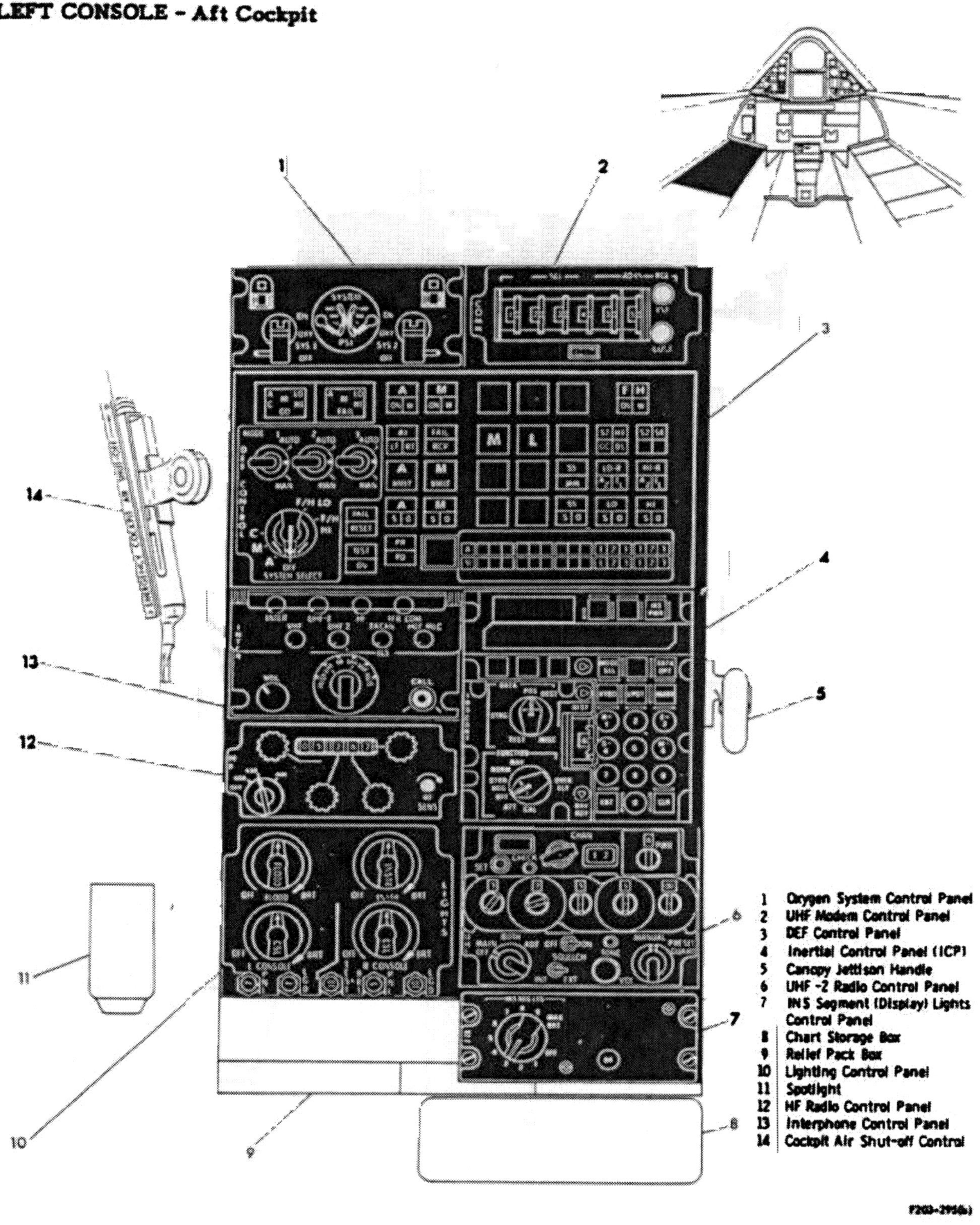
1 Oxygen System Control Panel
2 UHF Modem Control Panel
3 DEF Control Panel
4 Inertial Control Panel (ICP)
5 Canopy Jettison Handle
6 UHF -2 Radio Control Panel
7 INS Segment (Display) Lights Control Panel
8 Chart Storage Box
9 Relief Pack Box
10 Lighting Control Panel
11 Spotlight
12 HF Radio Control Panel
13 Interphone Control Panel
14 Cockpit Air Shut-off Control

Figure 1-18

RIGHT CONSOLE - Aft Cockpit

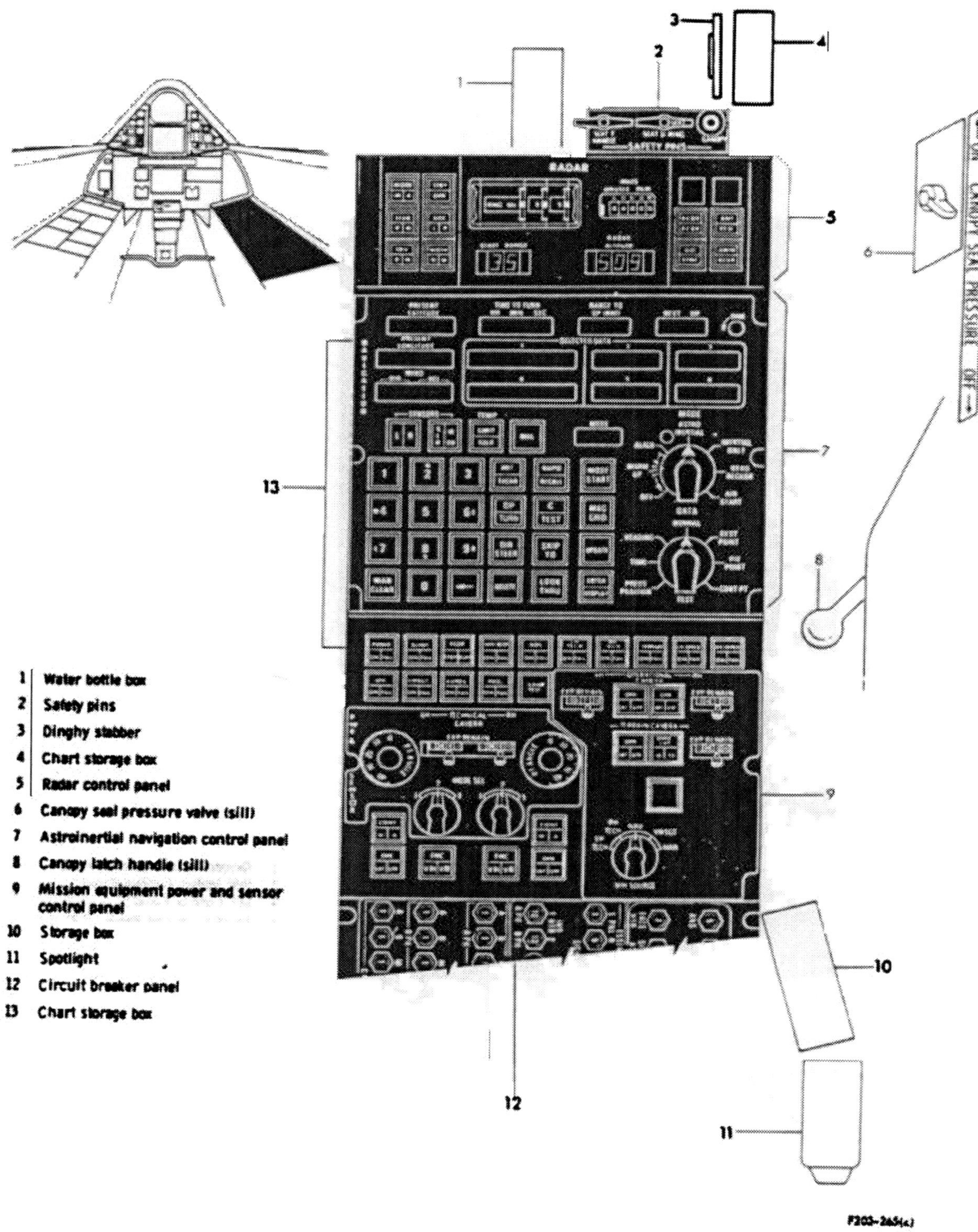

RADAR
SAFETY PINS
ON CANOPY SEAL PRESSURE OFF
1 Water bottle box
2 Safety pins
3 Dinghy stabber
4 Chart storage box
5 Radar control panel
6 Canopy seal pressure valve (sill)
7 Astroinertial navigation control panel
8 Canopy latch handle (sill)
9 Mission equipment power and sensor control panel
10 Storage box
11 Spotlight
12 Circuit breaker panel
13 Chart storage box

Figure 1-19

AIR INLET SYSTEM

The air inlet system in each nacelle includes the cowl structure, a moving spike to provide optimum internal airflow characteristics, variable forward and aft bypass openings, a spike porous centerbody bleed, and an internal shock trap bleed for internal shock wave positioning and boundary layer flow control. Each inlet is canted inboard and downward to align with the local airflow pattern. (See Figures 1-20 and 1-21.)

The forward and aft bypass openings control airflow characteristics within the inlet and mass flow to the engine. Normally, the spike and forward bypass are operated automatically by DAFICS and the aft bypass is scheduled manually. Overriding manual controls are provided for the spike and forward bypass. The forward bypass can be operated manually when the spike is in automatic operation; however, when the spike is controlled manually, the forward bypass is also in manual control. Manual operation of the spike alone while the forward bypass control is in the AUTO position will cause the forward bypass to open 100%.

INLET SPIKES

The spikes are automatically locked in the forward postion for ground operation and for flight below 30,000 feet. They are unlocked above this altitude, but remain in their forward positions until Mach 1.6. During automatic operation above Mach 1.6, the spikes retract approximately 1-5/8 inch per 0.1 Mach number. Total spike motion is approximately 26 inches. This increases the captured stream tube area 112%, from 8.7 square feet to 18.5 square feet. The throat closes down to 4.16 square feet, 54% of the area at Mach 1.6.

During automatic operation, DAFICS schedules spike position as a function of Mach number with biasing for vertical acceleration, angle of attack, and angle of sideslip. See Figures 1-23, 1-25 and 1-27. DAFICS air data (sensed at the nose-mounted pitot mast and measured by the PTAs) are used to compute Mach, angle of attack, and angle of sideslip for automatic spike control.

Spike position can also be set manually by cockpit controls.

The spike centerbody is equipped with small slots which remove spike boundary layer air and prevent flow separation. The air is ducted overboard through the nacelle louvers after passing through the spike and its supporting struts.

INLET FORWARD BYPASS

The forward bypass openings in each inlet provide overboard exhausts for inlet air which is not required by the engine. The openings are a rotating band of ports located a short distance aft of the inlet throat. The air exits overboard through louvers located forward of the louvers for the spike centerbody bleed. Bypass position is automatically modulated by DAFICS to control inlet pressure aft of the normal shock (major internal shock wave) to position this shock properly near the throat.

In automatic operation, the forward bypass remains closed until above Mach 1.4; then it is released to modulate in accordance with DAFICS schedules. In automatic operation, the forward bypass returns to closed when below Mach 1.3. The inlet usually "starts" between Mach 1.6 and Mach 1.8; that is, the normal shock moves from in front of the inlet to a position near the shock trap bleed in the

AIR INLET SYSTEM - Spike Forward

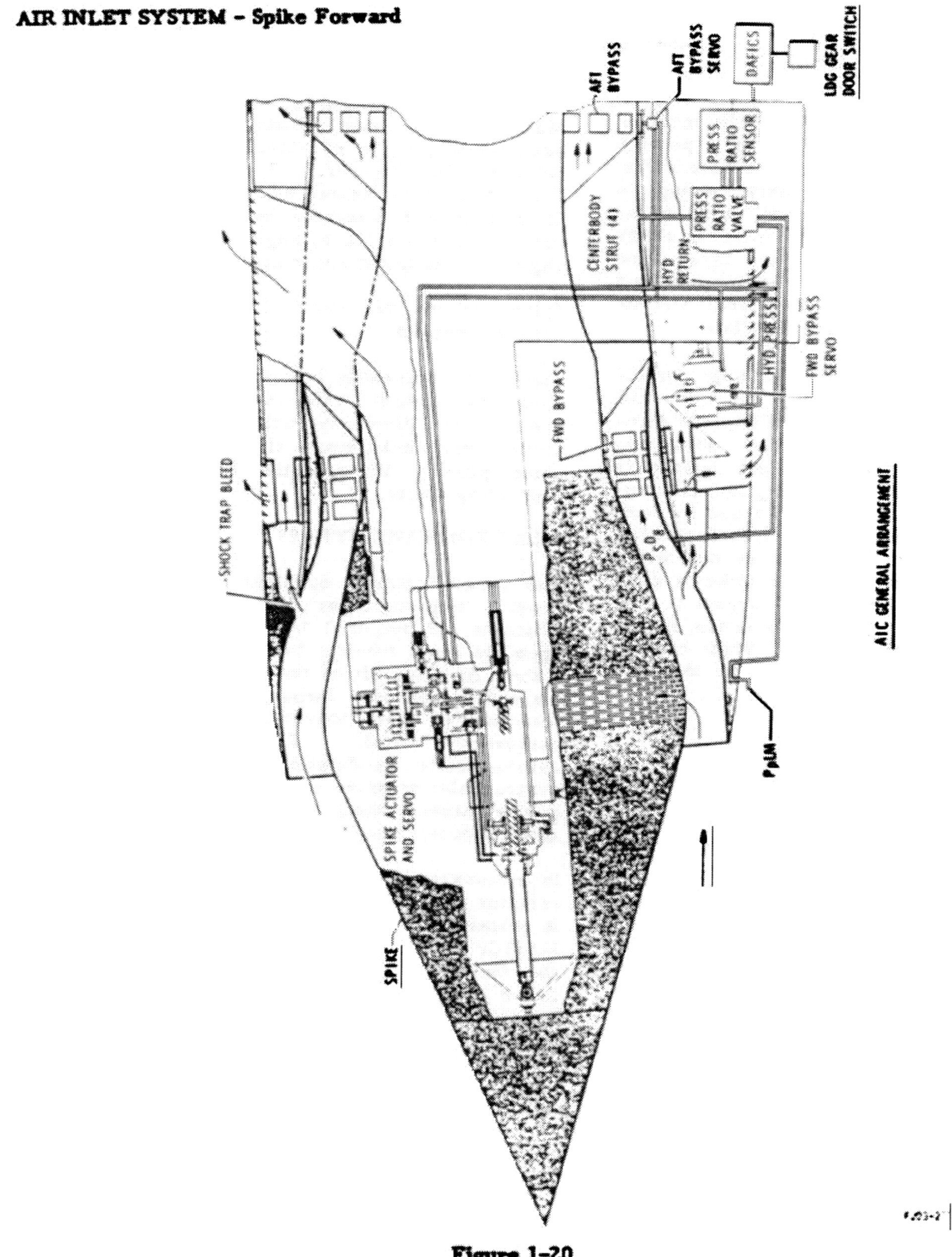

Figure 1-20

AIRFLOW PATTERNS

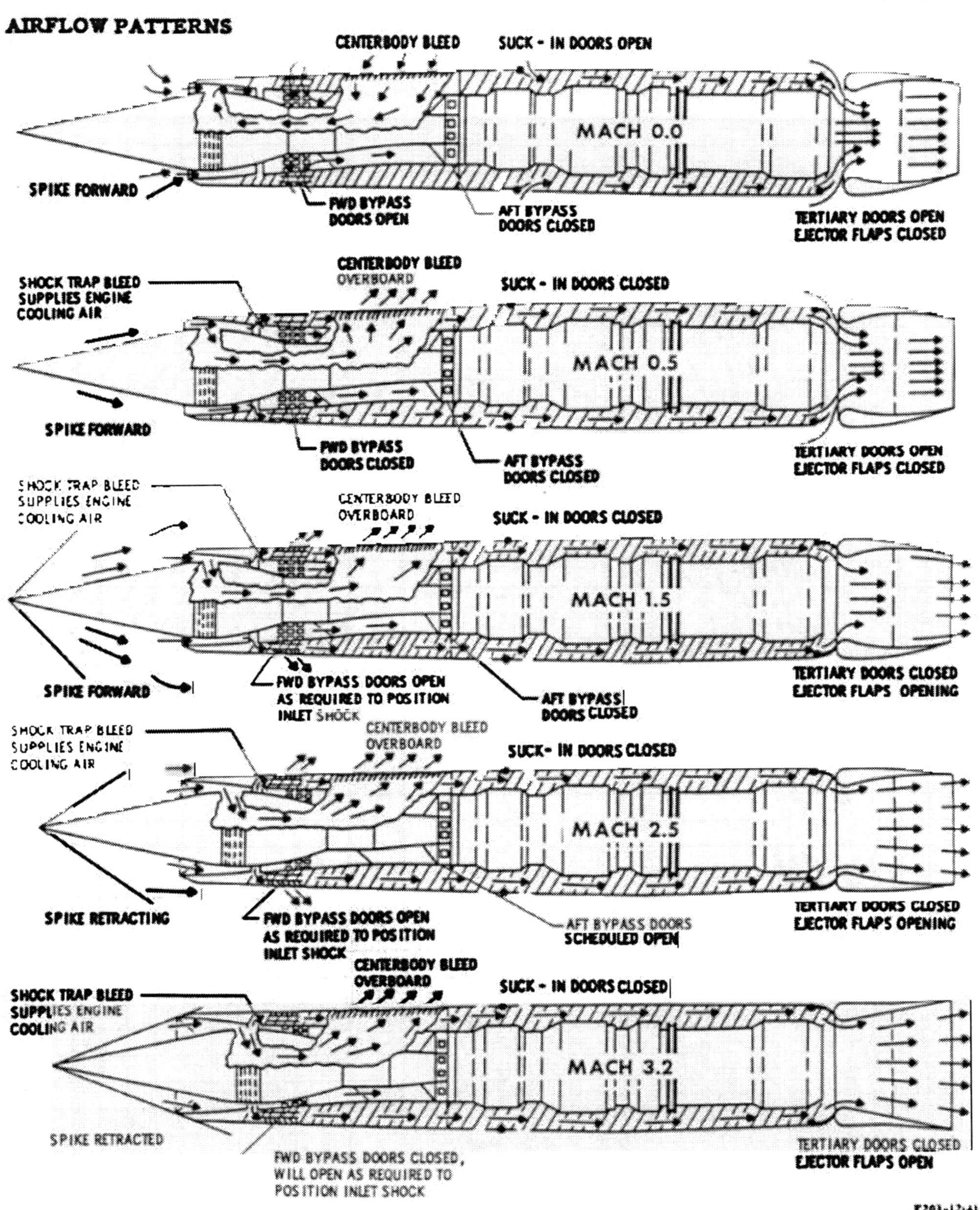

Figure 1-21

BARBER POLE C.I.P.

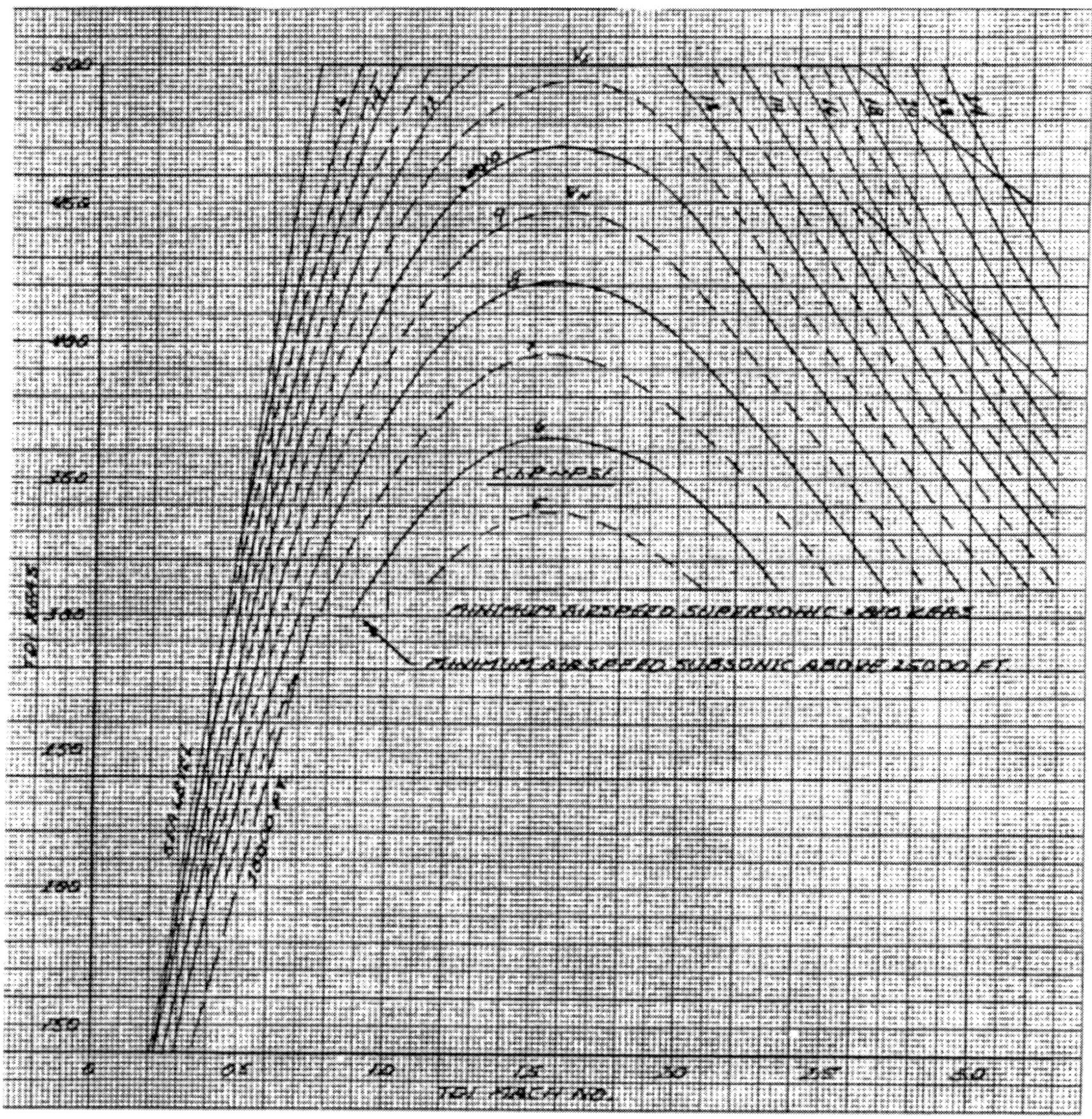

Figure 1-22 (Sheet 1 of 5)

AIR INLET SPIKE SCHEDULE

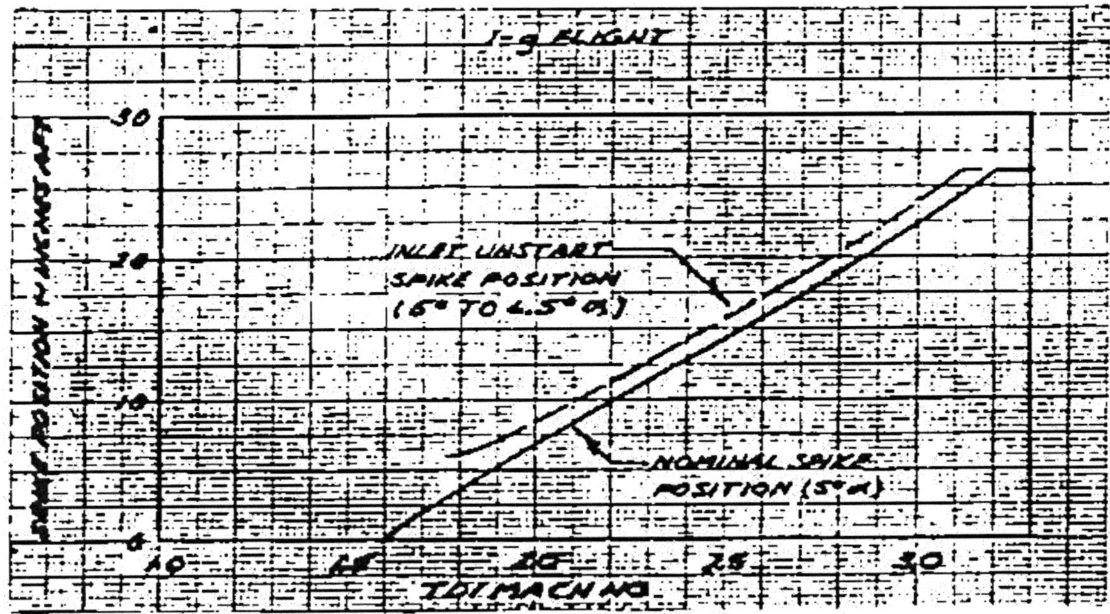

Figure 1-22 (Sheet 2 of 5)

throat. As Mach increases, the forward bypass position must modulate automatically, or by manual control, to keep the normal shock near the throat. When in manual control, the opening can be set in accordance with a Mach schedule or by referring to the position of the opposite inlet (if it is operating automatically). See Figure 3-5.

In automatic control, DAFICS positions the forward bypass to maintain a duct pressure ratio (DPR) schedule as a function of Mach number, with biasing for vertical acceleration, angle of attack, and angle of sideslip. (DPR = P_{sD8}/P_{pLM}, inlet duct static pressure immediately aft of the primary shock wave divided by total free stream pressure or pitot pressure outside the cowl.) (See Figure 1-20.) Four inlet duct static pressure taps measure P_{sD8}. These are located circumferentially on the inlet duct wall, aft of the shock trap bleed. The total free stream pressure, P_{pLM}, is sensed by two pitot probes located on the outside of each nacelle.

When the landing gear is down, the forward bypass is held 100% open by an override signal from a main gear door switch. Control of the forward bypass reverts to the mode selected by the pilot when the gear has retracted.

INLET AFT BYPASS

The inlet aft bypass consists of a rotating band of ports locâted just forward of the engine face. When rotated to an open position, inlet air is allowed to bypass the engine and join the flow from the shock trap bleed. The combined flow passes through the space between the engine case and the nacelle structure, cools these spaces, then augments the exhaust gas flow in the ejector area. See Figure 1-21.

FORWARD BYPASS POSITION - Mach 2.20 to 2.80

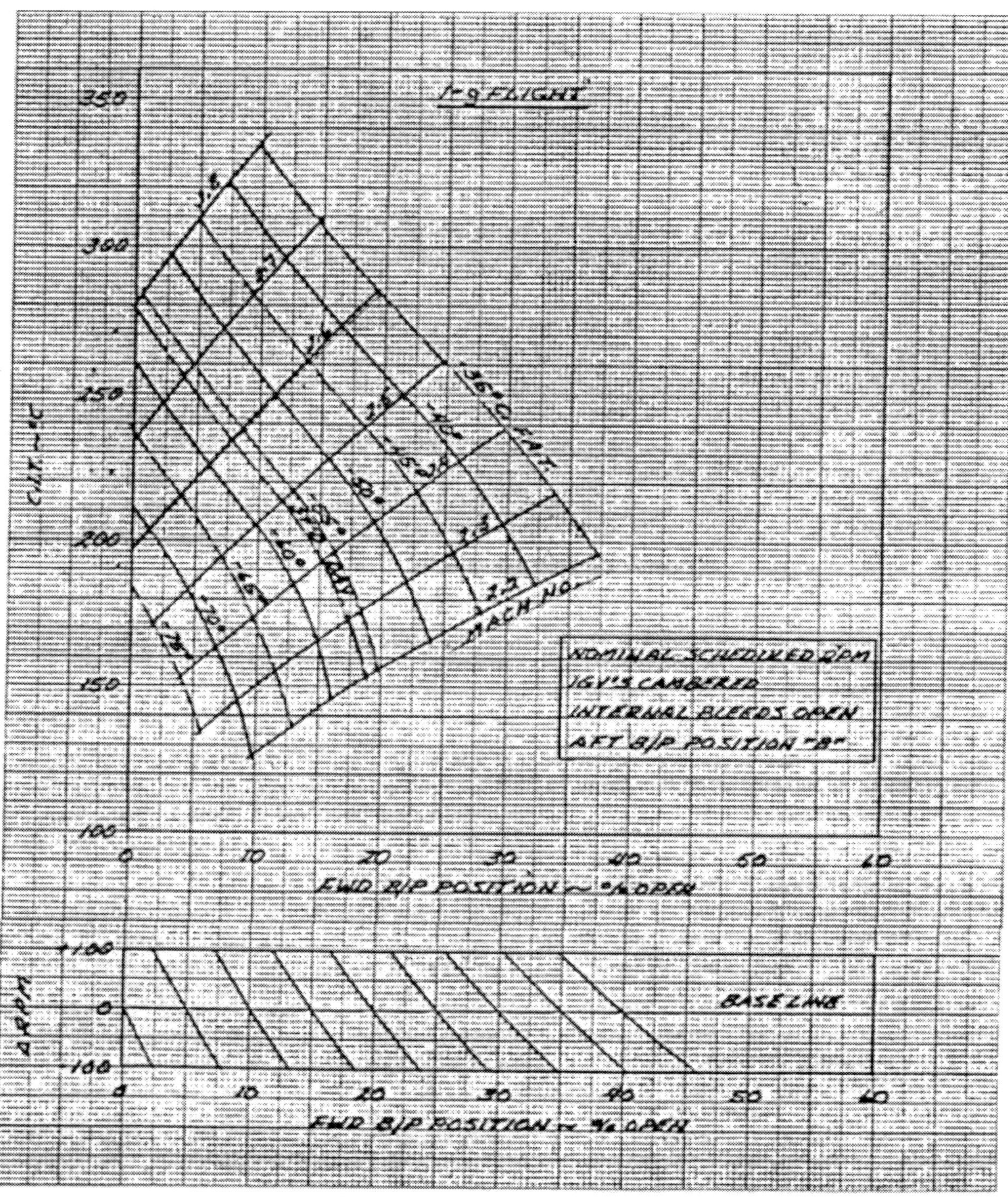

Figure 1-22 (Sheet 3 of 5)

FORWARD BYPASS POSITION - Mach 2.20 to 3.20

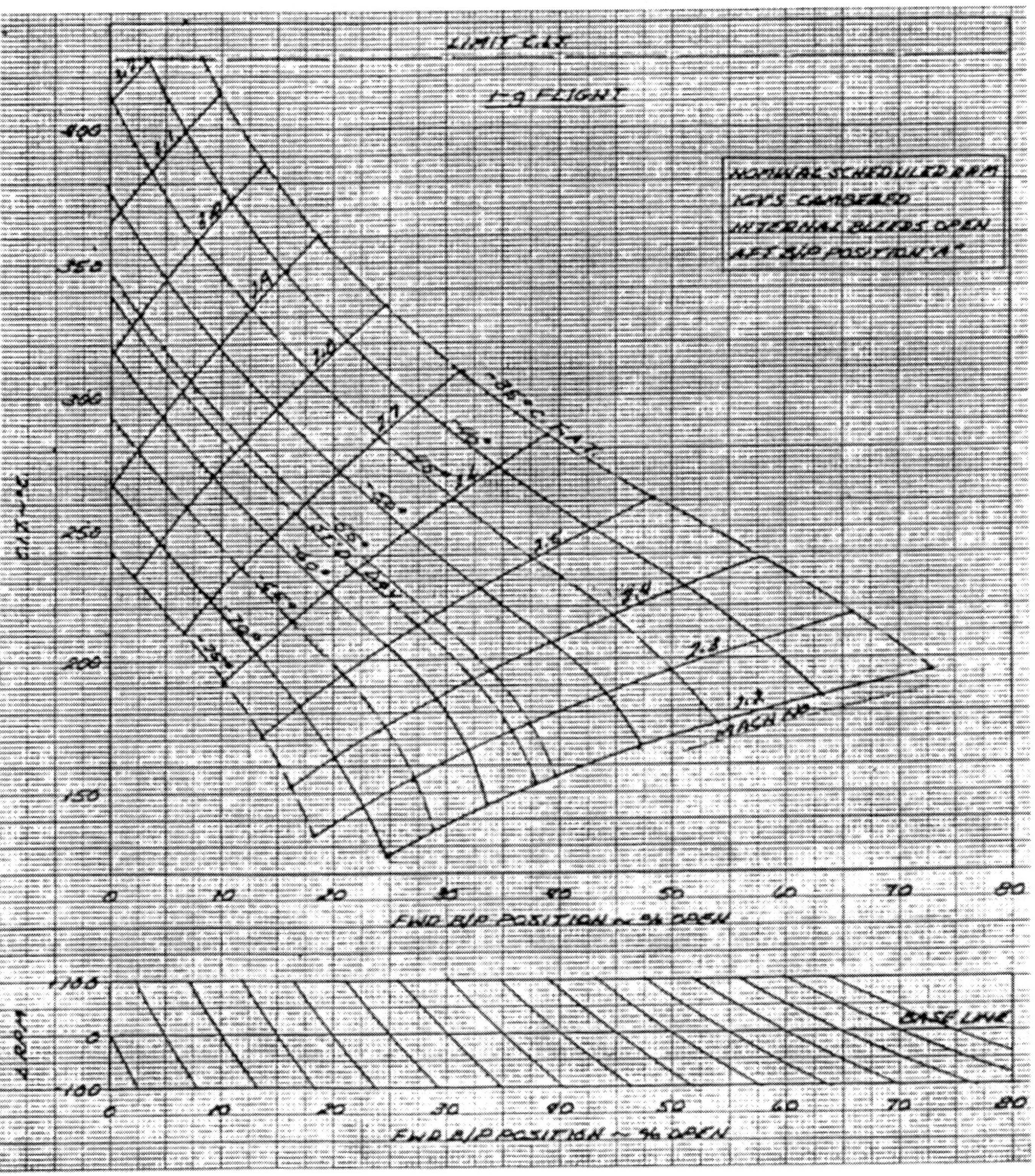

Figure 1-22 (Sheet 4 of 5)

FORWARD BYPASS POSITION - Mach 2.9

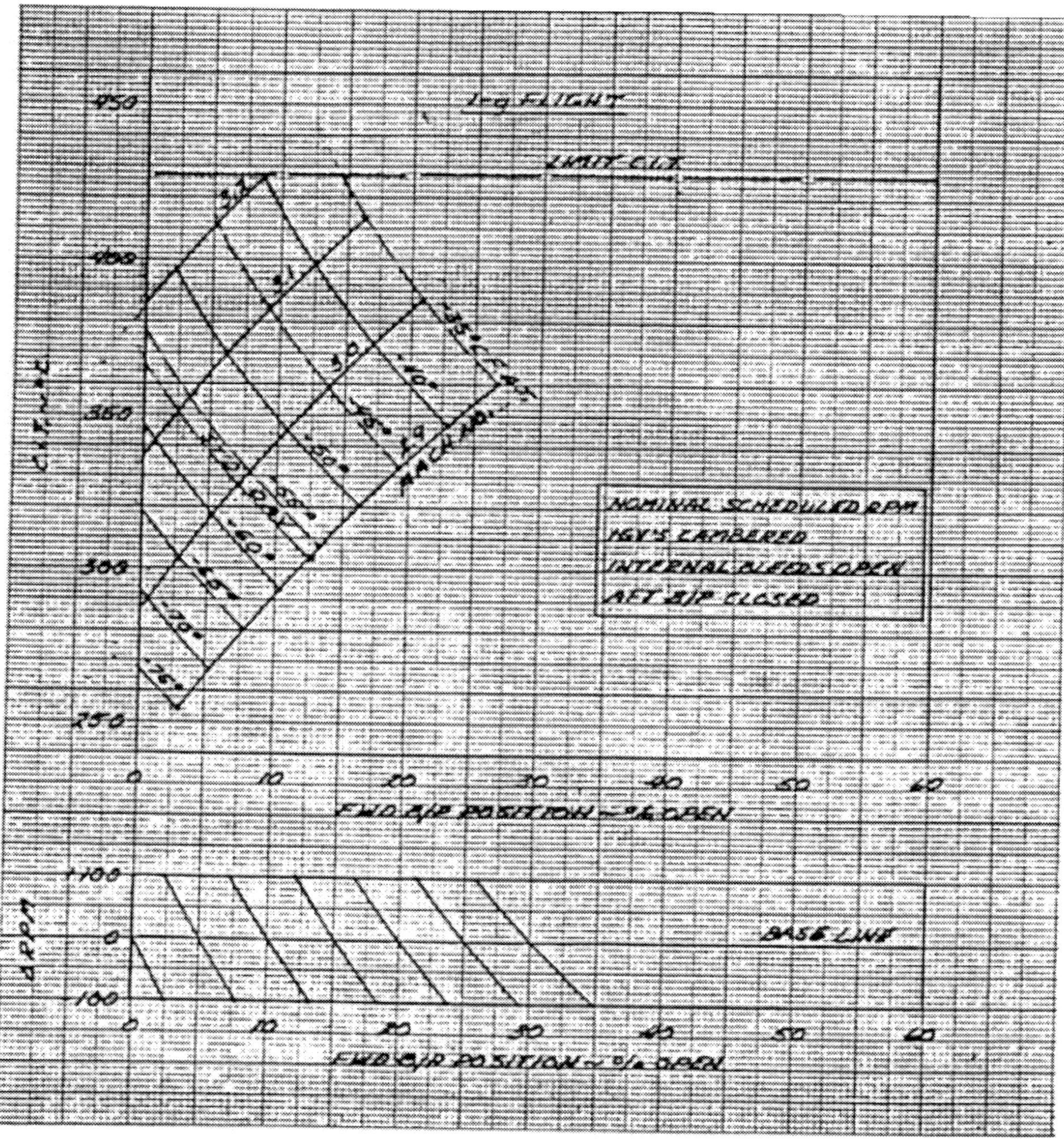

Figure 1-22 (Sheet 5 of 5)

The position of the aft bypass in each inlet is controlled by rotary selector switches in the pilot's cockpit. When the gear is extended, the aft bypass ports are held closed by an override signal from the nose gear downlock switch.

INLET CONTROL PARAMETERS

When the aircraft is supersonic, inlet airflow is controlled by positioning the spike and forward bypass so that the locations of shock waves ahead of the inlet and at the inlet throat produce maximum practical pressure recovery at the engine face, and supply the proper amount of air to the engine. Refer to spike position vs Mach number and forward bypass position vs Mach number and CIT, Figure 1-22. Manual operation of the aft bypass provides for those conditions where additional bypass area is required or when a reduction in forward bypass flow is desired. The forward and aft bypass openings and the spikes for the left and right inlets are operated by the L and R hydraulic systems, respectively. The inlet control also includes a shock expulsion sensor and restart system.

In automatic control, DAFICS schedules the spike and forward bypass positions as a function of Mach, with biasing for load factor, angle of attack, and angle of sideslip. Bias does not override a manually operated spike or forward bypass.

Load Factor Bias

The g-bias function in DAFICS schedules the spike forward and the forward bypass more open when load factor is greater than 1.5 g or less than 0.7 g during automatic inlet operation. The biasing noticeably decreases CIP and reduces the possiblity of unstarts. Figures 1-23 and 1-24 illustrate spike and door biasing, respectively.

SPIKE POSITION BIAS DUE TO LOAD FACTOR

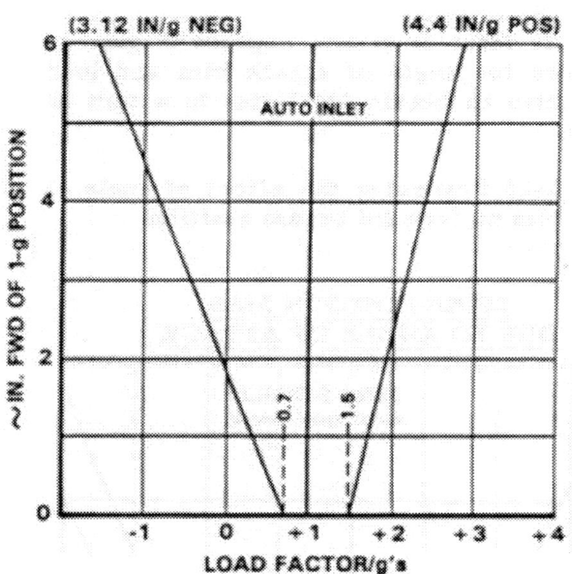

Figure 1-23

FWD BYPASS POSITION BIAS DUE TO LOAD FACTOR AND DPR

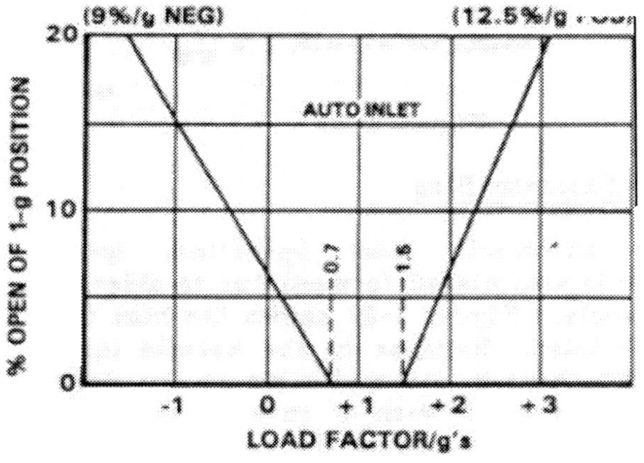

Figure 1-24

Angle of Attack Bias

During automatic inlet operation above Mach 1.6, spikes are biased forward when angle of attack deviates from five degrees (Figure 1-25). Add the angle of attack bias and load factor bias to obtain total bias in a turn or pull-up.

Figure 1-26 illustrates the effect of angle of attack bias on forward bypass position.

SPIKE POSITION BIAS DUE TO ANGLE OF ATTACK

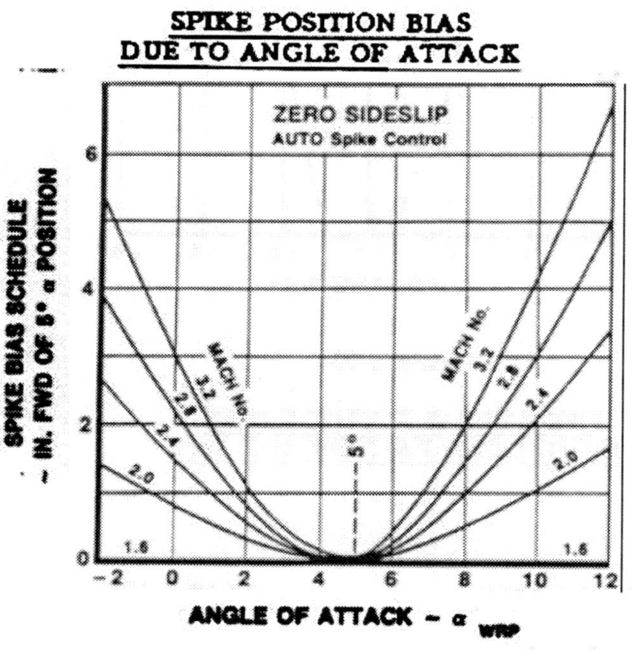

Figure 1-25

Angle of Sideslip Bias

During automatic inlet operation, spike position is also biased forward due to sideslip (yaw) angle. Figure 1-27 shows the bias for the left inlet. Because of the nacelle inlet cant, the chart is not applicable to the right inlet unless read with a reverse sign for sideslip angle, i.e., enter the chart with $-\beta$ for a nose-left condition.

EFFECT OF ANGLE OF ATTACK ON FORWARD BYPASS POSITION

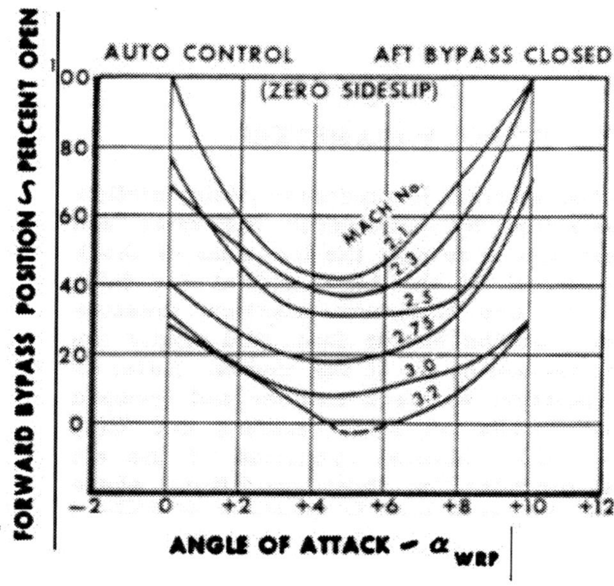

Figure 1-26

Sideslip angles of up to one degree are common. The result can be a "split" spike position indication. Because the spike control positioning tolerance is $\pm$ 0.2 inches, this could result in a split indication of up to 0.9 inches with a sideslip angle of 1° (with no indicator instrument error).

In automatic operation, the duct pressure ratio (DPR) is also biased lower due to sideslip angle. Figure 1-28 shows forward bypass position changes that result from spike and DPR biasing when flying with a sideslip at typical cruise speeds. Note that a 1° sideslip can result in a 3% difference between the two inlets. An additional 3% difference is possible due to the tolerance in DPR scheduling.

SPIKE POSITION BIAS DUE TO SIDESLIP (YAW)

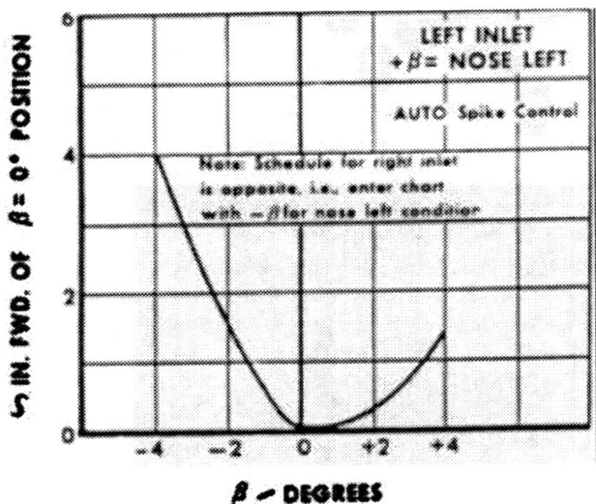

Figure 1-27

EFFECT OF SIDESLIP (YAW) ON FORWARD BYPASS POSITION

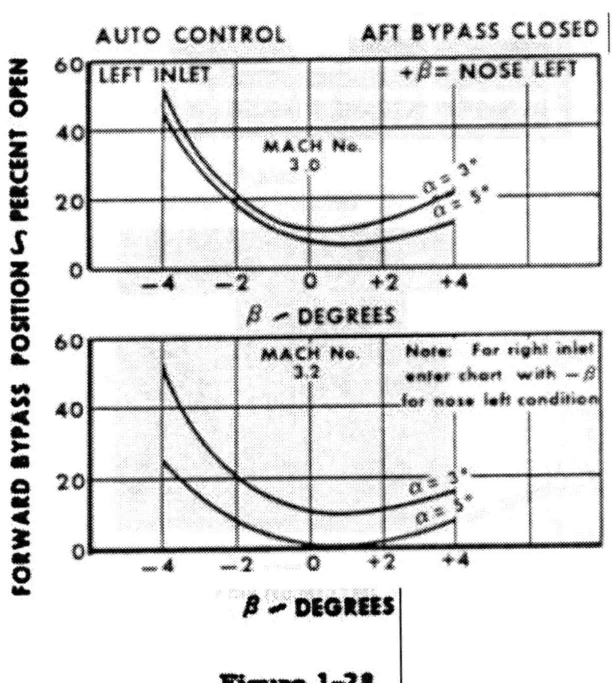

Figure 1-28

Effect of RPM on Forward Bypass Position

Engine speed also affects bypass position, as rpm directly affects engine airflow requirements. This amounts to approximately 4% increase in bypass opening per 100 rpm decrease in engine speed when at typical cruise conditions and with the bypass nearly closed. The influence of rpm change increases significantly as the forward bypass condition approaches 50% open, where a 10% difference in forward bypass indication per 100 rpm change is not unusual.

Malfunction Biases

During automatic inlet operation, the forward bypass door duct pressure ratio (DPR) schedule is biased slightly lower for: loss of DAFICS computer(s), loss of reliable M PTA air data for sideslip computations, or unstart(s).

The DPR schedules of both inlets are biased 40 mpr (milli-pressure ratios) lower if a DAFICS A, B, or M computer fails (appropriate A, B, or M CMPTR OUT caution light on the pilot's annunciator panel illuminated).

If DAFICS is not receiving sideslip angle data from the M PTA, or the sideslip angle data from the M PTA is unreasonably high (as determined by comparison with the yaw SAS lateral accelerometers), the angle of sideslip used for automatic inlet operation computations is zero and the DPR schedules for both inlets are biased 20 mpr lower.

Each time the shock expulsion sensor (SES) detects an unstart (as indicated by illumination of the L or R UNST light), the DPR schedule for the inlet that unstarted is reduced 10 mpr. The DPR schedule for the other inlet is not changed.

The total reduction in DPR schedule on each inlet will not exceed 40 mpr regardless of how many malfunctions occur. For comparison, a 40 mpr differential is the range of adjustment provided to maintenance for scheduling inlet DPR.

AIR INLET SYSTEM CONTROLS

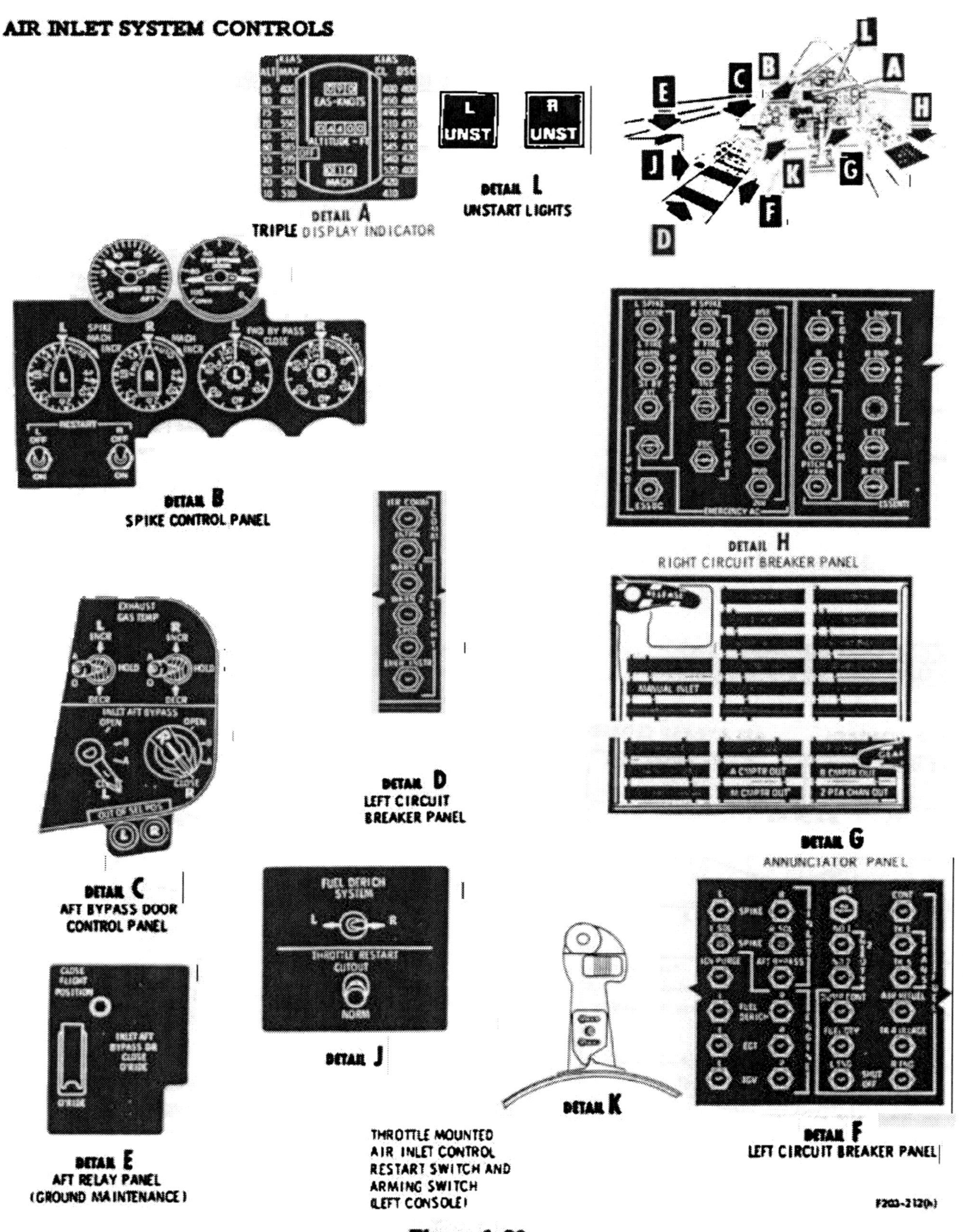

Figure 1-29

AUTOMATIC RESTART

The inlet control includes a shock expulsion sensor (SES) and an auto restart feature which operate automatically above Mach 1.6 (normally, SES is effective above Mach 2.0). If the inlet normal shock wave is expelled, the SES overrides the auto spike and forward bypass schedule (for both inlets above Mach 2.3). This "cross tie" feature keeps the other inlet from unstarting, reduces asymmetric thrust, and minimizes undesirable sideslip angles. During the automatic restart cycle, the forward bypasses open fully and the spikes move forward as much as 15 inches. Spike retraction starts 3.75 seconds after the expulsion is sensed. After the spikes return to their scheduled position, the forward bypasses return to automatic operation. In automatic operation, the forward bypass door duct pressure ratio schedule of the inlet that unstarted is biased slightly lower after each unstart, for up to 4 unstarts. The "cross tie" function is locked out below Mach 2.3 so that only the inlet which has unstarted will perform the automatic restart cycle.

Compressor inlet pressure (CIP) is the SES reference. The SES system actuates when a momentary CIP decrease of more than 23% occurs. Rapid CIP decrease is characteristic of an unstart; however, the SES can also be actuated by compressor stalls if CIP decreases rapidly more than 23%. Successive unstarts or compressor stalls may cause the SES reference pressure (CIP) to decay. The SES cannot operate if the momentary pressure drop is less than 23% of the existing reference pressure. In this event, manual restart will be necessary.

If an inlet remains unstarted (i.e, the shock is not recaptured) after automatic restart, the SES will not actuate since CIP will be low

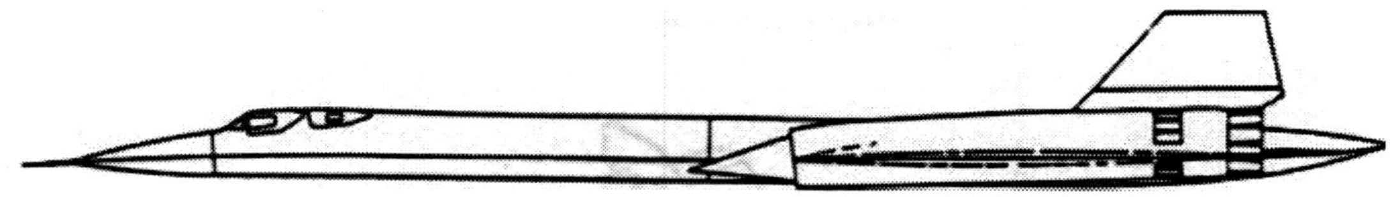

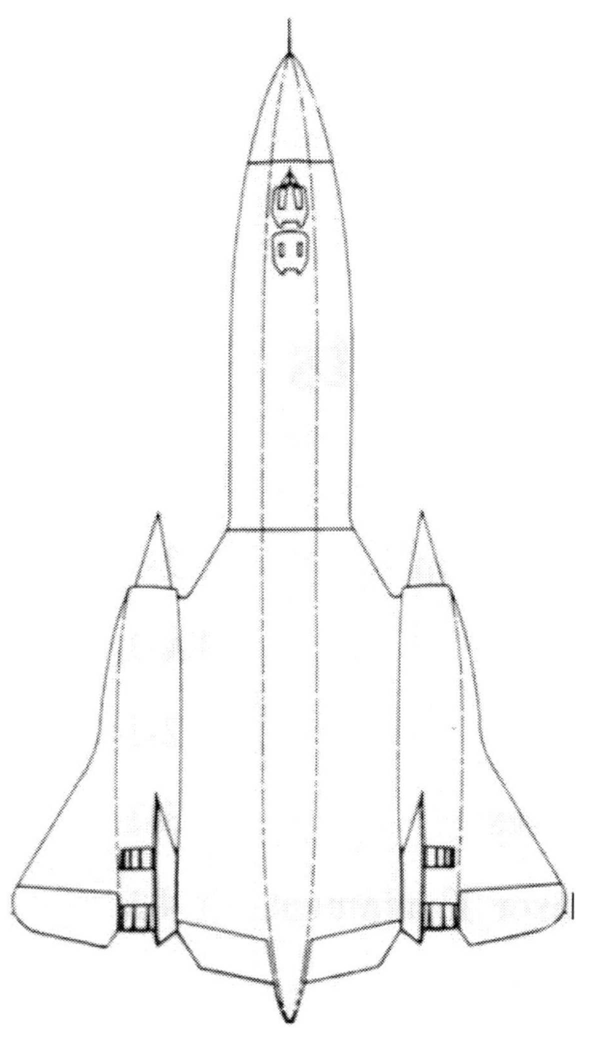

THIS PAGE
INTENTIONALLY
LEFT BLANK
OR
STILL
CLASSIFIED.

and will not change. In this event, manual restart will be necessary.

The automatic restart and cross tie features do not override any manually positioned inlet control. If an unstart occurs on the inlet in manual spike operation, neither that inlet nor the opposite inlet will respond. With only one forward bypass door in manual control, an unstart on either inlet will result in automatic actuation of both spikes and the door in automatic control. If the manually controlled door is not opened manually, an unstart and possible engine stall can be expected on that inlet, even if the opposite inlet unstarts. Manual operation of the air inlet restart switch overrides the SES automatic restart cycle and manual position settings of that inlet.

INLET CONTROLS, INSTRUMENTS AND INDICATOR LIGHTS

Spike Switches

Individual rotary control knobs for each spike are on the left side of the pilot's instrument panel. Each switch has two detent positions, AUTO and FWD, indexed to the instrument panel lubber line engraved at the 12 o'clock position. The knobs are graduated counterclockwise in Mach from 1.4 through 3.2 to represent the manual override range of control. With the knob in the AUTO detent, the corresponding spike is positioned automatically. The FWD setting moves the spike to its forward stop, bypassing the automatic circuitry. The manual override settings schedule the spike to any desired position from full forward at the Mach 1.4 - 1.6 setting to full aft (26-inch position) at the Mach 3.2 setting. There is no automatic bias for load factor or angles of attack and sideslip during manual control.

NOTE

Inlet restart switches override all other inlet controls (the forward bypass door opens and the spike is driven forward).

The emergency ac bus provides power for manual and automatic control of the spikes and forward bypass doors through the L and R SPIKE AND DOOR circuit breakers on the pilot's right console. Power for the automatic control circuit relays is furnished from the essential dc bus through the L and R SPIKE circuit breakers on the pilot's left console. Power for spike unlock above 30,000 feet and Mach 1.4 is furnished from the essential dc bus through the L and R SPIKE SOL circuit breakers on the pilot's left console.

Spike Position Indicator

A dual position indicator for the left and right spikes is located on the left side of the pilot's instrument panel. This instrument is marked in one inch increments from 0 (forward) through 26 (full aft). Instrument power is provided from the essential dc bus through the R SPIKE circuit breaker on the pilot's left console.

Inlet Forward Bypass Switches

Two rotary control switches, one for each forward bypass system, are provided on the pilot's instrument panel. Each switch has two detent positions, AUTO and OPEN, indexed to the instrument panel lubber line engraved at the 12 o'clock position. The knobs may be rotated clockwise from the OPEN position to schedule the bypass between full open and full closed. The 100% OPEN detent positions the bypass full open. The manual settings allow the pilot to schedule forward bypass position, overriding the automatic system. Power for the circuit is furnished by the essential dc bus through the L and R SPIKE circuit breakers on the pilot's left console. Power for inlet forward bypass unlock above Mach 1.4 is furnished from the essential dc bus through the L and R SPIKE SOL circuit breakers on the pilot's left console.

MANUAL INLET FORWARD BYPASS BIAS VS. SPIKE KNOB MACH NUMBER

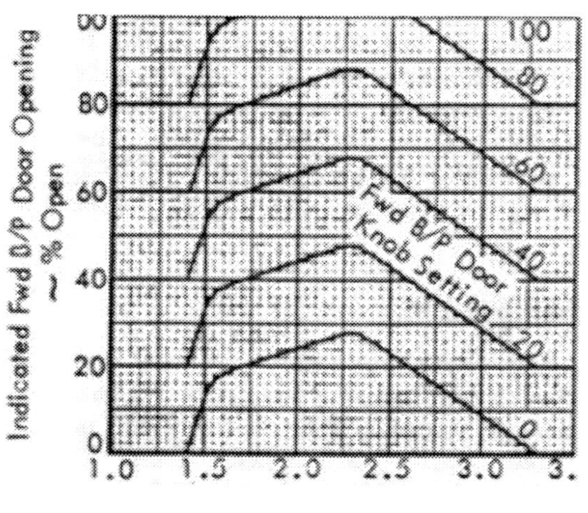

Figure 1-30

Forward Bypass Position Indicator

A dual position indicator for the left and right forward bypass is located on the left side of the pilot's instrument panel. The indicator shows the positions of the forward bypass from 0% (fully closed) to 100% (full open). The dial is marked in 20% increments, with additional marks at each 10% position. Instrument power is provided from the essential dc bus through the L SPIKE circuit breaker on the pilot's left console.

Forward Bypass Bias for Manual Spike Operation

A manual forward bypass bias schedule is applied to the forward bypass whenever the spike and the forward bypass are operated manually. This schedule prevents inadvertent closure of the forward bypass into a range where unstarts are likely. The indicator will show the forward bypass as being more open than the setting selected by the forward bypass switch. With the spike in manual:

a. With the forward bypass switch in AUTO or 100% position - The forward bypass will open fully.

b. With the forward bypass switch in any manual position, other than 100%, the forward bypass will close as far as the bias schedule will permit.

Inlet Aft Bypass Switches and Indicators

Two rotary INLET AFT BYPASS switches, one for each aft bypass system, are located left of the throttle quadrant. The switch positions are: CLOSE, A (15% open), B (50% open), and OPEN (100%). Two rotate-to-dim,amber lights, labeled L and R, are located below the switches. The lights illuminate when an aft bypass position and switch setting do not correspond and extinguish when the aft bypass reaches the setting selected by the aft bypass switches. Approximately 5 seconds is required for the aft bypass ring to move from full closed to full open. The aft bypass actuator control circuits and indicator lights are powered by the essential dc bus through the AFT BYPASS circuit breaker on the pilot's left console.

NOTE

Occasional momentary illumination of either or both aft bypass out-of-position lights may occur with any switch position selected. This is caused by door actuator drift away from the commanded position limit switch. The door is then commanded back to the selected position and the door out-of-position light extinguishes. Such uncommanded movement of the aft bypass doors are generally too small to be detected by changes in auto inlet position indications, and will not affect aircraft performance.

A ground maintenance switch, labeled INLET AFT BYPASS DR CLOSE O'RIDE, is located on the left console relay panel in the forward cockpit. It overrides the aft bypass close signal when the landing gear is extended.

Inlet Restart Switches

Two inlet restart switches are installed on the lower left edge of the pilot's instrument panel. The two positions are off (center), and RESTART ON (down). When the switch is moved to RESTART ON, the respective inlet spike is moved forward and the forward bypass is positioned full open. Selection of RESTART ON overrides all other spike and forward bypass control settings for that inlet. Power for the respective inlet restart switches is furnished by the essential dc bus through the L and R SPIKE SOL circuit breakers on the pilot's left console.

NOTE

The forward bypass door linear voltage differential transformer (LVDT) is not temperature compensated. With RESTART ON, the forward bypass is 100% open but the gauge may indicate as low as 80% open.

Throttle Restart Switch

A throttle restart switch, located on the inboard side of the right throttle, restarts both inlets simultaneously. The three-position slide switch can be moved fore and aft by the pilot's left thumb. In the forward (off) position, the inlet spikes and forward bypass are controlled by settings of the inlet spike and forward bypass control switches, or by the individual inlet restart switches. The center position opens the forward bypass of both inlets but does not affect the position of the spikes. The aft position causes the forward bypass to open and the spikes to move full forward in both inlets. This switch is operative only when the throttle restart cutout switch is in NORM. Power for throttle restart switch control of the respective spike and forward bypass positions is from the essential dc bus through the L and R SPIKE SOL circuit breakers on the pilot's left console.

NOTE

The forward bypass door linear voltage differential transformer (LVDT) is not temperature compensated. With the throttle restart switch in the center or aft position, the forward bypass is 100% open but the gauge may indicate as low as 80% open.

Throttle Restart Cutout Switch

The THROTTLE RESTART CUTOUT switch is on the pilot's left console. It arms the circuit to the throttle restart switch. In the NORM position, the throttle restart switch is operable. If the throttle restart switch malfunctions, the throttle restart cutout switch should be placed in CUTOUT.

Inlet Unstart Lights

The inlet unstart lights, labeled L UNST and R UNST, are located on the pilot's instrument panel. The L UNST or R UNST light illuminates when the shock expulsion sensor (SES) detects an unstart in the corresponding inlet and extinguishes after 3.75 seconds when the SES resets. Power for the lights is furnished by the essential dc bus through the L and R SPIKE circuit breakers on the pilot's left console.

WARNING

Near design Mach, the inlet may **remain unstarted after the L or R UNST light extinguishes.**

NOTE

For multiple unstarts or an inlet which has not restarted, the unstart light(s) will not illuminate unless the subsequent unstart(s) is sensed by the SES.

INLET CONTROL SYSTEM

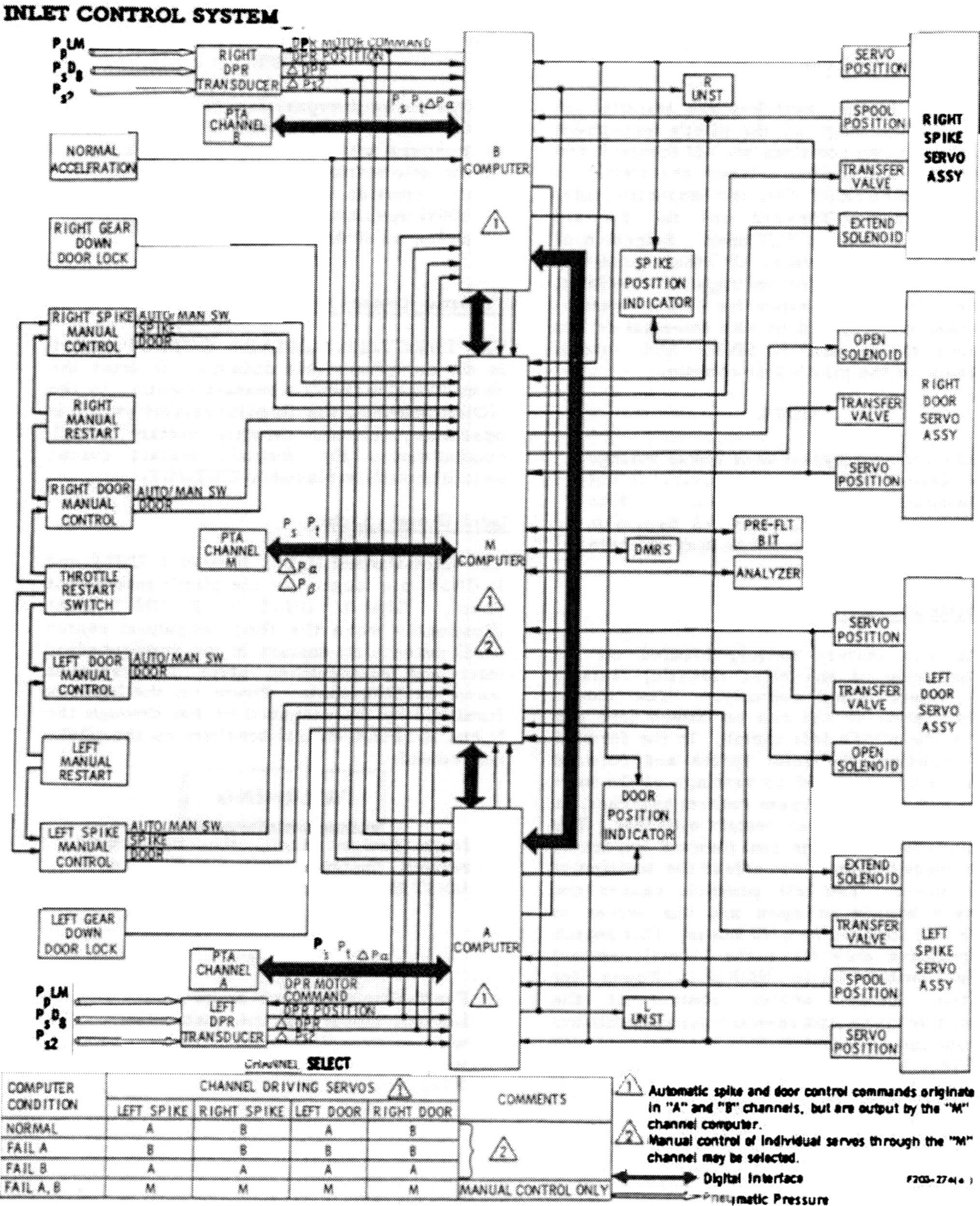

COMPUTER CONDITION	CHANNEL DRIVING SERVOS △1				COMMENTS
	LEFT SPIKE	RIGHT SPIKE	LEFT DOOR	RIGHT DOOR	
NORMAL	A	B	A	B	△2
FAIL A	B	B	B	B	
FAIL B	A	A	A	A	
FAIL A, B	M	M	M	M	MANUAL CONTROL ONLY

△1 Automatic spike and door control commands originate in "A" and "B" channels, but are output by the "M" channel computer.

△2 Manual control of individual servos through the "M" channel may be selected.

⟷ Digital Interface

Pneumatic Pressure

F203-27 4(a)

Figure 1-31

Manual Inlet Indicator Light

The MANUAL INLET caution light on the pilot's annunciator panel illuminates when one or more of the four rotary spike and/or forward bypass control knobs is not in AUTO, or an inlet restart switch is not in off.

INLET CONTROL SYSTEM

When in AUTO, the left and right spikes and forward bypass doors are controlled by the A and B computers, respectively, output through the M computer to the appropriate servo assemblies. Each computer (A and B) receives nonredundant inputs from its own duct pressure ratio (DPR) transducer assembly, but they share a common normal acceleration signal. Left and right DPR signals are crossfed, so that either A or B computer can control both inlets if the other computer fails. Left inlet auto control requires at least one phase of emergency ac power to the A computer through the COMPUTER circuit breakers in the aft cockpit. Right inlet auto control requires at least one phase of emergency ac power to the B computer through the COMPUTER circuit breakers in the aft cockpit.

Control of inlet unstart lights is direct from each computer (A and B), with cross control if either computer fails.

Manual spike and door control (both inlets) requires M computer A and C phase or B and C phase emergency ac power through the COMPUTER circuit breakers in the aft cockpit.

Left and right spike and door position signals are supplied to forward cockpit indicators. DMRS, PREFLIGHT BIT, and DAFICS Analyzer interfaces are via the M computer. Refer to Figure 1-31 for channel select table.

FUEL SYSTEM

There are five individual fuselage tanks, tanks 1A, 1, 2, 4, and 5, and two wing-fuselage tank groups, tanks 3 and 6. (See Figure 1-32.) Tank 6 is further divided into 6A and 6B. Interconnecting plumbing and electrically driven boost pumps are utilized for fuel feed, transfer, and dumping. Other components of the system include pump controls, nitrogen inerting, scavenging, pressurization and venting, single-point refueling receptacles, and a fuel quantity indicating system. The fuel heat sink system cools cockpit air, engine oil, accessory drive system oil, and hydraulic fluid.

The fuel system moves the aircraft center of gravity by:

a) automatic tank sequencing to the engine feed manifolds to control the fuel level in the various tanks;

b) early depletion of tank 1 fuel to a pre-selected shut-off level, which is dependent upon the zero fuel weight and C.G. position, and whether the flight is subsonic or supersonic;

c) automatic aft fuel transfer into tank 5 when tank 5 pumps are energized and tank 2 contains fuel. In addition, the manual aft transfer system is automatically energized if tank 1 has less than 1500 pounds of fuel, the tank 2 boost pumps are off, and the boost pumps in tanks 3 and 5 are on;

d) manual aft transfer into tank 5 when it is less than full;

e) manual forward transfer to tank 1. See Section 5, Use of Forward Transfer.

For automatic center of gravity scheduling to be effective, the fuel loading distribution given in T.M. SR-71-5 Weight and Balance Manual must be followed.

The automatic fuel feed system is designed to maintain the c.g. within the band depicted in Figure 1-33.

FUEL QUANTITY

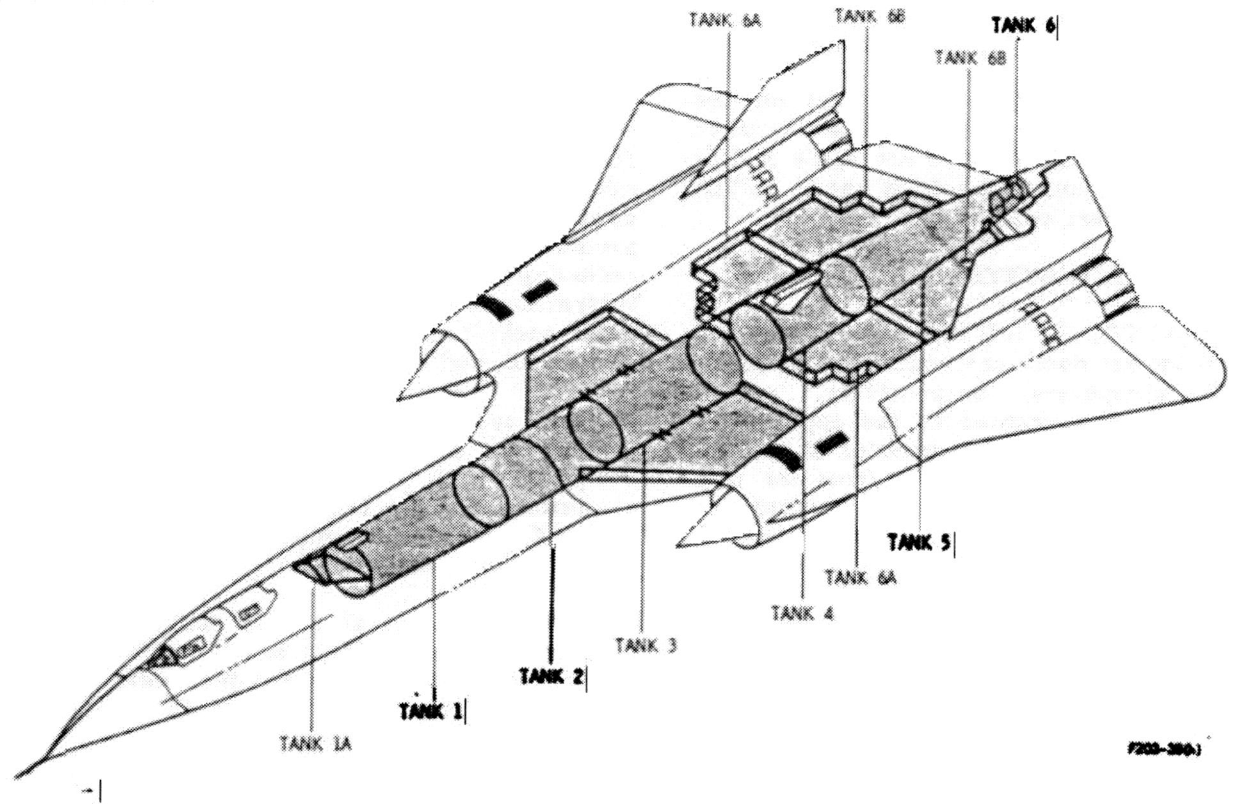

FUEL TANK CAPACITIES
Normal Flight Attitude

Tank	Fuel/Gal	Fuel (JP-7)
1A	251.1	1650 lb.
1	2095.9	13770 lb.
2	1974.1	12970 lb.
3	2459.7	16160 lb.
4	1453.6	9550 lb.
5	1758.0	11550 lb.
6A (forward)	1158.3	7610 lb.
6B (Aft)	1068.5	7020 lb.
Total	12219.2	80280 lb. *

* At average fuel density of 6.57 lb./gal.
(46.2° API, Fuel temperature = 78°F)

Figure 1-32

C.G. VS. GROSS WEIGHT
OFF TANKER WITH FULL FUEL LOAD - SUPERSONIC CRUISE - AUTOMATIC FUEL SEQUENCING

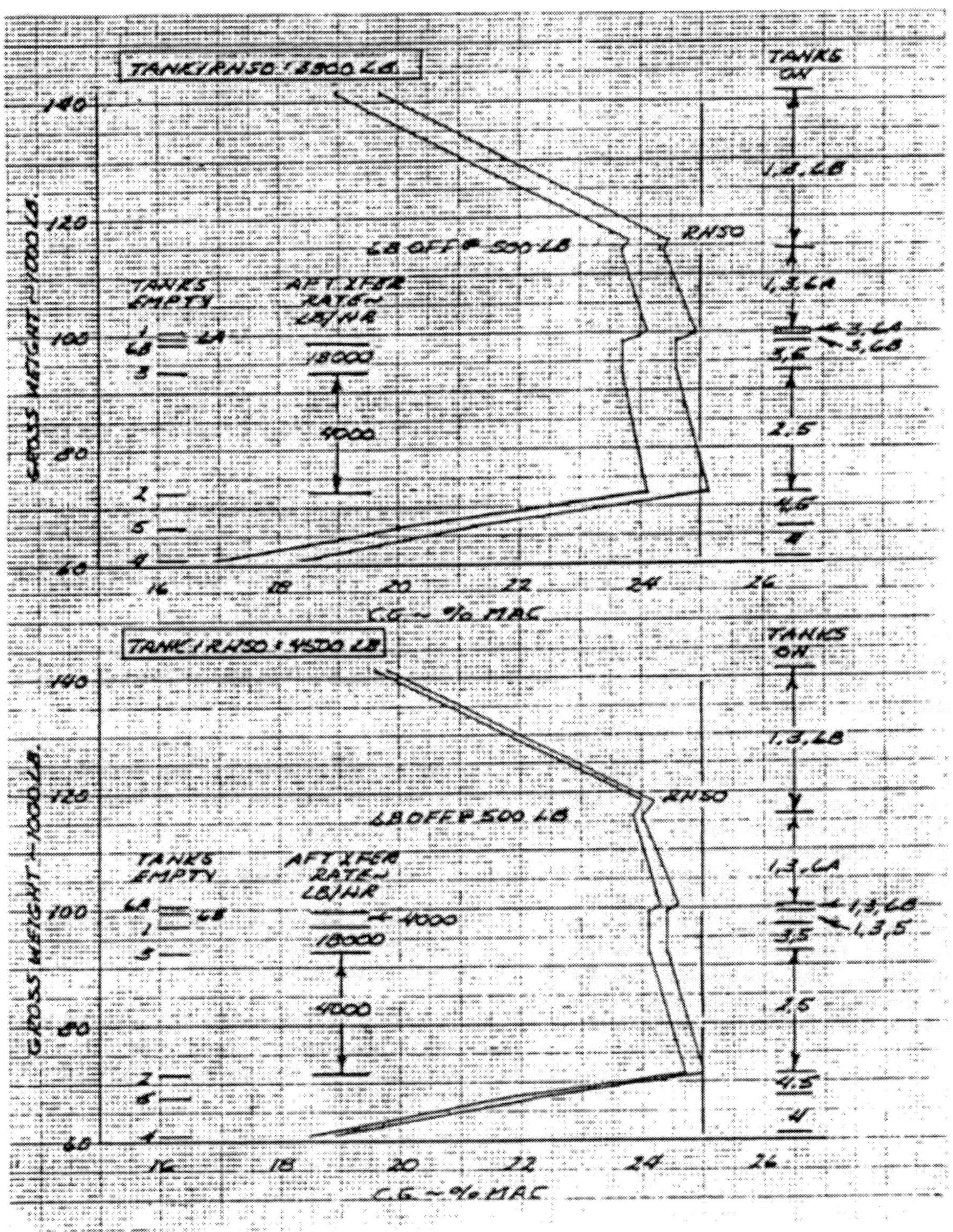

Figure 1-33

ZERO FUEL WEIGHT C.G. BANDWIDTH SUPERSONIC FLIGHT	
RHSO SETTING TANK 1 -LB	ZERO FUEL WEIGHT C. G. RANGE -% MAC
3300	17.0 to 18.4
4500	18.5 to 18.9
5000	19.0 to 19.5
5500	19.6 to 20.1
6000	20.2 to 20.8
6800	20.9 to 21.6
7500	21.7 to 22.3

Figure 1-34

FUEL TANKS

Tanks 1, 1A, 2, 4, and 5 are entirely contained in the fuselage. Tank 1A is a small tank located immediately forward of and feeding into tank 1. Tanks 3 and 6 consist of three and five tank groups, respectively. The No. 3 tank group is comprised of the forward section of each wing and a fuselage tank. The No. 6 tank group is located in the wings on either side of tanks 4 and 5 and includes a small sump tank (approximately 12 gallons) at the extreme aft end of the fuselage which contains the boost pumps for the group. All fuselage tanks are interconnected by a single vent line. Each wing tank is vented to its nearest fuselage tank. Submerged boost pumps are contained in the fuselage tanks. The pumps supply fuel through the left and right manifolds and transfer fuel for center of gravity control. The manifolds can be connected by opening a crossfeed valve. A fuel dump valve is installed in each manifold.

FUEL FEEDING AND SEQUENCING

The left engine is supplied from the left fuel manifold which is normally fed from tanks 1, 2, 3, and 4. The right engine is supplied from the right fuel manifold which is normally fed from tanks 1, 4, 5, and 6. Although crossfeed can be used to feed either engine from any tank, the normal automatic fuel sequencing schedule is:

Left engine supply tanks	Right engine supply tanks	Until
1, 3, 4*	1**, 6, 4*	6 empty
1, 3	1**, 5	1 empty
3	5	3 empty
2	5	2 or 5
2 or 4***	5 or 4***	empty
4	4	See note.

* If tank 4 is full initially, pumps 4-1 and 4-2 operate briefly to provide approximately 850 pounds ullage space. Tank 4 indicator light will not illuminate.

** Termination of tank 1 supply to the right engine depends on the right-hand shutoff float switch selection.

*** Depending on the right-hand shutoff float switch selection for tank 1, tank 2 may or may not shut off before tank 5 is depleted.

NOTE

Forward transfer is necessary during subsonic operation to maintain c.g. ahead of the aft limit.

Tank sequence is controlled automatically by float switches, see Figure 1-35. These switches optimize the center of gravity for supersonic cruise.

FUEL BOOST PUMP AND FLOAT ARRANGEMENT

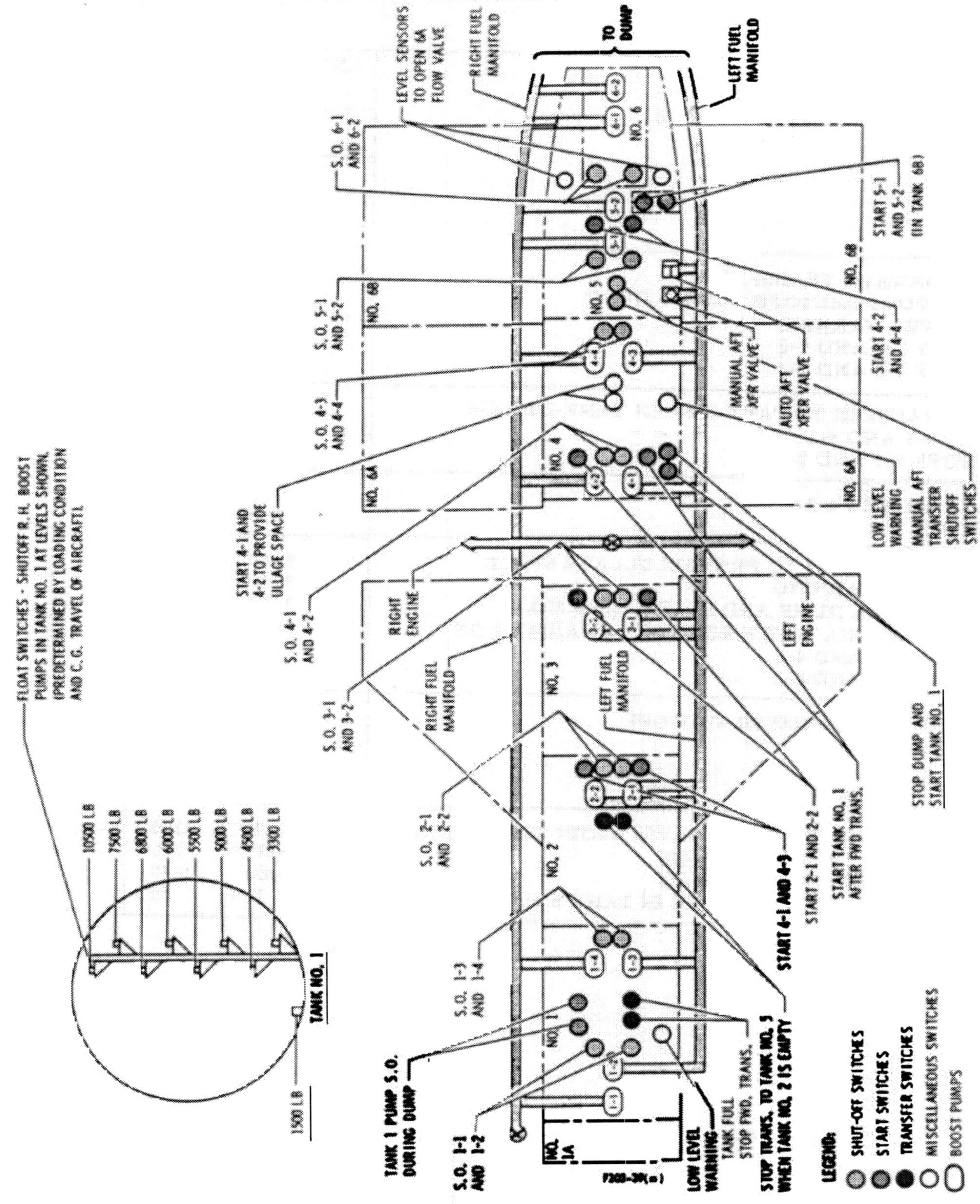

Figure 1-35

ATTITUDE EFFECTS ON FUEL TANK FLOAT SWITCH ACTUATION

TANK NO.	SWITCH	ANGLE OF ATTACK DEG 0 — Tank Fuel Remaining - Lb (6.7 lb/gal.)	ANGLE OF ATTACK DEG +6.2 — Tank Fuel Remaining - Lb (6.7 lb/gal.)
1	STOP FORWARD TRANSFER	13300	13700
	TANK 1 PUMP SHUTOFF DURING DUMP	4550	4850
	LOW LEVEL WARNING	4200	5400
	SHUTOFF 1-1 AND 1-2	1650	2850
	SHUTOFF 1-3 AND 1-4	400	65
2	STOP TRANSFER TO TANK 5 WHEN TANK 2 IS LOW	600	400
	START 4-1 AND 4-3	500	100
	SHUTOFF 2-1 AND 2-2	300	65
3	START 2-1 AND 2-2	500	175
	SHUTOFF 3-1 and 3-2	250	75
4	START 4-1 AND 4-2 TO PROVIDE ULLAGE SPACE	9550	9400
	LOW LEVEL WARNING	3750	4050
	STOP NORMAL DUMP AND START TANK NO. 1	3500	3950
	START TANK NO. 1 WITH FORWARD TRANSFER ON	400	1250
	SHUTOFF 4-3 AND 4-4	300	75
	SHUTOFF 4-1 AND 4-2	250	1100
5	MANUAL AFT TRANSFER SHUTOFF	10950	10950
	START 4-4	1700	350
	START 4-2	1600	300
	SHUTOFF 5-1 AND 5-2	1400	200
6	6A TO 6B FLOW CONTROL VALVES BEGIN SCHEDULING	10900	8100
	START 5-1 (SWITCH IN TANK 6B)	6475	300
	START 5-2 (SWITCH IN TANK 6B)	5200	150
	SHUTOFF 6-1 AND 6-2 (SWITCH IN TANK 6 SUMP TANK)	10	20

Figure 1-36

FUEL FEED SYSTEM

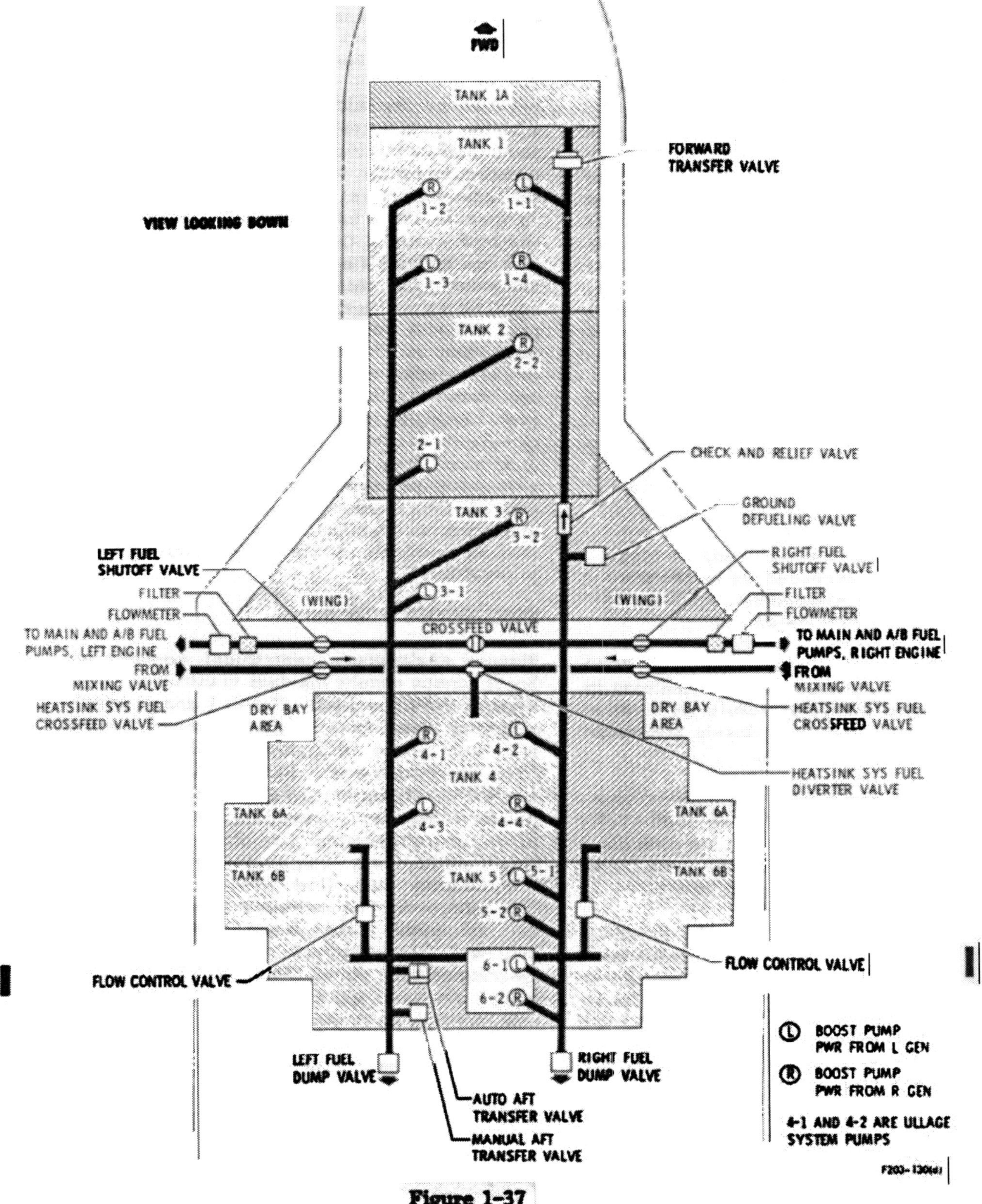

Figure 1-37

A stack of float switches, called right-hand shutoff switches, is installed in tank 1. These switches allow the automatic c.g. control system to compensate for variations in aircraft weight and c.g. due to sensor loading. The selection of a float switch determines the level of fuel remaining in Tank 1 at which boost pumps 1-1 and 1-4 (which supply the right fuel manifold) are shut off. Since boost pumps 1-2 and 1-3 continue to supply the left fuel manifold, the rate of fuel flow from tank 1 is reduced approximately 50% when shutoff occurs. This stops the rapid aft shift of c.g. experienced when tank 1 is supplying both fuel manifolds. Setting of the right-hand shutoff switches is accomplished by maintenance personnel prior to flight. Float switch selection is based on aircraft zero fuel weight and c.g. position as scheduled in the SR-71-5 Handbook of Weight and Balance Data.

NOTE

Recommended weight and/or c.g. limits can be exceeded by seemingly normal loading arrangements. Check loading documents carefully.

Eight settings are provided between 3300 pounds and 10,500 pounds of fuel remaining in tank 1. Proper selection results in a schedule for c.g. in a band that closely approaches the aft limit during cruise.

The preselected float switch setting is overridden if tank 4 contains less than 3600 $\pm$1200 lb of fuel. In this case, tank 1 supplies fuel to both manifolds until the tank 1 fuel low level switches shut off the boost pumps.

The tank 6 boost pump sump tank is fed by tank 6B. Two flow control valves are located between tanks 6A and 6B. When the valves are open, fuel in tank 6A can flow into tank 6B. The tank 6 flow control valves are held open by fuel pressure provided by the tank 6 boost pumps or by pump 5-1. Fuel pressure from the tank 6 boost pumps is provided to the valves only when the quantity of fuel in tank 6B is below approximately 500 pounds at cruise attitude. If pump 5-1 is not on, the valves function to maintain approximately 500 pounds of fuel in tank 6B until tank 6A is empty. If pump 5-1 is on, the valves are open regardless of the fuel quantity in tank 6B.

Because of the aft position of the tank 6 outlets, tank 6 may stop feeding if a nose down flight attitude is established while fuel remains in tank 6 wing tanks. A nose-down attitude and/or a deceleration shifts fuel forward in tanks 6A and/or 6B and uncovers the tank outlets. (Normally, the tank 6 group empties during climb and level cruise). With the wing tank outlets uncovered, fuel does not flow to the sump tank which contains the tank 6 boost pumps. As a result, tank 5 is sequenced on by the low level float switch in the aft portion of tank 6B. The tank 6 pumps are shut off when the sump tank empties, even though a substantial amount of fuel may remain in the wing tanks. An unusual c.g. condition can result if the incorrect sequencing persists. The c.g. may also move out of the desired range if a substantial mismatch of fuel flow to the engines exists (see Figure 2-8). Refer to Fuel Management in Section II.

FUEL BOOST PUMPS

Sixteen single-stage, centrifugal, ac-powered boost pumps supply the fuel manifolds. (See Figures 1-35 and 1-37). Tanks 1 and 4, which normally feed both engines, are equipped with four pumps each and tanks 2, 3, 5, and 6 have two pumps each. Either pump of a pair is capable of supplying sufficient pressure to permit engine operation at reduced afterburning thrust if the other pump fails. Two pumps are required for maximum afterburning fuel flow at lower altitudes. The pumps may be manually operated by use of individual tank boost pump control switches, located on the pilot's right instrument side panel. Manual control of the tank pumps supplements but does not terminate automatic tank sequencing. Manual selection of any tank pumps will change the programmed c.g. schedule and may cause a serious c.g. condition to develop. The boost pumps are cooled and lubricated by the fuel; therefore, manual activation of pumps should be terminated (by pressing the pump release switch) when the tank is empty.

In automatic operation, each pump is protected by a float switch that deactivates the pump when the tank is empty. Individual circuit breakers for each pump are loacted in the E-bay and are not accessible in flight. Three-phase ac power for the pumps is furnished by the generator buses. Odd-numbered pumps (except 4-1) are powered by the left generator ac bus and even numbered pumps (except 4-2) are powered by the right generator ac bus.

NOTE

- In tank 4, pump 4-2 and 4-3 are powered by the left generator bus and pump 4-1 and 4-4 are powered by the right generator bus. This precludes flameout due to fuel starvation if a single generator fails while the bus tie is split.
- If the left engine generator fails and the bus tie is split, fuel in tank 6A may become trapped due to insufficient tank 6 pump pressure and the unavailability of pump 5-1. Pump 5-1 pressure is ported directly to the level control and flow valves controlling tank 6A fuel feed.

FUEL TRANSFER SYSTEM

Fuel transfer systems control aircraft c.g. A manual forward transfer system and an automatic and a manual aft transfer system are provided.

Forward Transfer

Fuel may be transferred forward through the right fuel feed manifold from tanks 4, 5, and 6 into tank 1 at a maximum rate of approximately 950 lb/min. Fuel may be transferred from the left fuel feed manifold (tanks 2 and 3) by opening the crossfeed valve. Fuel transfer is initiated by the forward transfer valve in the forward end of the right fuel feed manifold opening into tank 1. This valve is controlled by the FWD TRANS fuel transfer switch. With RHSO settings of 6000 lbs or higher, forward transfer may be required more often.

Automatic Aft Transfer

Fuel can be transferred from the left fuel feed manifold into tank 5 either by automatic or manual operation. The automatic aft transfer system is started when the tank 5 boost pumps are energized if tank 2 contains fuel above the level of its low fuel quantity float switch. The automatic aft transfer stops when the tank 5 boost pumps are de-energized or when tank 2 fuel level reaches the low fuel quantity float switch. The rate of fuel transfer is determined by throttle position. The fuel automatic aft transfer rates are: 65 lb/min (4000 lb/hr) when both throttles are in afterburner, 23 lb/min (1400 lb/hr) otherwise. In addition, the 233 lb/min (14,000 lb/hr) manual aft transfer rate is automatically added when:

1. Tank 1 contains less than 1500 lbs, and
2. Tank 2 contains fuel, but it's boost pumps are not on, and
3. Tank 3 and 5 boost pumps are on.

Manual Aft Transfer

The manual aft transfer system permits the pilot to transfer fuel into tank 5 from any tank(s) whose pumps are supplying the left fuel manifold.

The manual aft transfer rate is approximately 233 lb/min (14,000 lb/hr).

NOTE

Forward transfer operates at a rate of approximately 950 lb/min, a rate more than sufficient to overcome the effects of automatic and manual aft transfer.

FUEL SYSTEM - Pressurization

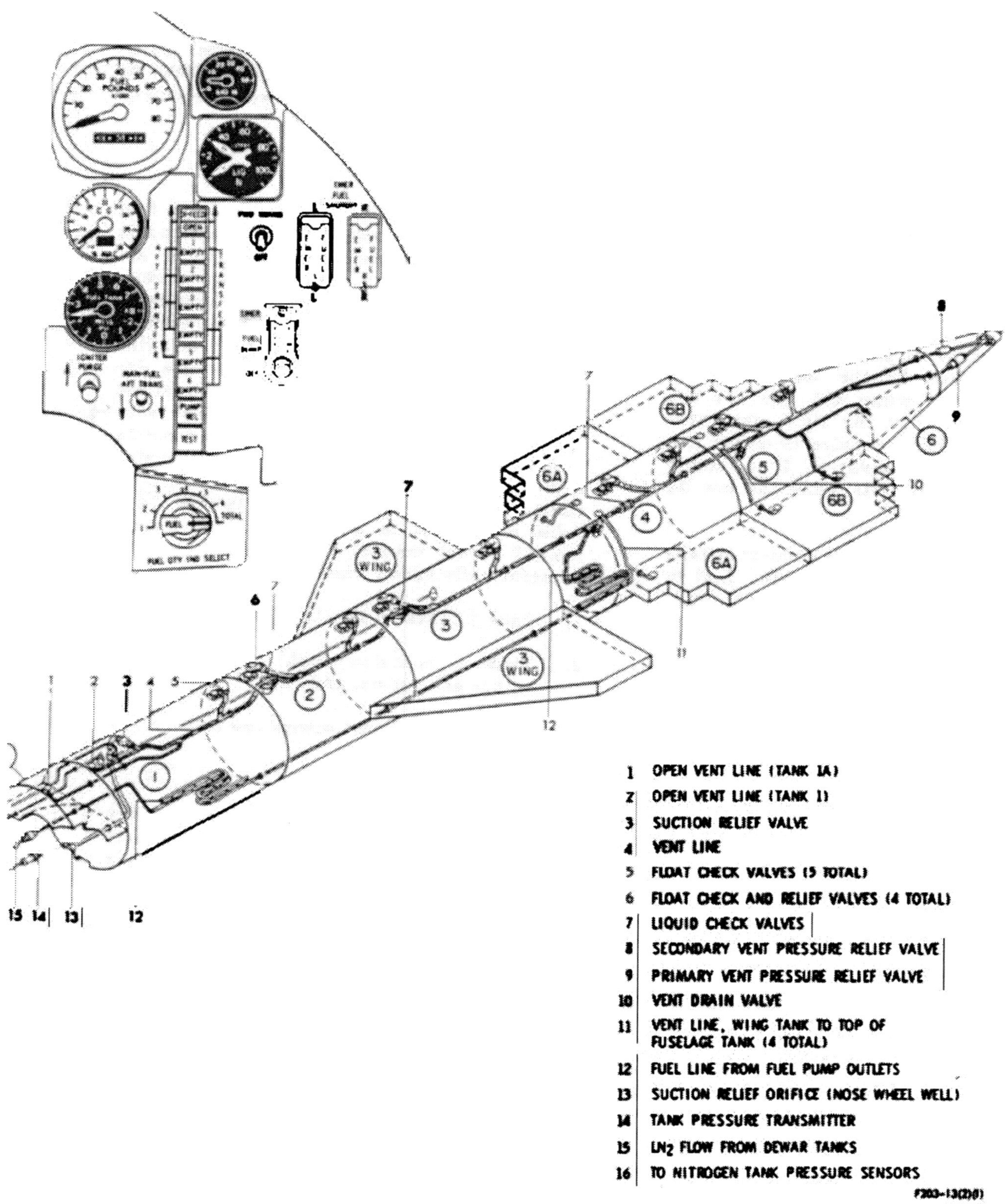
1 OPEN VENT LINE (TANK 1A)
2 OPEN VENT LINE (TANK 1)
3 SUCTION RELIEF VALVE
4 VENT LINE
5 FLOAT CHECK VALVES (5 TOTAL)
6 FLOAT CHECK AND RELIEF VALVES (4 TOTAL)
7 LIQUID CHECK VALVES
8 SECONDARY VENT PRESSURE RELIEF VALVE
9 PRIMARY VENT PRESSURE RELIEF VALVE
10 VENT DRAIN VALVE
11 VENT LINE, WING TANK TO TOP OF FUSELAGE TANK (4 TOTAL)
12 FUEL LINE FROM FUEL PUMP OUTLETS
13 SUCTION RELIEF ORIFICE (NOSE WHEEL WELL)
14 TANK PRESSURE TRANSMITTER
15 LN2 FLOW FROM DEWAR TANKS
16 TO NITROGEN TANK PRESSURE SENSORS
3 WING
6A
6B
FUEL QTY IND SELECT
F203-13(2)(8)

Figure 1-38

FUEL HEAT SINK SYSTEM

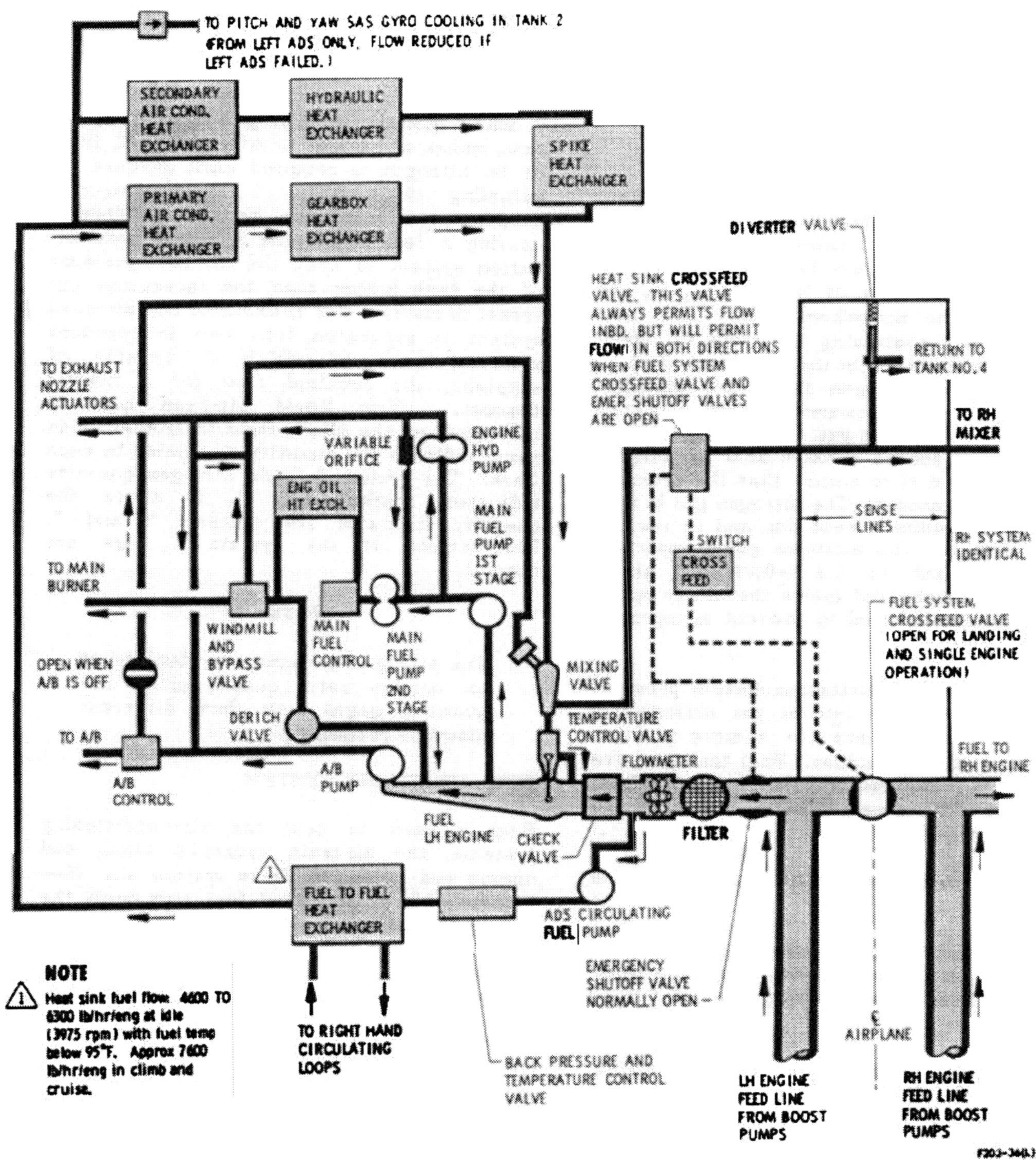

Figure 1-39

Manual aft transfer is controlled by the MAN-FUEL AFT TRANS switch on the pilot's right instrument side panel. Two ullage float switches automatically terminate manual aft transfer if the switch is held on while tank 5 is full; however, automatic aft transfer is not stopped.

FUEL TANK PRESSURIZATION SYSTEM

The fuel tank pressurization system (Figure 1-38) consists of indicators, three Dewar flasks and associated valves and plumbing to the fuel tanks. Two Dewar flasks, each containing 106 liters of liquid nitrogen, are located in the nosewheel well. The third Dewar flask, containing 50 liters of liquid nitrogen, is installed in the left forward chine (B bay). The nitrogen flasks are equipped with automatic ac-powered heaters to change the liquid nitrogen to gas. The nitrogen from the flasks is routed through heat exchangers (in tanks 1 and 4) to ensure that the nitrogen has become gaseous. The nitrogen gas is then ported to a common vent line and to the top of all tanks. The nitrogen gas pressurizes each fuel tank to 1.5 (±0.25) psi above ambient pressure and inerts the ullage space above the heated fuel to prevent autogenous ignition.

The fuel tank pressurization system provides 1.5 (±0.25) to 3.25 (±0.25) psi differential pressure to fuel tanks for a more positive fuel supply to the engines. Fuel tank positive pressure is prevented from exceeding 4.15 psi by a secondary pressure relief valve in the tail cone vent line.

Positive tank pressure is maintained in three ways:

1. The inerting system allows nitrogen flow into the tanks whenever fuel tank pressure differential drops below 1.5 (±0.25) psi.
2. Fuel vapor pressure tends to maintain 1.5 (±0.25) to 3.25 (±0.25) differential during supersonic cruise when fuel warms up.
3. Ascent into less dense atmosphere causes the pressure differential in the tanks to increase to the setting of the pressure relief valve of 3.25 (±0.25) psi at which time the tank will continue venting until level-off.

A small quantity of nitrogen is required for taxi, runup, and takeoff. After takeoff, little or no nitrogen is required until descent for refueling or landing. As the aircraft descends, atmospheric pressure increases causing a demand on the nitrogen pressurization system to keep the internal pressure of the tank higher than the increasing external pressure. For reliability, the nitrogen system is separated into two independent systems, each of which is capable of supplying the required flow for a normal descent. Two liquid nitrogen quantity indicators on the pilot's right instrument side panel indicate the quantity remaining in each flask. The system 3 liquid nitrogen quantity indicator, marked LIQ N, is above the quantity indicator for systems 1 and 2. Fluctuations of the system 3 gage are normal.

NOTE

The nitrogen systems may deplete at an uneven rate; consequently, the quantity gages may show different amounts remaining.

FUEL HEAT SINK SYSTEM

Fuel is used to cool the air-conditioning systems, the aircraft hydraulic fluid, and engine and accessory drive system oil. (See Figure 1-39.) Circulated fuel also cools the TEB tank and the control lines which actuate the afterburner nozzle. Engine oil is cooled by main engine fuel flow through an oil cooler, located between the main fuel control and the windmill bypass valve. This fuel is then directed to the main burner section. The other cooling is accomplished by fuel circulation through several cooling loops. If within engine consumption requirements, the

FUEL SYSTEM - Refueling

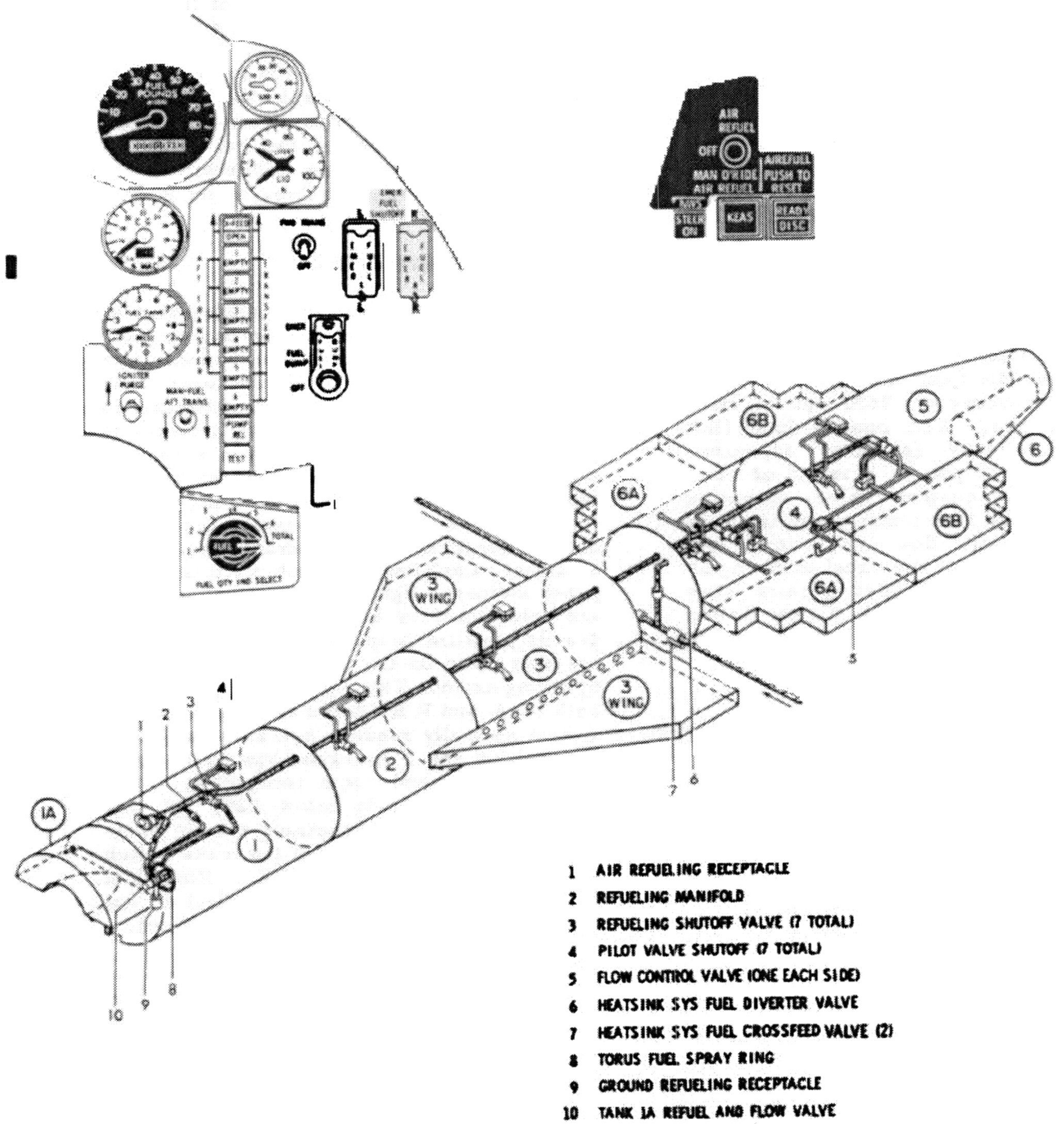

1 AIR REFUELING RECEPTACLE
2 REFUELING MANIFOLD
3 REFUELING SHUTOFF VALVE (7 TOTAL)
4 PILOT VALVE SHUTOFF (7 TOTAL)
5 FLOW CONTROL VALVE (ONE EACH SIDE)
6 HEATSINK SYS FUEL DIVERTER VALVE
7 HEATSINK SYS FUEL CROSSFEED VALVE (2)
8 TORUS FUEL SPRAY RING
9 GROUND REFUELING RECEPTACLE
10 TANK 1A REFUEL AND FLOW VALVE

F203-13(5)(F)

Figure 1-40

hot fuel returning from the accessory drive system heat exchanger, the primary and secondary air-conditioning heat exchangers, the hydraulic fluid heat exchanger, the spike heat exchanger, and the exhaust nozzle actuators is circulated through a mixing valve and temperature limiting valve and returned to the main engine and afterburner fuel manifold. The quantity in excess of engine requirements is diverted to tank 4.

NOTE

> When cooling loop fuel temperature is less than 96°F, flow through the loop should be between 4600 and 6300 pph at idle rpm. Loop flow increases to approximately 7600 pph at military rpm. The cooling loop flow automatically increases approximately 3600 pph as a result of temperature control valve operation when loop fuel temperature is above 96°F. Loop flow and cockpit fuel flow indication are equal until engine consumption becomes greater than flow through the loop. Excess flow is returned to the fuel tanks when engine consumption is less than cooling loop flow.

If the temperature of the mixed cooling loop and incoming engine fuel exceeds 290°F, the temperature control valve starts to close and some of the cooling loop fuel is prevented from mixing with the incoming engine fuel. A pressure-operated valve routes the hot fuel to tank 4. The temperature control valve is completely closed at 300°F and all cooling loop fuel is returned to tank 4. If tank 4 is full, the return fuel will be diverted to the next tank that has space for it. During single engine operation with the inoperative engine throttle in OFF, actuation of the fuel crossfeed valve allows the hot recirculated fuel from the windmilling engine to cross over and mix with the cooling loop and incoming fuel for the operating engine. If within engine consumption requirements and if the mixed fuel temperature is below 290°F, all of the hot fuel will be burned by the operating engine and afterburner. If the mixed fuel temperature is above 300°F, all hot fuel from both engines is returned to tank 4 or to the next tank with space available. Placing the emergency fuel shutoff switch to the shutoff (up) position terminates heat sink fuel to the windmilling engine. When either of the fuel shutoff valves is closed, the corresponding heat sink crossfeed valve is deenergized to prevent fuel circulation through the inoperative engine.

AIR REFUELING SYSTEM

The air-refueling system can receive fuel at approximately 6000 pounds per minute from KC-10 or KC-135 boom-equipped tanker aircraft. Calibrated orifices for each tank allows all tanks to be filled simultaneously in 12 to 15 minutes with a refueling pressure of 65-70 psi. A shutoff pilot valve in each tank terminates refueling flow when the tank or tank group is full. The air refueling system consists of a boom receptacle, receptacle doors, hydraulic valves, hydraulic actuators, a signal amplifier, control switches, and panel indicator lights. The refueling doors are held closed by hydraulic pressure; if hydraulic pressure is lost, as when the engines are shut down on the ground, the doors open by spring action. This enables air refueling if both the L and R hydraulic systems fail. The system normally requires hydraulic actuating power from the L hydraulic system to operate the doors and boom receptacle. If L hydraulic pressure is below 2200 psi, the refueling doors and receptacle can be operated by R system pressure if the brake switch is in ALT STEER & BRAKE. Electrical power is required from the essential dc bus for operation of the controls and indicators.

Unless the SR-71 receiver is using the manual boom latching procedure for refueling (air refueling switch in MAN O'RIDE) the boom will automatically disconnect if fuel pressure exceeds 70 psi. Pressure disconnect is normal when tanks reach full if refueling with a KC-135 tanker. Because the KC-10 refueling system automatically reduces flow to maintain normal refueling pressure, pressure disconnect does not normally occur if refueling with a KC-10 tanker.

FUEL SYSTEM CONTROLS AND INDICATORS

Crossfeed Switch

The push-button crossfeed switch is mounted above the column of fuel boost pump switches. When depressed, a motor operated valve between the left and right fuel manifolds opens to join the two fuel manifolds, so either or both manifolds can feed either or both engines. The fuel heat sink systems are also interconnected. The crossfeed switch must be depressed a second time to terminate crossfeed operation. The legend XFEED illuminates as the valve starts to open and an OPEN legend illuminates when the valve is fully opened. The OPEN legend extinguishes when the valve starts to close, but the XFEED legend remains on until the valve is fully closed. Circuit control power is furnished by the essential dc bus through the fuel CONT circuit breaker on the pilot's left console. Three phase power for the crossfeed valve is furnished by the essential ac bus through circuit breakers in the E bay.

Fuel Boost Pump Switches and Indicator Lights

Six self-illuminated, square plastic fuel boost pump pushbutton switches are installed in a vertical line on the pilot's right instrument side panel. The switches control manual operation of the fuel boost pumps. The switches have an electrical hold and bail arrangement that allows manual selection of only one tank of tank group 1, 2, 3 and one tank of tank group 4, 5, 6 at the same time.

NOTE

> Manual operation supplements, but does not terminate, automatic fuel tank sequencing.

When a set of boost pump relays are actuated, either automatically or manually, a clear numeral on an illuminated green background in the upper half of the pushbutton illuminates. When a tank is empty, an amber EMPTY light in the lower half of the push button illuminates and that pump group stops, unless manually selected.

NOTE

> If all the fuel is used from a manually selected tank, its EMPTY light will not illuminate until the tank which normally turns on that tank is also empty.

Automatic operation of the ullage system pumps 4-1 and 4-2 does not affect the Tank 4 indicator light. The Tank 4 indicator light only illuminates if pump 4-3 or 4-4 (or both) is on. In automatic operation, pumps 4-3 and 4-1 are turned on by individual float switches in Tank 2; and pumps 4-4 and 4-2 are turned on by individual float switches in Tank 5.

When manually depressed, the boost pump switch will hold down electrically and the pumps continue to operate until the pump release switch is pressed. Power for the boost pump switch circuits is furnished by the essential dc bus through the fuel CONT circuit breaker on the pilot's left console. Power for the indicator lights is furnished by the ac hot bus through the INSTR light circuit breaker on the pilot's right console and the FUEL CONT circuit breaker on the light control panel on the pilot's left console.

NOTE

> Pulling the TK 5 TRANS circuit breaker will disable tank 2, 3, and 5 fuel boost pump indicator lights. The fuel boost pumps are not disabled.

Pump Release Switch

A push-button PUMP REL switch is located below the fuel boost pump switches. When the PUMP REL switch is depressed, any boost pump switch that has been manually actuated is released. Power for the circuit is furnished by the essential dc bus through the fuel CONT circuit breaker on the pilot's left console.

CAUTION

A manually selected boost pump should be released when the tank indicates EMPTY.

Tank Lights Test Switch

A push-button tank lights TEST switch, is below the pump release switch. When the switch is depressed, the fuel boost pump lights, crossfeed, pump release, and test lights illuminate. Power is furnished by the ac hot bus through the INSTR light circuit breaker on the pilot's right console and the FUEL CONT TEST circuit breaker on the pilot's light control panel.

Fuel Forward Transfer Switch

The two-position forward transfer switch, labeled FWD TRANS (up) and OFF (down), is located on the pilot's right instrument side panel. If tank 1 is not full, moving the switch to FWD TRANS shuts off all tank 1 boost pumps and opens the forward transfer valve, allowing fuel to transfer into tank 1. Forward transfer is stopped automatically, when tank 1 is full, by the tank 1 "full" float switches; however, tank 1 boost pumps will not resume operation until the forward transfer switch is OFF or tank 4 is almost empty. Power for the circuit is furnished by the essential dc bus through the TK 1 TRANS circuit breaker on the pilot's left console.

Fuel Forward Transfer Light

With S/B R-2691, a FWD TRANSFER light on the RSO's instrument panel illuminates when the fuel forward transfer switch in the forward cockpit is in FWD TRANS. Power for the light is from the TRANS TK1 circuit breaker on the pilot's left circuit breaker panel.

Fuel Aft Transfer Switch

The fuel aft transfer switch, located on the pilot's right instrument side panel, is spring-loaded to the off (up) position. Holding the switch in the down position operates a solenoid controlled valve in the left fuel manifold, if tank 5 is not full, and allows fuel to enter tank 5. If tank 5 is full, the two float switches prevent the solenoid valve from operating. Control power for the manual aft transfer valve solenoid is from the essential dc bus through the TK 5 TRANS circuit breaker on the pilot's left console.

NOTE

Pulling the TK 5 TRANS circuit breaker will disable tank 2, 3, and 5 fuel boost pump indicator lights. The fuel boost pumps are not disabled.

Fuel Dump Switch

A guarded, three-position, lift-lock FUEL DUMP switch is installed on the pilot's right instrument side panel. The switch positions are: EMER (up), FUEL DUMP (center), and a guarded OFF (down) position. The switch must be pulled out and up to move to the EMER position. In the FUEL DUMP position, dual solenoid dump valves in each fuel manifold open to commence dumping. All fuel tanks continue to feed in automatic sequence until tank 1 reaches approximately 4700 pounds, depending on aircraft attitude (see Figure 1-36), then tank 1 boost pumps stop. Fuel pumps in all other tanks continue to operate in automatic sequence until tank 4 reaches approximately 3700 pounds (again, a function of aircraft attitude). At 3700 pounds, fuel dump ceases and, if there is any fuel in tank 1, tank 1 boost pumps will start. With the switch in the EMER position, dumping is identical except that at the 3700 pound level in tank 4, fuel dump does not cease and dumping will continue to empty tanks.

WARNING

Emergency fuel dumping must be terminated by positioning the dump switch to OFF (or FUEL DUMP) or all tanks will empty.

The nominal dump rate is 2500 pounds per minute for both FUEL DUMP and EMER switch positions, but the rate varies with the amount of fuel remaining and the number of boost pumps operating. (Refer to Section II, Fuel Dumping). Power for the circuit is furnished by the essential dc bus through the fuel DUMP CONT circuit breaker on the pilot's left console.

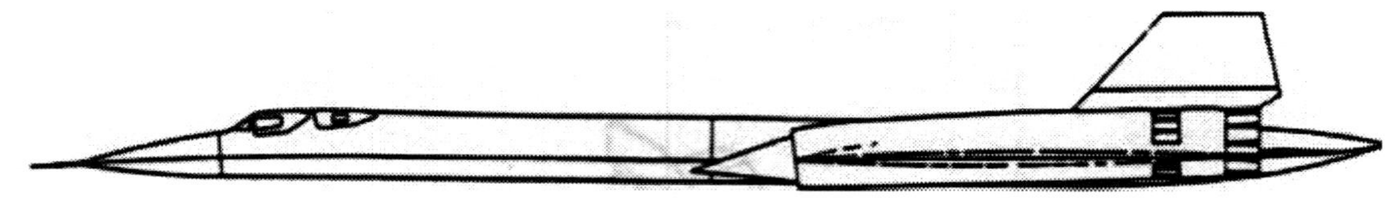

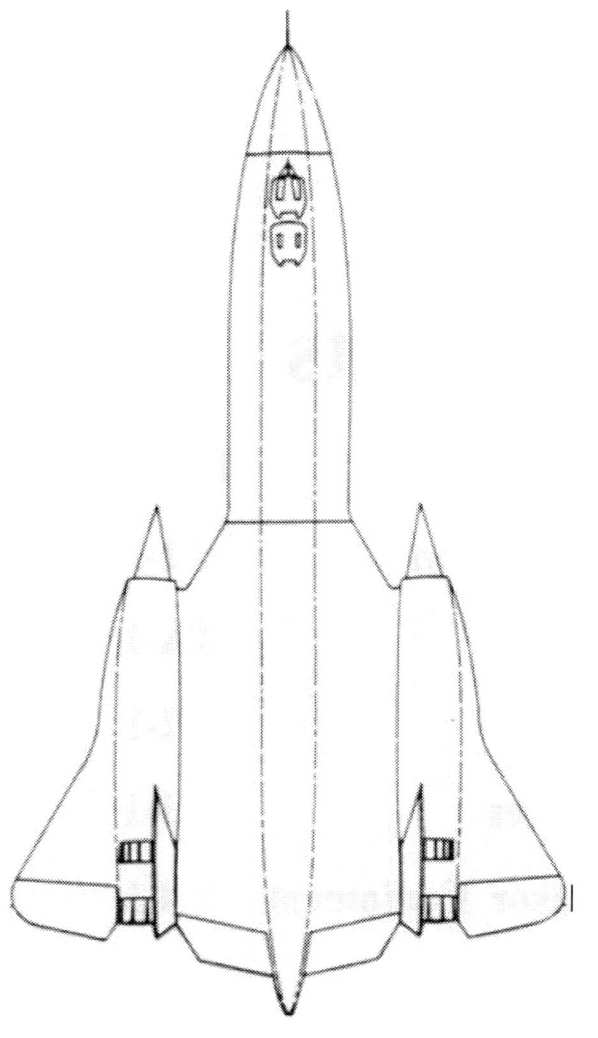

THIS PAGE INTENTIONALLY LEFT BLANK OR STILL CLASSIFIED.

Emergency Fuel Shutoff Switches

Independent emergency fuel shutoff switches for each engine are located on the pilot's right instrument side panel. The switches operate dc powered relays which control ac motor driven gate valves in the engine fuel lines. (See Figure 1-37). When the switches are in the guarded (fuel on) down position, the relays are deenergized and the gate valves are open and held open by the gate valve motors. When a shutoff switch is moved to the up (fuel off) position, its relay is energized and the corresponding gate valve motor closes the valve. Allow three to five seconds for the valve to close and shut off the fuel supply to that engine. This also isolates the fuel cooling loop for that side. (See Figure 1-39.) The valve remains closed as long as the relay is energized. The valve opens if the dc relay is deenergized and ac power is available.

Control power for the emergency fuel shutoff switch relays is provided from the essential dc bus through the L and R ENG SHUT OFF circuit breakers on the pilot's left console. Operating power for the shutoff valve motors is supplied from the essential ac bus through circuit breakers in the E-Bay.

Fuel Quantity Selector Switch

A fuel quantity selector switch is installed on the pilot's right instrument side panel and on the right side of the RSO's instrument panel. The switch has seven positions: TOTAL and six individual tank positions. The position selected determines the indication presented by the quantity gage. Operation of each fuel quantity selector switch and its respective quantity indicator is independent of the other cockpit.

Fuel Quantity Indicator

A fuel quantity indicator is located on the pilot's right instrument side panel and on the RSO's instrument panel. The indicator has a circular scale and a pointer that displays fuel quantity from 0 to 85,000 pounds. A five-digit window indicates fuel quantity to the nearest 100 pounds.

Power for both indicators is furnished by the ac hot bus through individual FUEL QTY circuit breakers in each cockpit.

Fuel Quantity Indicator Characteristics

When the aircraft is on the ground or in stabilized flight and the fuel quantity indicator selector switch is in TOTAL, the indicator should read within 780 lb of the sum of the individual tank readings. When the selector switch is rotated to another position, the indicator may require 8 to 10 seconds to reach a new reading. The normal time required to stabilize for individual tank readings is 1 to 3 seconds.

During forward acceleration or deceleration, or climbs and descents at relatively steep angles, the quantity indicator readings become inaccurate. This becomes apparent when either the forward or aft averaging probe in any tank becomes completely submerged in fuel and cannot compensate for response of the opposite probe. Because no sideslip compensation is provided, uncoordinated turns cause a lower quantity reading. Most malfunctions that affect an individual fuel tank quantity indication also affect total indication, even when the individual tank is empty, but do not influence the readings of the other individual tanks.

NOTE

> Fluctuations of the fuel quantity indications of about 5-percent of full scale can be expected with keying or modulation of the HF transmitter.

Center of Gravity (CG) Indicator

A cg indicator is located on the pilot's right instrument side panel and on the left side of the RSO's instrument panel. The pilot's indicator is connected to the tank fuel quantity sensors and displays aircraft center of gravity location in percent of reference chord (although indicator is marked % MAC). The indicator dial face has 1% scale divisions from 14% to 30% and labels in 2% increments. The RSO's indicator is a repeater.

FUEL QUANTITY AND C.G. INDICATION SYSTEMS

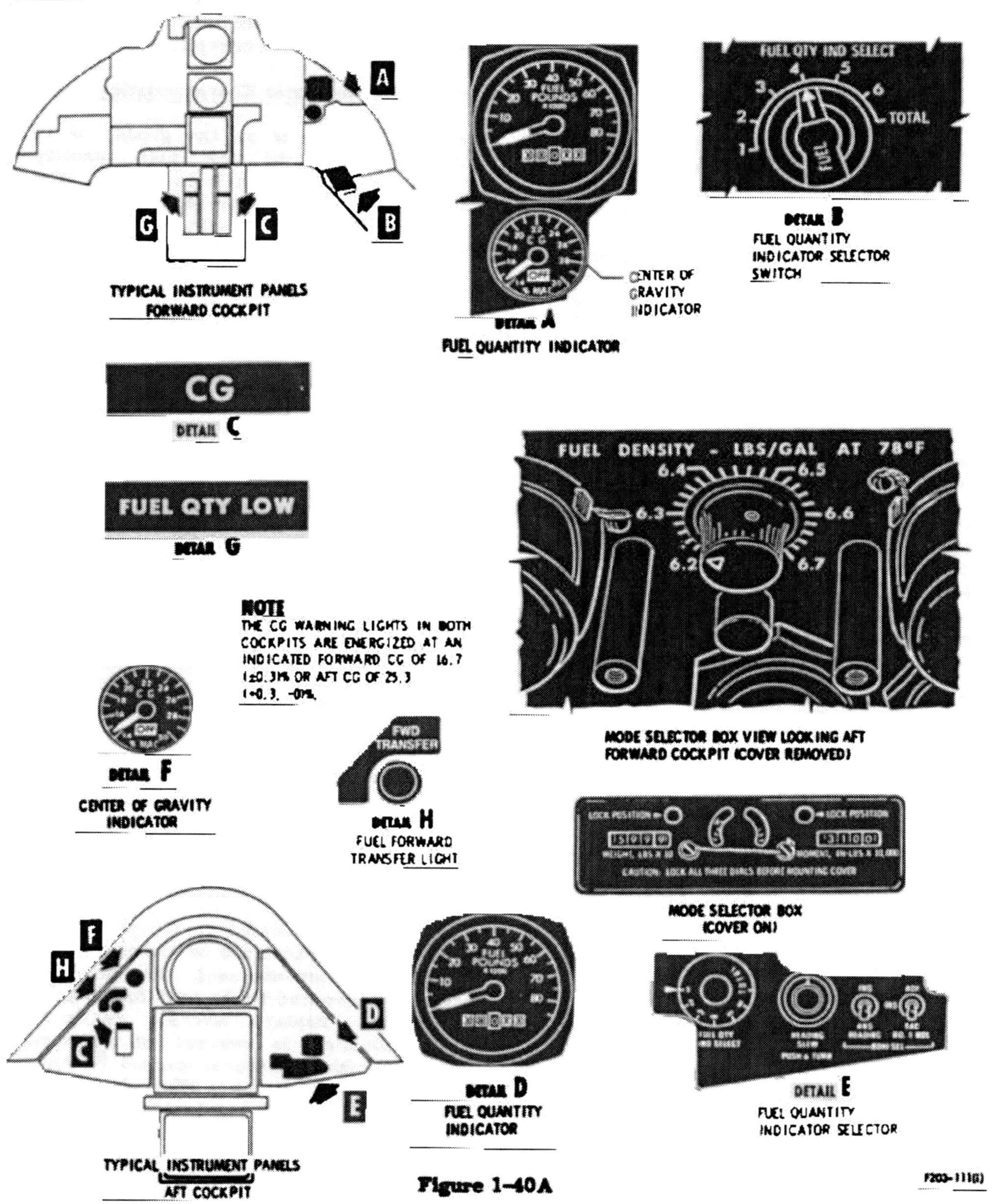

Figure 1-40A

A switch in the forward cockpit indicator causes the CG annunciator warning light in both cockpits to illuminate when cg reaches 25.3% to 25.6% aft cg or 16.4% to 17.0% forward cg. Power is furnished to the indicators by the ac hot bus through a FUEL QTY circuit breaker in each cockpit. An OFF warning flag is displayed only on the affected indicator if power is interrupted. If power to the forward cockpit indicator is lost, cg indication on both indicators will remain at the cg shown at the time of power interruption, but the aft cockpit indicator will not display an OFF warning flag.

NOTE

- An erroneous cg indication can be expected during steady sideslip, as during single-engine operation, when fuel remains in tank 6.
- Fluctuations of the cg indications of about 5 percent can be expected with keying or modulation of the HF transmitter.

CG Indicator Mode Selector

A mode selector for the cg indicator is located on a bulkhead aft and left of the seat in the forward cockpit. Values representing reference fuel density, aircraft weight without fuel, and the corresponding moment value must be set with the mode selector knobs. The mode selector control settings can be viewed by the pilot when entering the aircaft, but re not accessible in flight.

NOTE

The correct cg indicator mode selector settings must be set on the ground to obtain a proper cg indication in-flight. If the correct settings are not made, the cg indicator will appear to operate properly, but will be erroneous, and manual cg computations will be necessary.

Fuel Low Pressure Lights

Warning lights for each engine, labled L and R FUEL PRESS, are located on the pilot's annunciator panel. The light illuminates when fuel pressure in the respective main fuel manifold decreases to less than 7 (± 1/2) psi. The light extinguishes when fuel pressure rises above 10 psi.

NOTE

The L and/or R FUEL PRESS warning light(s) may illuminate sporadically during fuel dumping. The light for the left manifold may illuminate during manual aft transfer when tank sequencing occurs. The light for the right side may illuminate during forward transfer.

Fuel Quantity Low Light

A FUEL QTY LOW caution light on the pilot's annunciator panel is illuminated by the closing of low-level float switches in tanks 1 and 4. The switches are connected in series and both must close to illuminate the caution light. The switch in tank 1 closes when the fuel level has dropped below 4200 pounds at 0 degrees pitch angle and below 5400 pounds at +6.2 degrees pitch angle. The tank 4 switch closes when the fuel level has dropped below 3750 pounds at 0 degrees pitch angle and below 4050 pounds at +6.2 degrees pitch angle. Therefore, the light may illuminate at any condition of tanks 1 and 4 between 3750 pounds (0-degree pitch angle, fuel in tank 4 only) and 9450 pounds (+6.2 degree pitch angle and fuel in both tanks). This light will stay on after initial sensing and may be reset by moving the refuel switch to AIR REFUEL, then to OFF.

Fuel Tank Pressure Indicator

A fuel tank pressure indicator is installed on the pilot's right instrument side panel. The gage senses tank 3 pressure, which is common to all tanks, and is marked from -2 to +8 in increments of 1 psi. Power is supplied by the essential ac bus 26 volt instrument transformer through the FUEL TK PRESS circuit breaker on the pilot's annunciator panel.

Fuel Tank Low Pressure Light

A TANK PRESS warning light is located on the pilot's annunciator panel. The light illuminates when tank pressure is less than +0.25 psi.

Liquid Nitrogen Quantity Indicators

Two liquid nitrogen quantity indicators are installed on the pilot's right instrument side panel.

The dual needle lower indicator displays the quantity of liquid nitrogen remaining in the system 1 and system 2 Dewar flasks. The indicator is marked in 5-liter increments from 0 to 110 liters. When the indicators and warning lights test button (IND & LT TEST) is depressed, the quantity indicator needles move toward zero. Power for the indicator is furnished by the essential dc and ac busses through four circuit breakers. Two, labeled N2 QUAN NO 1 and NO 2, are located on the pilot's left console, and two labeled N QTY NO 1 and NO 2 are located on the pilot's right console.

NOTE

The Dewar flasks may deplete at an unequal rate.

A second liquid nitrogen quantity gage is installed above the dual needle indicator. The gage is marked LIQ N and displays the quantity of liquid nitrogen remaining in the system 3 Dewar flask. The dial is marked in 10 liter increments from 0 to 50 liters. Fluctuations of the gage are normal. The instrument indication is not affected by operation of the indicators and warning lights test button. Power is furnished from the essential dc bus through a circuit breaker in the C-Bay.

Nitrogen Quantity Low Indicator Lights

Two nitrogen quantity low caution lights, one for each system, are located on the pilot's annunciator panel. The lights are labeled SYS 1 N QTY LOW and SYS 2 N QTY LOW. When illuminated, the respective nitrogen quantity is less than 3 liters. Operation of the lights can be checked by depressing the indicators and warning lights test button until the quantity gauge indicates below 3 liters. When the test button is released, the nitrogen quantity low lights will remain illuminated momentarily.

NOTE

- If necessary, loiter in accordance with emergency procedures to cool the fuel tanks if LN_2 has been depleted. Cooling is not required if speed has not exceeded Mach 2.6.
- There is no caution light to indicate depletion of the system 3 liquid nitrogen supply.

Air Refuel Switch

An air refuel switch, located at the top of the pilot's instrument panel, has three positions: AIR REFUEL (up), OFF (center), and MAN O'RIDE (down). When the switch is placed in the AIR REFUEL or MAN O'RIDE position, the refueling door opens, the receptacle lights illuminate, and the READY light in the air refuel reset switch illuminates. In AIR REFUEL, the boom latches are automatically armed. In MAN O'RIDE, opening and closing of the boom latches must be controlled manually by the air refuel disconnect trigger switch.

CAUTION

Before opening or closing the refueling door, ensure that the probe is clear.

Air Refuel Reset Switch and Indicator Light

A dual-indicating, self-illuminating push-button switch, labeled PUSH TO RESET, is located on the top of the pilot's instrument panel. The upper half of the push-button switch illuminates green and displays the word READY when the air refuel ready switch is in either AIR REFUEL or MAN O'RIDE and the refueling system signal amplifier is on. The READY light extinguishes when the boom is seated and latched. If the boom disconnects from the refueling receptacle, the lower half of the push-button switch illuminates amber and displays DISC. The push-button switch must then be depressed (or the air refuel switch recycled) to illuminate READY before the boom can be reengaged. The READY light illuminates and the DISC light does not illuminate when a disconnect occurs while refueling in MAN O'RIDE.

Air Refuel Disconnect Trigger Switch

The trigger switch on the forward side of the control stick grip can be used to disconnect the refueling boom.

When refueling with the air refuel switch in MAN O'RIDE, the trigger switch is used to open and close the boom latches. Depressing the trigger switch opens the boom latches and holding it depressed keeps the boom latches open. When the boom is seated, the READY light extinguishes and the switch can be released to close the boom latches, locking the boom in the receptacle.

ELECTRICAL SYSTEM

Electrical power is normally supplied by two ac generators, rated at 60 KVA each. The generators are mechanically driven by their respective engines through constant speed drive (CSD) units and operate in parallel. They provide 115/200 volt, 400-cycle, three-phase power to five ac buses and to two, two-hundred ampere transformer-rectifiers (T-Rs). Three 28-volt dc buses are normally energized by these T-Rs. Either generator is capable of supplying the normal ac and dc power requirements of the aircraft. Figure 1-44 shows the ac and dc power supplies, control circuits, and power distribution system.

The electrical system operates automatically and the protective features and the emergency dc system are automatically available after the generators have been set and the batteries switched on. An emergency ac generating system is provided which may be available after some types of electrical system failures. It must be selected manually, using the generator control switches.

Electrical system back-up controls provided in the forward cockpit for emergency conditions include the generator bus tie, instrument inverter, and emergency ac bus switches. Seven caution lights indicate: generator(s) out, generator bus tie open, transformer-rectifier(s) out, instrument inverter on, and emergency battery on.

BATTERIES

Two 25 ampere-hour batteries are provided for emergency service. If both generators are off or inoperative, or both transformer-rectifiers fail, each battery individually supplies one of the two essential dc buses. The battery relays will not engage unless sufficient charge remains in the No. 1 battery.

The No. 1 battery energizes the No. 1 essential dc bus, which supplies the SAS pilot valves. It also energizes the emergency ac bus through an instrument inverter, rated at 1 KVA. See Emergency DC Power Supply, Figure 1-46.

The No. 2 battery energizes the No.2 essential dc bus which supplies SAS control power, DAFICS computers, and all other essential dc system loads. The No. 2 battery is always connected to the BAILOUT and PILOT EJECTED warning lights, regardless of the battery switch position.

The maximum duration of the dual-battery power system is approximately 40 minutes if unnecessary equipment is turned off. Figure 1-43 lists power requirements of equipment energized from the essential dc buses.

EMERGENCY AC POWER SYSTEM

Each main generator has an emergency operating mode which may be available if neither generator will function normally. To be usable, the generator(s) selected must still be rotating with intact windings. Then, setting the system to the emergency mode may generate usable, but unregulated, ac power. There is no control of voltage or frequency in the emergency mode since generator speed is not governed and the No. 2 battery provides direct excitation of the generator field. If either generator is operating in EMER, the bus tie will open. Emergency ac power is applied directly to the ac hot bus and to the generator bus associated with that engine. See Figure 1-45.

CAUTION

In-flight, do not operate either generator in EMER unless both generators have failed.

Primarily, the emergency ac system will supply power for the fuel boost pumps and will power the ac hot bus to provide fuel transfer, cross-feed, and pitch axis trim capability. It also supplies power to the corresponding T-R unit and, if sufficient voltage is generated, the dc buses will be powered by the T-R. With neither generator operating in NORM, the emergency ac bus is energized by the No. 1 essential dc bus through the instrument inverter and the essential ac bus is dead; the monitored dc bus is also dead.

If sufficient ac generator voltage is not available to the T-Rs, the essential dc buses and the instrument inverter will be powered by the batteries.

WARNING

During emergency ac operation, the normal automatic fuel sequencing system is disabled and the pilot must manually select tanks. The automatic aft transfer and ullage systems are inoperative. Normal generator fault protection is not provided.

EXTERNAL POWER

An external power receptacle in the nosewheel well accepts a six-wire type cable connection from an MD-3 or MD-4 (or equivalent) ground support unit. The supply must provide 115-120/200-208 volt, 400-cycle, three-phase ac power with A-B-C phase rotation, and 28 volt dc power. (The 28 volt dc supply only energizes the aircraft external power relay. Aircraft dc power is obtained from the external ac supply through the ship's T-R units.)

Transfer to Internal Power

The aircraft generator switches are ineffective when the engines are stopped. Normally, the generator switches are set to NORM after the engines start. The generators remain disengaged until the right engine CSD reaches a speed which allows its generator to synchronize with the frequency and phasing of the external supply. However, if the generator switches are set to NORM <u>before</u> the engines are started, the resulting paralleling transient may cause the ANS to trip off. The external power contactor is automatically opened when the right engine reaches a parallel condition with the external supply and the generator line contactor closes. The external power connector can be removed after this occurs, but its removal is normally delayed until both generators are on-line and the system operating normally - a condition indicated by all seven electrical system caution lights being off.

CIRCUIT BREAKER PANELS – Fwd Cockpit

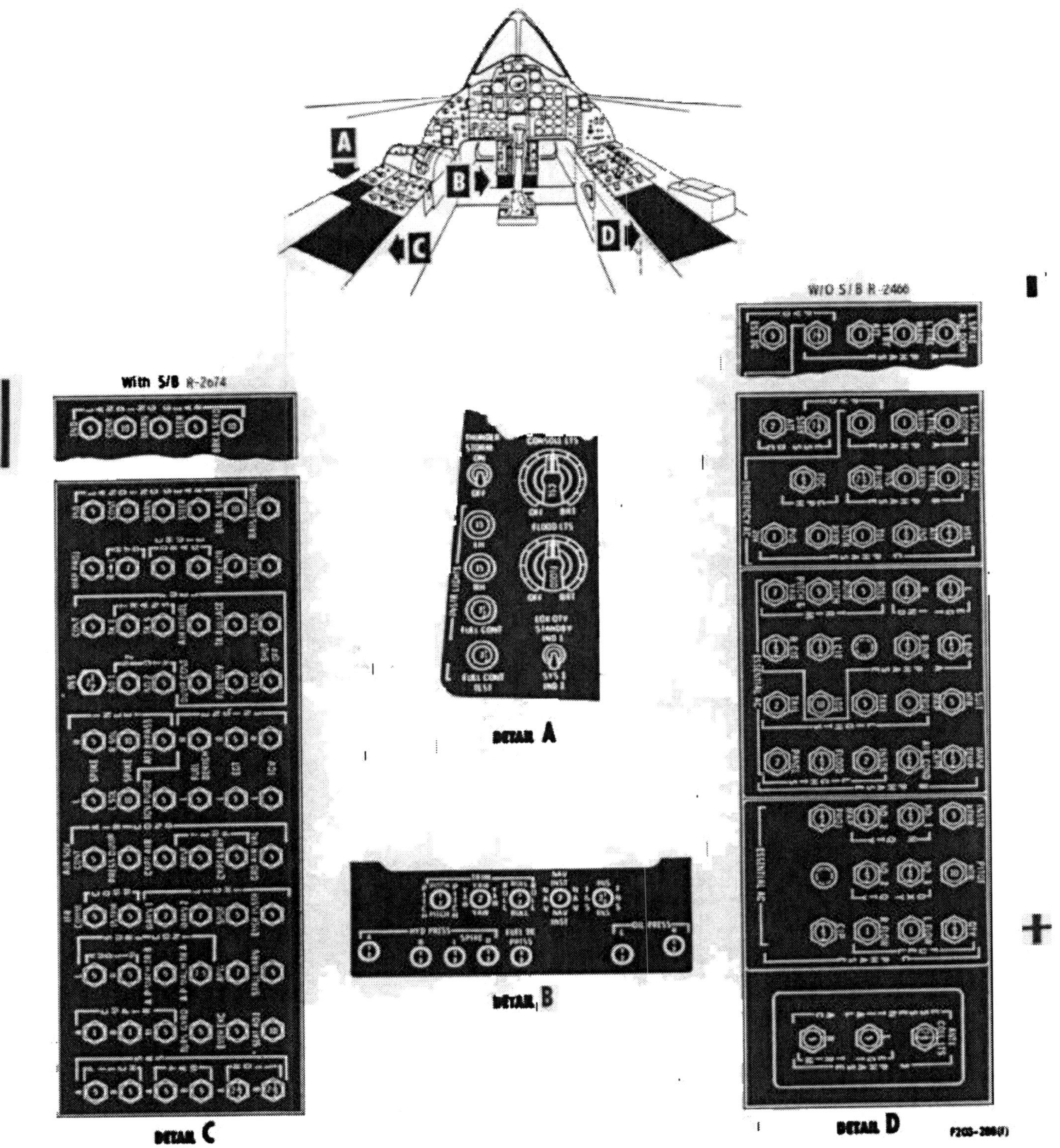

Figure 1-41

CIRCUIT BREAKER PANEL
Aft Cockpit

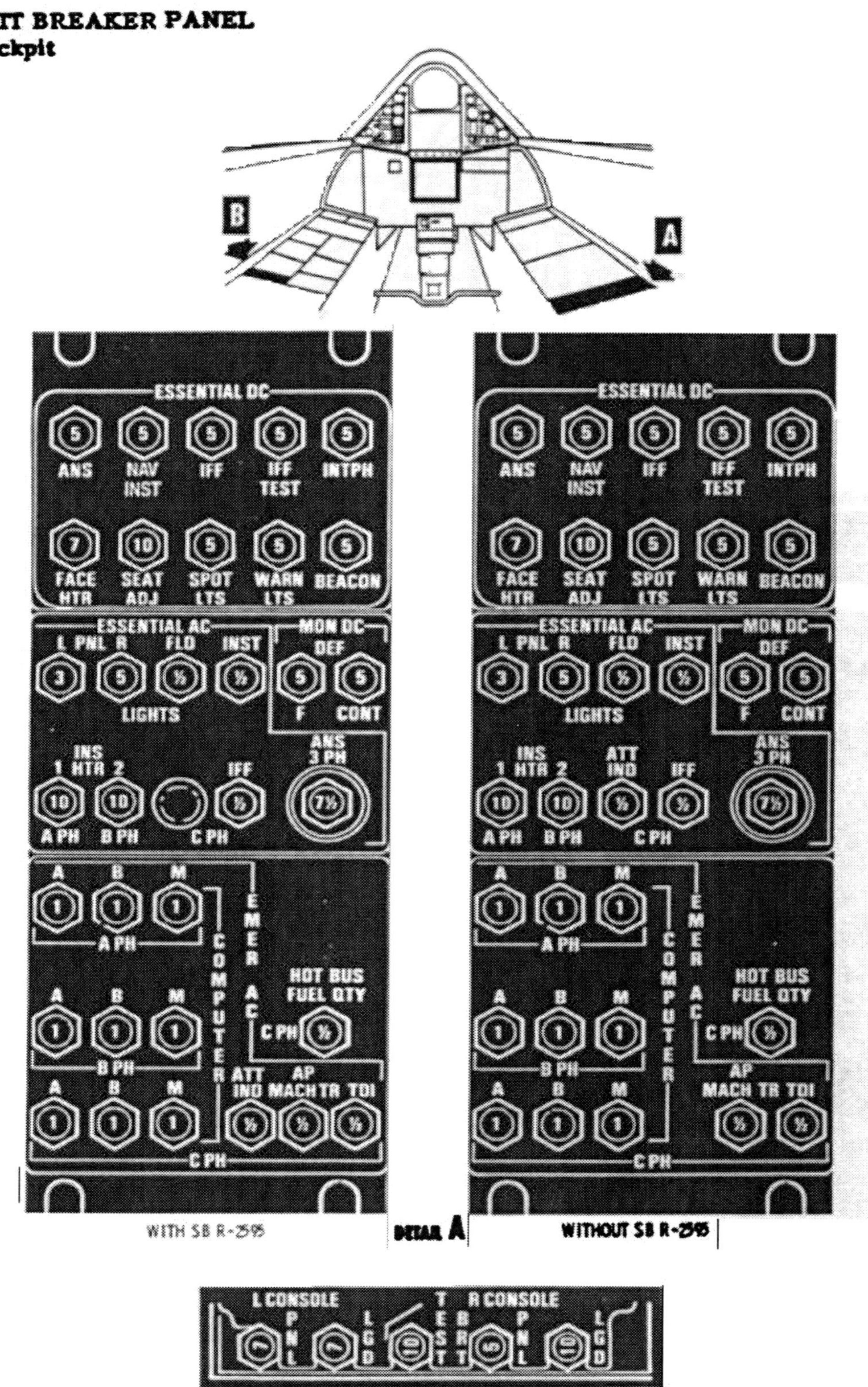

Figure 1-42.

The left engine generator automatically synchronizes its output with the right generator, then its contactor closes and the generators operate in parallel. If the right generator is not on line, the left generator can be set in NORM, but it may not parallel the external power supply and it will not come on line until external power is disconnected or shut off. A momentary power surge can result if the phasing is not synchoronized. The L GEN OUT caution light will remain on until power is transferred.

Engine starting without external electrical power is possible, but not recommended. Instrument indications, including EGT, require essential ac bus power and would not be available until the generators are turned on.

CIRCUIT BREAKERS

Circuit breaker panels, located in the forward and aft cockpits, contain pullout/push-to-reset breakers for certain ac and dc circuits. See Figures 1-41 and 1-42. Services interrupted by opening these circuit breakers are listed by Figure 1-47. Other circuit breaker panels which are not accessible during flight are located in the C and E bays.

Differential Protection Relays

A differential protection relay (DPR) is a part of each generator system. The DPRs provide automatic protection by disconnecting the associated generator for a significant fault within the generator, in the generator feeder lines (to the buses), and/or in the generator line contactors.

ELECTRICAL SYSTEM CONTROLS AND INDICATOR LIGHTS

Generator Switches

A 3-position lift-loc control switch for each generator is located on the pilot's right instrument side panel. The switch positions are NORM (down), OFF (center), and EMER (up). Placing either switch in NORM will return the respective generator to normal operation if it has been removed from the bus for any reason other than generator or system failure. The NORM position is locked to prevent accidental actuation to OFF or EMER. In OFF, the corresponding generator is removed from service.

NOTE

> If its protective circuits trip a generator and it does not reset automatically, the generator switch must be moved to OFF and then back to NORM to attempt a manual reset.

In EMER (used when both generators have failed), a 28 volt dc excitation current from the essential dc bus is applied to the respective generator exciter fields through a single 5 amp EMER EXC circuit breaker. If the generator is rotating and the windings are operative, an unregulated voltage and frequency ac current is developed which will power the hot bus and corresponding generator bus. The essential ac bus will be dead. Extinction of the corresponding L and/or R GEN OUT light(s), after selecting EMER, indicates successful emergency system operation. The corresponding L and/or R XFMR RECT OUT light may extinguish; if this occurs, the EMER BAT ON light will extinguish if either the battery voltage is exceeded by the transformer-rectifier output, or the battery output is less than 10 amperes.

Bus Tie Switch

A push-button bus tie switch is located on the pilot's right instrument side panel. If the GEN BUS TIE OPEN light illuminates simultaneously with indication of generator failure, depressing the switch should retie the L and R generator buses and extinguish the light, if the failure was in the generator or its control system.

ELECTRICAL LOAD ANALYSIS WHEN NO AC GENERATOR AVAILABLE

DC ESSENTIAL BUS ITEM	POSSIBLE LOAD IN AMPERES	CAN BE USED OR TURNED OFF BY SWITCH	CAN BE USED OR TURNED OFF BY CB	CANNOT BE USED-TURNED OFF BY SWITCH	CANNOT BE USED-TURNED OFF BY CB
AIR REFUEL	1.20	X			
AIR SHUTOFF CONTROL	2.80	X			
APW	2.50	X			
AUTOPILOT/MACH TRIM A	0.30		X		
AUTOPILOT/MACH TRIM B	0.30		X		
BRAKE AND SKID	1.60	X			
COCKPIT AIR	0.10		X		
COCKPIT AND BAY TEMP	0.30	X			
COMNAV-50	2.75			X	
COMPUTER A	0.20		X		
COMPUTER B	0.20		X		
COMPUTER M	0.20		X		
DRAG CHUTE	3.00	X			
EGT TRIM, L AND R	1.00		X		
EMER FUEL S/O, L AND R	0.40			X	
EMER GEN EXCITER	1.00	X			
FACE HEAT	0.88	X			
FUEL CONTROL	0.40				X
FUEL DERICH, L AND R	2.40		X		
FUEL DUMP	2.30				
FUEL FWD TRANSFER	1.50	X	X		
FUEL TANK 5 TRANSFER	2.40				X
GROUND AIR VALVES	0.70				X
IFF	0.18			X	
IFR INTERCOMM	0.07		X		
IGNITER PURGE, L AND R	2.00	X			
ILS	1.83	X			
INLET AFT BYPASS, L AND R	4.00		X		
INLET GUIDE VANES	0.08				
INS	2.00	X	X		
INSTRUMENT INVERTER	28.50	X			
INTERPHONE	0.32		X		
LANDING GEAR CONTROL			X		
LANDING GEAR INDICATORS			X		
LANDING GEAR WARNING			X		
LN_2 QUANTITY, NO. 1 AND NO. 2	2.00				X
MANIFOLD TEMP	1.30		X		
NAVIGATION INSTRUMENTS	0.86		X		
NAVIGATION SYSTEM					X
NOSE STEERING	2.15				X
PILOT VALVES	0.00				

Figure 1-43 (Sheet 1 of 2)

ELECTRICAL LOAD ANALYSIS WHEN NO AC GENERATOR AVAILABLE (CONT.)

DC ESSENTIAL BUS ITEM	POSSIBLE LOAD IN AMPERES	CAN BE USED OR TURNED OFF BY SWITCH	CAN BE USED OR TURNED OFF BY CB	CANNOT BE USED-TURNED OFF BY SWITCH	CANNOT BE USED-TURNED OFF BY CB
PITOT HEAT	0.00			X	
PRESSURE DUMP	0.40	X			
RAIN REMOVAL (W/O S/B 2674)	1.50	X			
RUDDER LIMITER, L AND R	3.30	HANDLE			
SAS A	8.80	X	X		
SAS B	8.80	X	X		
SEAT ADJUST	5.00	X			
SPIKE AND DOOR POS. IND.	0.80		X		
SPIKE CONTROL L	1.00		X		
SPIKE CONTROL R	1.00		X		
SPIKE SOL L	3.50		X		
SPIKE SOL R	3.50		X		
SPOT LIGHTS	0.36	X			
STALL WARNING	0.15		X		
STANDBY ATT. IND., 3-INCH	2.00		X		
TACAN (ARN-118)	2.00			X	
TANK 4 ULLAGE	0.15				X
T-STORM LIGHTS	1.43	X			
TURN AND SLIP INDICATOR	0.25		X		
UHF ANTENNA ACTUATOR	3.50	X			
VHF (ARC-186)	2.96	X			
WARNING LIGHTS	1.41		X		
WINDSHIELD DEICE	0.70	X			

EMERGENCY AC BUS ITEM	POSSIBLE LOAD IN VA	CAN BE USED OR TURNED OFF BY SWITCH	CAN BE USED OR TURNED OFF BY CB	CANNOT BE USED-TURNED OFF BY SWITCH	CANNOT BE USED-TURNED OFF BY CB
ANGLE OF ATTACK IND.	5.00				
ATTITUDE DIRECTOR IND.	20.00		X		
ATTITUDE IND-RSO (S/B 2595)	20.00		X		
AUTOPILOT/MACH TRIM	30.00		X		
COMPUTER A	124.00		X		
COMPUTER B	124.00		X		
COMPUTER M	124.00		X		
EMER INSTR TRANSFORMER	18.20		X		
FIRE WARNING, L AND R	2.20		X		
HSI	50.00		X		
INLET, L	20.00		X		
INLET, R	20.00		X		
INS	220.00	X			
PITCH INDICATOR	0.21		X		
STANDBY ATT. IND., 2-INCH	13.20		X		
TDI	39.00		X		

Fig [illegible] 2)

ELECTRICAL POWER DISTRIBUTION

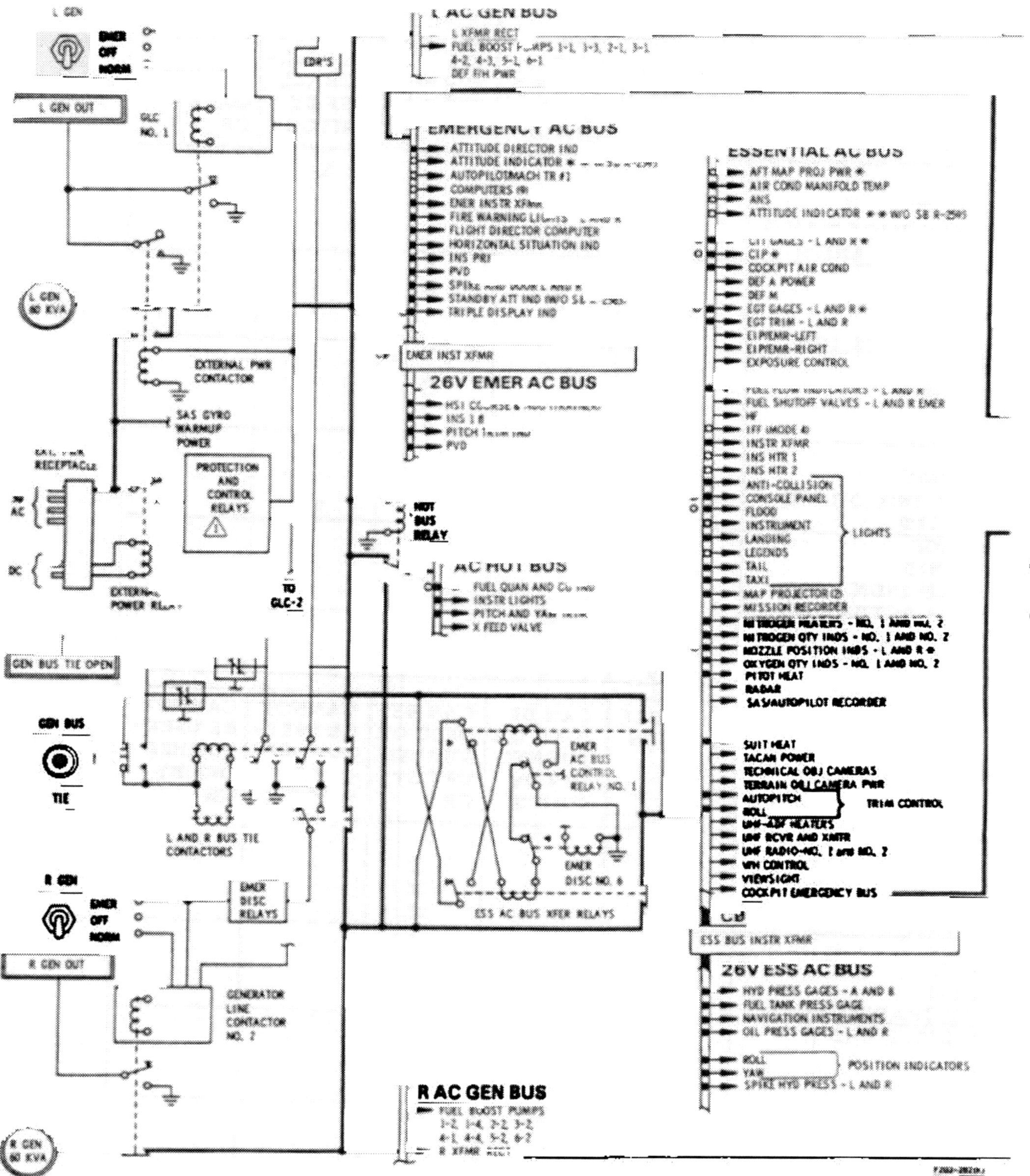

Figure 1-44 (Sheet 1 of 2)

ELECTRICAL POWER DISTRIBUTION

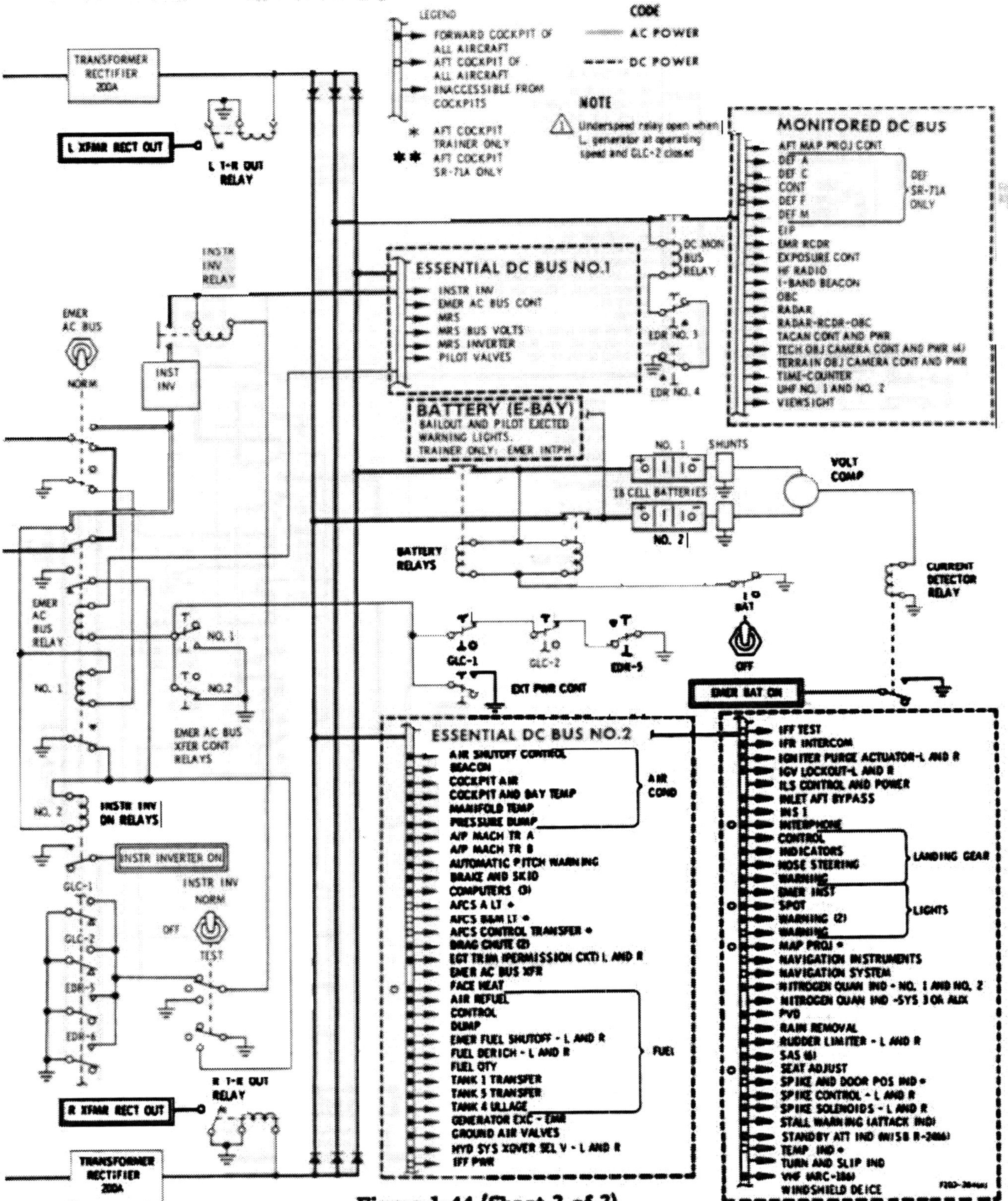

LEGEND
- FORWARD COCKPIT OF ALL AIRCRAFT
- AFT COCKPIT OF ALL AIRCRAFT
- INACCESSIBLE FROM COCKPITS
- * AFT COCKPIT TRAINER ONLY
- ** AFT COCKPIT SR-71A ONLY

CODE
- AC POWER
- DC POWER

NOTE

Underspeed relay open when L generator at operating speed and GLC-2 closed

TRANSFORMER RECTIFIER 200A

L XFMR RECT OUT

L T-R OUT RELAY

INSTR INV RELAY

EMER AC BUS

NORM

INST INV

ESSENTIAL DC BUS NO.1
- INSTR INV
- EMER AC BUS CONT
- MRS
- MRS BUS VOLTS
- MRS INVERTER
- PILOT VALVES

DC MON BUS RELAY

EDR NO. 3

EDR NO. 4

MONITORED DC BUS
- AFT MAP PROJ CONT
- DEF A
- DEF C
- CONT
- DEF F
- DEF M
- EIP
- EMR RCDR
- EXPOSURE CONT
- HF RADIO
- I-BAND BEACON
- OBC
- RADAR
- RADAR-RCDR-OBC
- TACAN CONT AND PWR
- TECH OBJ CAMERA CONT AND PWR (4)
- TERRAIN OBJ CAMERA CONT AND PWR
- TIME-COUNTER
- UHF NO. 1 AND NO. 2
- VIEWSIGHT

DEF SR-71A ONLY

BATTERY (E-BAY)
BAILOUT AND PILOT EJECTED WARNING LIGHTS.
TRAINER ONLY: EMER INTPH

NO. 1 SHUNTS

18 CELL BATTERIES

NO. 2

VOLT COMP

BATTERY RELAYS

CURRENT DETECTOR RELAY

BAT

OFF

EMER AC BUS RELAY

NO. 1

NO. 2

GLC-1

GLC-2

EDR-5

EXT PWR CONT

EMER BAT ON

EMER AC BUS XFER CONT RELAYS

INSTR INV ON RELAYS

INSTR INVERTER ON

INSTR INV

NORM

OFF

TEST

GLC-1

GLC-2

EDR-5

EDR-6

R T-R OUT RELAY

R XFMR RECT OUT

TRANSFORMER RECTIFIER 200A

ESSENTIAL DC BUS NO.2
- AIR SHUTOFF CONTROL
- BEACON
- COCKPIT AIR
- COCKPIT AND BAY TEMP
- MANIFOLD TEMP
- PRESSURE DUMP
- A/P MACH TR A
- A/P MACH TR B
- AUTOMATIC PITCH WARNING
- BRAKE AND SKID
- COMPUTERS (3)
- AFCS A LT *
- AFCS B&M LT *
- AFCS CONTROL TRANSFER *
- DRAG CHUTE (2)
- EGT TRIM (PERMISSION CKT) L AND R
- EMER AC BUS XFR
- FACE HEAT
- AIR REFUEL
- CONTROL
- DUMP
- EMER FUEL SHUTOFF - L AND R
- FUEL DERICH - L AND R
- FUEL QTY
- TANK 1 TRANSFER
- TANK 5 TRANSFER
- TANK 4 ULLAGE
- GENERATOR EXC - EMR
- GROUND AIR VALVES
- HYD SYS XOVER SEL V - L AND R
- IFF PWR

AIR COND

FUEL

- IFF TEST
- IFR INTERCOM
- IGNITER PURGE ACTUATOR-L AND R
- IGV LOCKOUT-L AND R
- ILS CONTROL AND POWER
- INLET AFT BYPASS
- INS 1
- INTERPHONE
- CONTROL
- INDICATORS
- NOSE STEERING
- WARNING
- EMER INST
- SPOT
- WARNING (2)
- WARNING
- MAP PROJ *
- NAVIGATION INSTRUMENTS
- NAVIGATION SYSTEM
- NITROGEN QUAN IND - NO. 1 AND NO. 2
- NITROGEN QUAN IND -SYS 3 OR AUX
- PVD
- RAIN REMOVAL
- RUDDER LIMITER - L AND R
- SAS (6)
- SEAT ADJUST
- SPIKE AND DOOR POS IND *
- SPIKE CONTROL - L AND R
- SPIKE SOLENOIDS - L AND R
- STALL WARNING (ATTACK IND)
- STANDBY ATT IND (MISB R-2086)
- TEMP IND *
- TURN AND SLIP IND
- VHF (ARC-186)
- WINDSHIELD DEICE

LANDING GEAR

LIGHTS

Figure 1-44 (Sheet 2 of 2)

EMERGENCY AC POWER SUPPLY

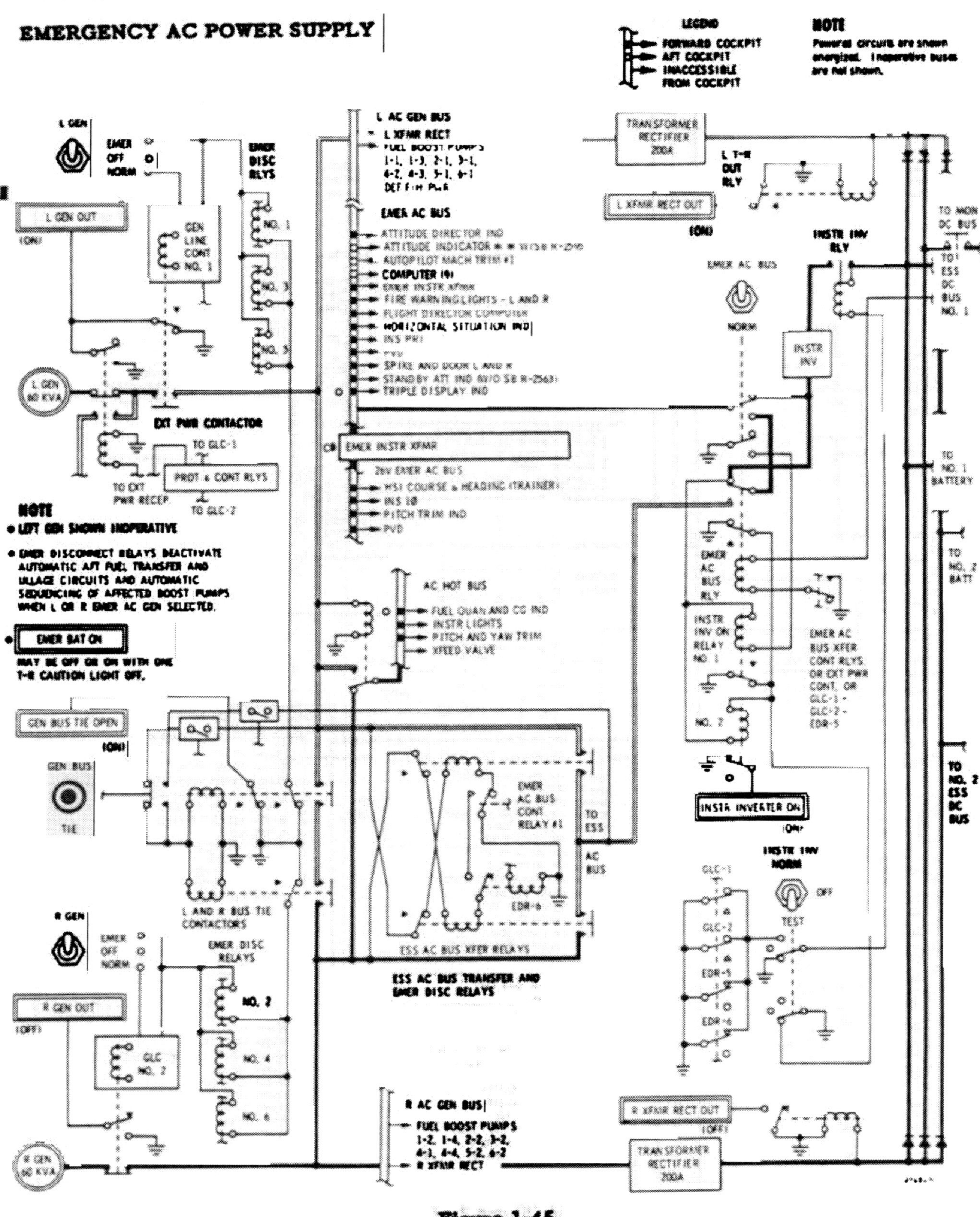

Figure 1-45

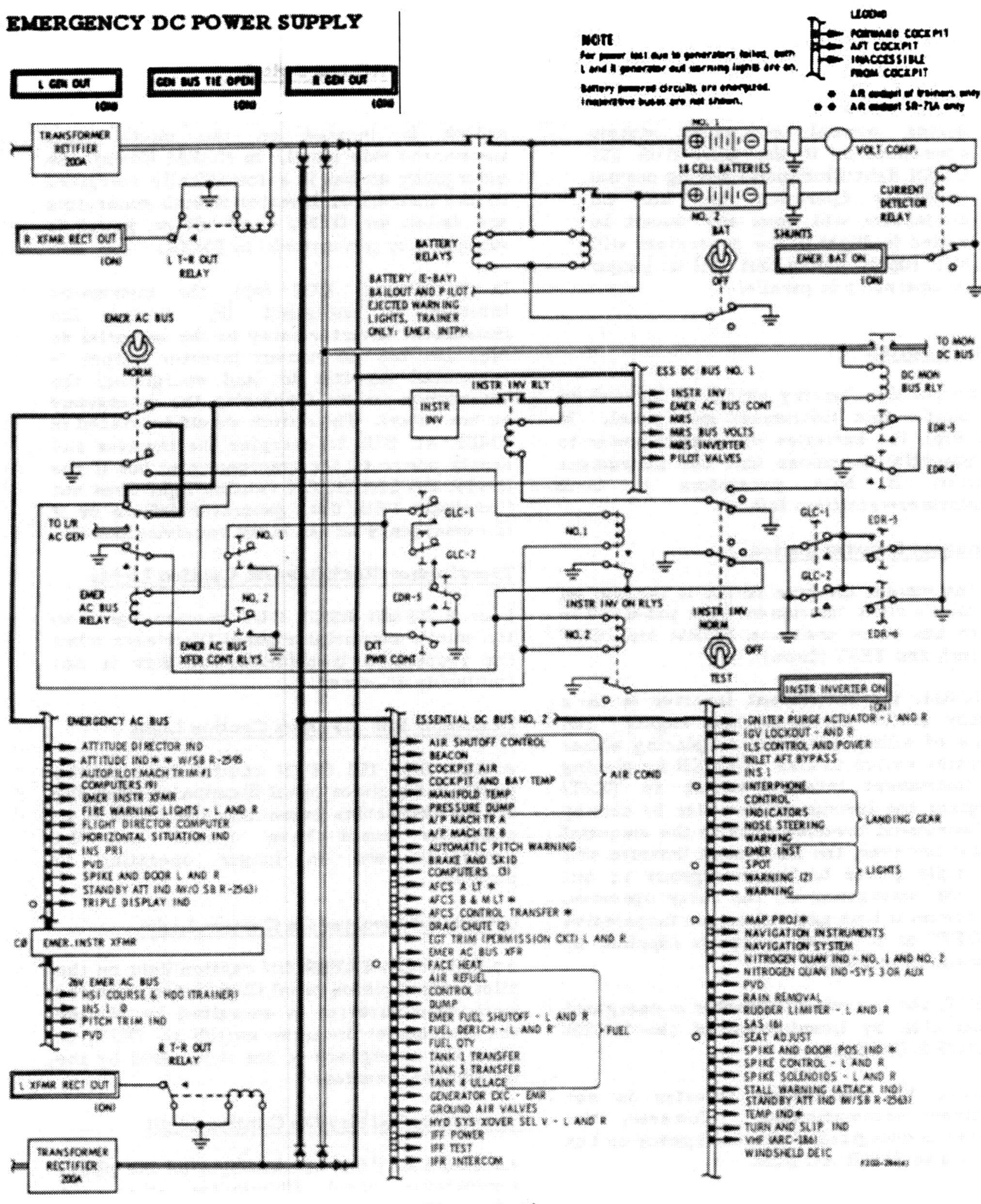
EMERGENCY DC POWER SUPPLY
L GEN OUT
(ON)
GEN BUS TIE OPEN
(ON)
R GEN OUT
(ON)
NOTE
For power lost due to generators failed, both L and R generator out warning lights are on.
Battery powered circuits are energized. Inoperative buses are not shown.
LEGEND
FORWARD COCKPIT
AFT COCKPIT
INACCESSIBLE FROM COCKPIT
AR cockpit of trainers only
AR cockpit SR-71A only
TRANSFORMER RECTIFIER 200A
R XFMR RECT OUT
(ON)
L T-R OUT RELAY
NO. 1
18 CELL BATTERIES
NO. 2
VOLT COMP.
CURRENT DETECTOR RELAY
SHUNTS
BAT
OFF
EMER BAT ON
(ON)
BATTERY RELAYS
BATTERY (E-BAY) BAILOUT AND PILOT EJECTED WARNING LIGHTS, TRAINER ONLY: EMER INTPH
EMER AC BUS
NORM
TO MON DC BUS
DC MON BUS RLY
EDR-5
EDR-4
ESS DC BUS NO. 1
INSTR INV
EMER AC BUS CONT
MRS
MRS BUS VOLTS
MRS INVERTER
PILOT VALVES
INSTR INV RLY
INSTR INV
TO L/R AC GEN
NO. 1
NO. 2
EMER AC BUS RELAY
EMER AC BUS XFER CONT RLYS
GLC-1
GLC-2
EDR-5
EXT PWR CONT
NO. 1
NO. 2
INSTR INV ON RLYS
INSTR INV NORM
OFF
TEST
INSTR INVERTER ON
GLC-1
GLC-2
EDR-5
EDR-4
EMERGENCY AC BUS
ATTITUDE DIRECTOR IND
ATTITUDE IND * * W/SB R-2505
AUTOPILOT MACH TRIM #1
COMPUTERS (9)
EMER INSTR XFMR
FIRE WARNING LIGHTS - L AND R
FLIGHT DIRECTOR COMPUTER
HORIZONTAL SITUATION IND
INS PRI
PVD
SPIKE AND DOOR L AND R
STANDBY ATT IND W/O SB R-2563
TRIPLE DISPLAY IND
CB
EMER. INSTR XFMR
28V EMER AC BUS
HSI COURSE & HDG (TRAINER)
INS 1
PITCH TRIM IND
PVD
R T-R OUT RELAY
L XFMR RECT OUT
(ON)
TRANSFORMER RECTIFIER 200A
ESSENTIAL DC BUS NO. 2
AIR SHUTOFF CONTROL
BEACON
COCKPIT AIR
COCKPIT AND BAY TEMP
MANIFOLD TEMP
PRESSURE DUMP
AIR COND
A/P MACH TR A
A/P MACH TR B
AUTOMATIC PITCH WARNING
BRAKE AND SKID
COMPUTERS (3)
AFCS A LT *
AFCS B & M LT *
AFCS CONTROL TRANSFER *
DRAG CHUTE (2)
EGT TRIM (PERMISSION CKT) L AND R
EMER AC BUS XFR
FACE HEAT
AIR REFUEL
CONTROL
DUMP
EMER FUEL SHUTOFF - L AND R
FUEL DERICH - L AND R
FUEL QTY
TANK 1 TRANSFER
TANK 3 TRANSFER
TANK 4 ULLAGE
FUEL
GENERATOR CKC - EMR
GROUND AIR VALVES
HYD SYS XOVER SEL V - L AND R
IFF POWER
IFF TEST
IFR INTERCOM
IGNITER PURGE ACTUATOR - L AND R
IGV LOCKOUT - AND R
ILS CONTROL AND POWER
INLET AFT BYPASS
INS 1
INTERPHONE
CONTROL
INDICATORS
NOSE STEERING
WARNING
LANDING GEAR
EMER INST
SPOT
WARNING (2)
WARNING
LIGHTS
MAP PROJ *
NAVIGATION INSTRUMENTS
NAVIGATION SYSTEM
NITROGEN QUAN IND - NO. 1 AND NO. 2
NITROGEN QUAN IND - SYS 3 OR AUX
PVD
RAIN REMOVAL
RUDDER LIMITER - L AND R
SAS (6)
SEAT ADJUST
SPIKE AND DOOR POS. IND *
SPIKE CONTROL - L AND R
SPIKE SOLENOIDS - L AND R
STALL WARNING (ATTACK IND)
STANDBY ATT IND W/SB R-2563
TEMP IND *
TURN AND SLIP IND
VHF (ARC-186)
WINDSHIELD DEIC

Figure 1-46

NOTE

If the bus tie switch is depressed during normal electrical system operation, or if the GEN BUS TIE OPEN light illuminates during normal generator operation, the bus tie contactors will open and cannot be retied in-flight. The generators will still supply power, but will no longer be operating in parallel.

Battery Switch

A two-position battery switch is located on the pilot's right instrument side panel. In BAT (up), the batteries will supply power to the essential dc busses (and the instrument inverter) if both generators or both transformer-rectifiers fail.

Instrument Inverter Switch

The instrument inverter switch is located on the pilot's right instrument side panel. The switch has three positions: NORM (up), OFF (center), and TEST (down).

In NORM, the instrument inverter is in a standby status during normal flight. The failure of either generator or placing either generator switch to OFF or EMER (or placing the instrument inverter switch to TEST) energizes the instrument inverter by closing the instrument inverter relay to the essential dc bus; however, the instrument inverter will not supply power to the emergency ac bus until the emergency ac bus relay operates. This occurs if both generators are inoperative (or OFF) or if all ac power is supplied by generator(s) in EMER.

In TEST, the instrument inverter is energized as indicated by illumination of the INSTR INVERTER ON light.

In OFF, the instrument inverter is not energized automatically. However, the inverter is energized if the emergency ac bus switch is in EMER AC BUS.

Emergency AC Bus Switch

A two-position lift-loc emergency ac bus switch is located on the pilot's right instrument side panel. In NORM (down), the emergency ac bus is automatically energized by the instrument inverter if both generators are failed (or OFF), or if all ac power is suppplied by generator(s) in EMER.

In EMER AC BUS (up), the instrument inverter is energized (by closing the instrument inverter relay to the essential dc bus) and the instrument inverter output is connected directly to (and energizing) the emergency ac bus (bypassing the emergency ac bus relay). The switch should be placed in EMER AC BUS to energize the inverter and supply power to the emergency ac bus if the INSTR INVERTER ON caution light does not illuminate with dual generator failure or if the emergency ac bus is not receiving power.

Transformer-Rectifier Out Caution Lights

L or R XFMR RECT OUT caution lights on the pilot's annunciator panel illuminate when the respective transformer-rectifier is not furnishing dc power.

Generator Bus Tie Open Caution Light

A GEN BUS TIE OPEN caution light on the pilot's annunciator panel illuminates when the bus tie contactors connecting the L and R ac generator buses have opened and the generators are no longer operating in parallel.

Instrument Inverter On Caution Light

An INSTR INVERTER ON caution light on the pilot's annunciator panel illuminates when the instrument inverter is energized by placing the instrument inverter switch to TEST, or when the emergency ac bus is powered by the instrument inverter.

Emergency Battery On Caution Light

An EMER BAT ON caution light on the pilot's annunciator panel illuminates when the

CIRCUIT BREAKER FUNCTION TABLE

CIRCUIT BREAKER	EFFECT OF POWER INTERRUPTION
	ESSENTIAL DC BUS (Forward Cockpit)
LANDING GEAR	
IND	Disabled: Landing gear indicator lights. Depress gear solenoid release button to move gear handle from DOWN to UP. Landing and tail lights will illuminate, if selected, while nose gear is up.
CONT	Disabled: Landing gear retraction, normal gear extension, normal nosewheel steering system.
WARN	Disabled: Lights in gear handle, pulse tone from gear warning horn.
STEER	Disabled: Nosewheel steering control switch (CSC/NWS) on control stick, and normal and alternate steering systems.
BRK & SKD	Disabled: High α system stick shaker, anti-skid brakes, alternate brake system, alternate nose wheel steering system.
RAIN REMOVAL	Disabled: Windshield rain removal system. Valve closes, if open.
MAP PROJ	Disabled: Pilot's map projector.
HYD R↔L	Disabled: Hydraulic system crossover for gear retraction, alternate nosewheel steering system, aerial refueling with power from R hydraulic system.
DRAG CHUTE (2)	Disabled: Drag chute system if both breakers open.
FACE HTR	Disabled: Helmet face heat, forward cockpit.
DEICE	Disabled: Windshield deice system.
FUEL	
CONT	Disabled: Manual boost pump selection. Fuel crossfeed valve will close, if open. Fuel system reverts to automatic sequencing.
TRANS	
TK 1	Disabled: Forward fuel transfer. Transfer valve will close, if open.
TK 5	Disabled: Aft fuel transfer, manual and automatic. Also tank 2, 3, and 5 boost pump indicator lights.
AIR REFUEL	Refuel door opens, as energized circuit required to close door.
TK 4 ULLAGE	System will not sense fuel high level condition in tank 4 to start ullage pumps in that tank.
DUMP CONT	Disabled: Fuel dump valves. Valve will close, if open.
FUEL QTY	Disabled: Forward and aft cockpit fuel quantity and c.g. indicators.
SHUTOFF - L ENG	Disabled: Left engine fuel shutoff. AC power will open the valve, if closed. Left heat sink crossfeed will open, if closed.
SHUTOFF - R ENG	Disabled: Right engine fuel shutoff. AC power will open the valve, if closed. Right heat sink crossfeed valve will open, if closed.
INS	Disabled: Holding power for warning flag at top of ADI, flag comes in view but displayed information is still valid.
N2 QTY	
NO 1	Disabled: SYS 1 N QTY LOW light, LN_2 crossfeed valve, No. 1 LN_2 Dewar heater.
NO 2	Disabled: SYS 2 N QTY LOW light, No. 2 LN_2 aft regulator solenoid (will not sense low tank pressure), No. 2 LN_2 Dewar heater.
INLET	
SPIKE - L	Disabled: Left and right inlet forward bypass position indications, left inlet automatic control and left inlet unstart light.
SPIKE - R	Disabled: Left and right inlet spike position indications, right inlet automatic control and right inlet unstart light.

Figure 1-47 (Sheet 1 of 7)

CIRCUIT BREAKER FUNCTION TABLE

CIRCUIT BREAKER	EFFECT OF POWER INTERRUPTION
	ESSENTIAL DC BUS (Forward Cockpit - Cont.)
INLET (cont.)	
SPIKE - L SOL	Disabled: Left spike and door override solenoids. Open fwd bypass manually when landing.
SPIKE - R SOL	Disabled: Right spike and door override solenoids. Open fwd bypass manually when landing.
	Note: For the respective inlet, if the SPIKE SOL circuit breaker is open: the spike only moves 15 inches forward of auto schedule when restart ON is selected or the throttle restart switch is set to the aft position; the center position of the throttle restart switch is inoperative; if manual spike is selected, all normal manual restart capability is lost; 30,000 foot and Mach 1.4 switch for spike and Mach 1.4 switch for forward bypass door are inoperative. Forward bypass remains closed when gear is extended.
AFT BYPASS	Disabled: Left and right inlet aft bypass position control. Doors remain as set.
ENGINE	
IGN PURGE	Disabled: Chemical ignition system (TEB) tank dump.
FUEL DERICH - L	Disabled: Left engine fuel derich system.
FUEL DERICH - R	Disabled: Right engine fuel derich system.
EGT - L	Disabled: Left engine automatic EGT trim system permission circuit, and EGT autotrim.
EGT - R	Disabled: Right engine automatic EGT trim system permission circuit, and EGT autotrim.
IGV - L	Disabled: Left engine IGV lockout solenoid. Goes to unlocked condition.
IGV - R	Disabled: Right engine IGV lockout solenoid. Goes to unlocked condition.
AIR COND	
AIR SOV CONT	Disabled: Shutoff controls for L and R air conditioning systems, left and right mission bays, and nose section. Valves remain in set positions. Air conditioning continues if refrigeration switches are on when c/b opens.
PRESS DUMP	Disabled: Cockpit pressure dump valve. Valve closes, if open.
CKPT AIR	Disabled: Cockpit automatic and manual temperature controls and defog control.
TEMP	
MANF	Disabled: Manifold temperature control valve, full cold selection, hot air bypass system, ANS flow limiting valve.
CKPT & BAY	Disabled: Temperature indication for cockpit, R-Bay and E-Bay.
GRD AIR VAL	Disabled: L and R forward mission bay ground air shutoff valves.
COMM	
IFR COMM	Disabled: Air refueling system boom communication amplifier.
INTPH	Disabled: All audio to forward cockpit headset (except trainer EMERG ICS).
LIGHT	
WARN 1	Disabled: All forward cockpit warning and caution lights except nacelle fire, inlet unstart, and landing gear warning lights and lights on WARN 2 circuit breaker.
WARN 2	Disabled: KEAS warning, RSO ejected, both IGV, both Derich and A/P off lights.
SPOT	Disabled: Forward cockpit spot lights and flex point lights, thunderstorm lights, altimeter vibrator.
EMER INSTR	Disabled: Instrument panel emergency lighting circuit.
RUD LIM	
L	Disabled: Full deflection of left rudder. 10° available.
R	Disabled: Full deflection of right rudder. 10° available.

Figure 1-47 (Sheet 2 of 7)

CIRCUIT BREAKER FUNCTION TABLE

CIRCUIT BREAKER	EFFECT OF POWER INTERRUPTION
	ESSENTIAL DC BUS (Forward Cockpit – Cont.)
A/P MACH TR A	Autopilot redundancy is lost and Mach Trim redundancy is reduced.
A/P MACH TR B	Autopilot redundancy is lost and Mach Trim redundancy is reduced.
	NOTE • To disable the Autopilot, both A/P MACH TR A & B circuit breakers must be opened. • To disable Mach Trim, both A/P MACH TR A & B circuit breakers and the CMPTR M circuit breaker must be opened.
APW	Disabled: APW pusher and shaker. High α system shaker not affected.
STALL WARN	Disabled: High α warning system.
CMPTR	
A	Disabled: A computer.
B	Disabled: B computer.
M	Disabled: M computer, Mach trim redundancy is reduced.
TURN GYRO	Disabled: ADI turn rate indication.
EMER EXC	Disabled: Emergency generator DC exciter system.
SEAT ADJ	Disabled: Forward cockpit seat adjustment. Seat remains as set.
SAS	
PITCH	
A	Disabled: SAS Pitch A servos (left and right engage solenoids).
B	Disabled: SAS Pitch B servos (left and right engage solenoids).
YAW	
A	Disabled: SAS Yaw A servos (left and right engage solenoids).
B	Disabled: SAS Yaw B servos (left and right engage solenoids).
ROLL	
A	Disabled: SAS Roll A servo (left engage solenoid).
B	Disabled: SAS Roll B servo (right engage solenoid).
PVD (on RH Console)	Disabled: PVD.
STBY ATT (on RH Console)	Disabled: 3-inch standby attitude indicator gyro.
	EMERGENCY AC BUS (Forward Cockpit)
L SPIKE AND DOOR (AØ)	Disabled: Left inlet spike and forward bypass controls and position indications. Mechanical bias within servos programs spike full forward and forward bypass door full open. Indications freeze.
L FIRE WARN (AØ)	Disabled: Left nacelle fire warning system.
STBY ATT (AØ)	Disabled: 2-inch standby attitude indicator gyro.
PVD (AØ)	Disabled: PVD.

Figure 1-47 (Sheet 3 of 7)

CIRCUIT BREAKER FUNCTION TABLE

CIRCUIT BREAKER		EFFECT OF POWER INTERRUPTION
		EMERGENCY AC BUS (Forward Cockpit - Cont.)
R SPIKE AND DOOR	BØ	Disabled: Right inlet spike and forward bypass controls and position indications. Mechanical bias within servos programs spike full forward and forward bypass door full open. Indications freeze.
R FIRE WARN	BØ	Disabled: Right nacelle fire warning system.
INS PRIME	BØ	Disabled: Primary INS operate power and INS segment (display) lights. INS operates until INU battery reaches 18 volts, then shuts down.
FDC	CØ	Disabled: Flight director computer.
HSI	CØ	Disabled: HSI instrument except for localizer signal during ILS approach.
ATT IND	CØ	Disabled: ADI instrument except for glide slope signal and off flag during ILS approach.
TDI	CØ	Disabled: Forward cockpit TDI.
INSTR XFMR	CØ	Disabled: 26 v Emergency AC Bus (All functions of 26 v Emergency AC bus circuit breakers.)
		26 VOLT EMERGENCY AC BUS (Forward Cockpit)*
TRIM		
└ PITCH	CØ	Disabled: Pitch trim indication.
INS	CØ	Disabled: INS synchro power. INS attitude and heading indications are invalid (INS platform is not affected).
PVD (on RH Console)	CØ	Disabled: PVD
		*NOTE: Emergency AC Bus INSTR XFMR circuit breaker must be in for power to 26 V Emergency AC Bus.
		AC HOT BUS (Forward Cockpit - C/Bs located on ESS AC Bus C/B panel)
TRIM		
└ PITCH AND YAW		Disabled: Manual pitch and yaw trim.
LIGHTS		
└ INSTR		Disabled: All left console INSTR LIGHTS C/B functions. (LH, RH, FUEL CONT and FUEL CONT TEST C/Bs)
FUEL		
└ QTY		Disabled: Fuel quantity indication, forward cockpit, & c.g. indications, both cockpits.
INSTR LIGHTS		Left Console Lighting Panel
LH		Disabled: Left instrument panel lights.
RH		Disabled: Right instrument panel lights.
FUEL CONT		Disabled: Fuel system control panel lights, including cross-feed light.
FUEL CONT TEST		Disabled: Fuel system control panel lights test function, and pilot's ADI light.

Figure 1-47 (Sheet 4 of 7)

CIRCUIT BREAKER FUNCTION TABLE

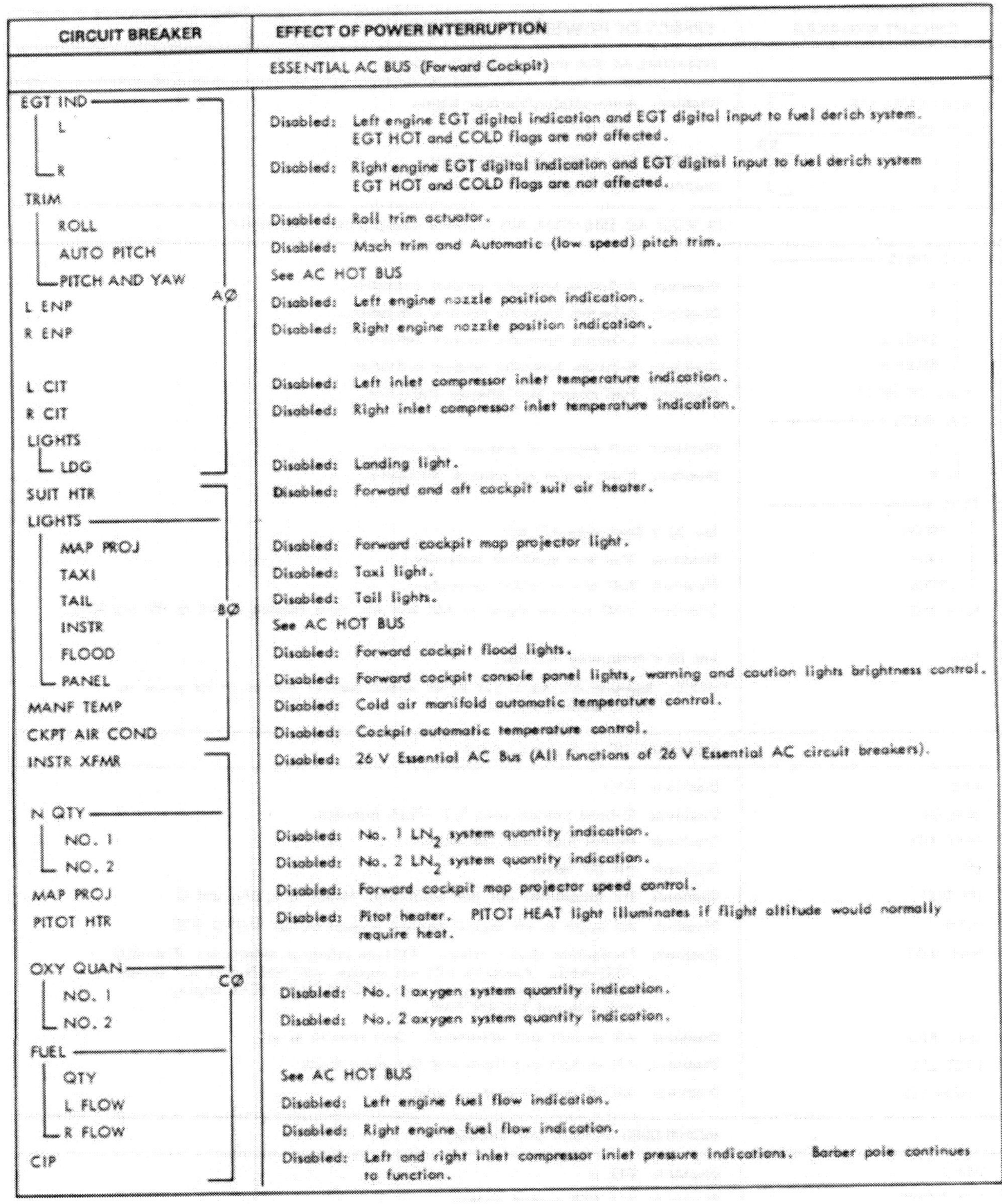

CIRCUIT BREAKER		EFFECT OF POWER INTERRUPTION
		ESSENTIAL AC BUS (Forward Cockpit)
EGT IND	AØ	
L	AØ	Disabled: Left engine EGT digital indication and EGT digital input to fuel derich system. EGT HOT and COLD flags are not affected.
R	AØ	Disabled: Right engine EGT digital indication and EGT digital input to fuel derich system EGT HOT and COLD flags are not affected.
TRIM	AØ	
ROLL	AØ	Disabled: Roll trim actuator.
AUTO PITCH	AØ	Disabled: Mach trim and Automatic (low speed) pitch trim.
PITCH AND YAW	AØ	See AC HOT BUS
L ENP	AØ	Disabled: Left engine nozzle position indication.
R ENP	AØ	Disabled: Right engine nozzle position indication.
L CIT	AØ	Disabled: Left inlet compressor inlet temperature indication.
R CIT	AØ	Disabled: Right inlet compressor inlet temperature indication.
LIGHTS	AØ	
LDG	AØ	Disabled: Landing light.
SUIT HTR	BØ	Disabled: Forward and aft cockpit suit air heater.
LIGHTS	BØ	
MAP PROJ	BØ	Disabled: Forward cockpit map projector light.
TAXI	BØ	Disabled: Taxi light.
TAIL	BØ	Disabled: Tail lights.
INSTR	BØ	See AC HOT BUS
FLOOD	BØ	Disabled: Forward cockpit flood lights.
PANEL	BØ	Disabled: Forward cockpit console panel lights, warning and caution lights brightness control.
MANF TEMP	BØ	Disabled: Cold air manifold automatic temperature control.
CKPT AIR COND	BØ	Disabled: Cockpit automatic temperature control.
INSTR XFMR	CØ	Disabled: 26 V Essential AC Bus (All functions of 26 V Essential AC circuit breakers).
N QTY	CØ	
NO. 1	CØ	Disabled: No. 1 LN_2 system quantity indication.
NO. 2	CØ	Disabled: No. 2 LN_2 system quantity indication.
MAP PROJ	CØ	Disabled: Forward cockpit map projector speed control.
PITOT HTR	CØ	Disabled: Pitot heater. PITOT HEAT light illuminates if flight altitude would normally require heat.
OXY QUAN	CØ	
NO. 1	CØ	Disabled: No. 1 oxygen system quantity indication.
NO. 2	CØ	Disabled: No. 2 oxygen system quantity indication.
FUEL	CØ	
QTY	CØ	See AC HOT BUS
L FLOW	CØ	Disabled: Left engine fuel flow indication.
R FLOW	CØ	Disabled: Right engine fuel flow indication.
CIP	CØ	Disabled: Left and right inlet compressor inlet pressure indications. Barber pole continues to function.

Figure 1-47 (Sheet 5 of 7)

CIRCUIT BREAKER FUNCTION TABLE

CIRCUIT BREAKER	EFFECT OF POWER INTERRUPTION
	ESSENTIAL AC BUS (Forward Cockpit - Cont.)
ANTI COLL LTS (3Ø)	Disabled: Anti-collision/fuselage lights.
EGT TRIM (3Ø)	
L	Disabled: Left engine EGT trim motor.
R	Disabled: Right engine EGT trim motor.
	26 VOLT AC ESSENTIAL BUS (Forward Cockpit/Center Pedestal)*
HYD PRESS	
A	Disabled: A-System hydraulic pressure indication.
B	Disabled: B-System hydraulic pressure indication.
SPIKE L	Disabled: L-System hydraulic pressure indication.
SPIKE R	Disabled: R-System hydraulic pressure indication.
FUEL TK PRESS	Disabled: Fuel system tank pressure indication.
OIL PRESS	
L	Disabled: Left engine oil pressure indication.
R	Disabled: Right engine oil pressure indication.
TRIM	
PITCH	See 26 V Emergency AC BUS
YAW	Disabled: Yaw trim condition indication.
ROLL	Disabled: Roll trim condition indication.
NAV INST	Disabled: ANS attitude signal to ADI and AI. ANS heading signal to HSI and BDHI.
INS	See 26 V Emergency AC BUS *NOTE: Essential AC Bus INSTR XFMR circuit breaker must be in for power to 26 V Essential AC BUS.
	ESSENTIAL DC BUS (Aft Cockpit)
ANS	Disabled: ANS
BEACON	Disabled: G-band beacon when S/B 1763K installed:
FACE HTR	Disabled: Helmet face heat, aft cockpit.
IFF	Disabled: All IFF modes.
IFF TEST	Disabled: IFF transponder self test capability, Modes 1, 2, 3A, and C.
INTPH	Disabled: All audio to aft cockpit headset (except trainer EMERG ICS).
NAV INST	Disabled: Navigation display relays. Attitude reference reverts to: Pilot-INS RSO-ANS. Autopilot will not engage with Pilot's ATT REF switch in ANS, HSI displays mag hdg and TACAN DME. BDHI displays true hdg and TACAN DME.
SEAT ADJ	Disabled: Aft cockpit seat adjustment. Seat remains as set.
SPOT LTS	Disabled: Aft cockpit spot lights and flex point lights.
WARN LTS	Disabled: All aft cockpit caution lights.
	MONITORED DC BUS (Aft Cockpit)
DEF F	Disabled: DEF H
DEF CONT	Disabled: ALL DEF control systems.

Figure 1-47 (Sheet 6 of 7)

CIRCUIT BREAKER FUNCTION TABL

CIRCUIT BREAKER	EFFECT OF POWER INTERRUPTION
	EMERGENCY AC BUS (Aft Cockpit)
COMPUTER	NOTE: • With one circuit breaker opened to each computer, no capability is lost. • With two circuit breakers opened to a computer, that computer is likely to shut down. The corresponding pitch, yaw and/or roll sensors and servos are disabled if the following pairs of circuit breakers are opened: A Computer: A & B or A & C phase; B Computer: A & B or B & C phase; M Computer: A & C or B & C phase
A, B, M — AØ; A, B, M — BØ; A, B, M — CØ	• With all three A computer circuit breakers open, left auto inlet is disabled (inlet goes to restart). Manual control of left spike and door required. • With all three B computer circuit breakers open, right auto inlet is disabled (inlet goes to restart). Manual control of right spike and door required. • Manual inlet control is disabled if the M computer A and C phase or B and C phase circuit breakers are opened.
ATT IND (W/SB R-2595)	Disabled: Aft cockpit attitude indicator.
AP MACH TR	Disabled: Autopilot and Mach Trim systems, Analytical Redundancy, AOA, TDI OFF flags in view (indications remain valid), Bank Command steering bar remains centered.
TDI	Disabled: Aft cockpit TDI. TAS to Pilot's & RSO's Map Projector – Automatic Map Rate.
	AC HOT BUS (Aft Cockpit)
FUEL QTY	Disabled: Fuel quantity and c.g. indications, aft cockpit.
	ESSENTIAL AC BUS (Aft Cockpit)
INS HTR 1 (AØ)	Disabled: No. 1 INS heater. INS performance will degrade.
LIGHTS — PNL L (AØ)	Disabled: All functions of L CONSOLE PNL and L CONSOLE LGD circuit breakers on left console c/b panel.
LIGHTS — PNL R (AØ)	Disabled: All functions of R CONSOLE PNL, R CONSOLE LGD AND TEST & BRT circuit breakers on left console c/b panel.
LIGHTS — FLD (AØ)	Disabled: Aft cockpit flood lights.
LIGHTS — INST (AØ)	Disabled: Aft cockpit instrument panel lights.
INST LIGHTS (AØ)	Left Console, aft cockpit
L CONSOLE — PNL	Disabled: Left console panel lights.
L CONSOLE — LGD	Disabled: Left console legend lights.
TEST & BRT	Disabled: Legend test function and, if right console rheostat switch is off, legend and panel lights associated with the right console rheostat.
R CONSOLE — PNL	Disabled: Right console panel lights.
R CONSOLE — LGD	Disabled: Right console legend lights.
INS HTR 2 (BØ)	Disabled: No. 2 INS heater. INS performance will degrade.
ATT IND (W/O SB R-2595) (CØ)	Disabled: Aft cockpit attitude indicator.
IFF (CØ)	Disabled: IFF Mode 4 capability.
ANS 3 PH (3Ø)	Disabled: ANS

Figure 1-47 (Sheet 7 of 7)

batteries are furnishing at least 10 amperes to the essential dc buses. The services in Figure 1-43 can be powered by the batteries and instrument inverter.

If dc power from the T-Rs is interrupted, a time delay of ten seconds occurs before the EMER BAT ON light illuminates. The EMER BAT ON light will not illuminate if T-R dc power is restored within ten seconds.

NOTE

Occasional flickering of the EMER BAT ON caution light can be disregarded if not accompanied by other indications of abnormal electrical system operation.

Generator Out Caution Lights

The L or R GEN OUT caution light on the pilot's annunciator panel illuminates when the corresponding generator is disconnected automatically or the respective generator control switch is OFF.

HYDRAULIC SYSTEMS

Four separate hydraulic systems are provided, each with its own pressurized reservoir and engine-driven pump. (See Figure 1-48). Normally, all the systems are independent. The pumps for the A and L system are driven from the left engine accessory drive system (ADS) and the B and R pumps from the right engine ADS. The A, B, and L system reservoirs are serviced to 2.8 gallons of hydraulic fluid. The R system reservoir is serviced to 4.5 gallons of hydraulic fluid. Hydraulic fluid is cooled by fuel-oil heat exchangers, using the aircraft fuel supply for cooling.

The A and B hydraulic systems power the flight controls. An accumulator is provided in each of these two systems. The A hydraulic system also powers the APW system stick pusher.

The L hydraulic system powers the left engine air inlet system and normally powers the landing gear (including uplocks and door cylinders), brakes, air refueling door and receptacle, and nosewheel steering. The R hydraulic system powers the right engine air inlet system. In addition, if L system pressure is less than 2200 psi, the R system will automatically power the landing gear retraction cycle, and the pilot may select the R system to power nosewheel steering and the air refueling system (by setting the brake switch to ALT STEER & BRAKE). Regardless of L system pressure, the pilot may always select the R system to power the brakes. Antiskid braking is available with either L or R hydraulic system.

Hydraulic System Pressure Gages

Two dual-indicating hydraulic gages are installed on the pilot's instrument panel. The bottom (SURF CONT) gage indicates A and B system hydraulic pressures, and the top (SPIKE) gage indicates L and R system hydraulic pressures. The gages are calibrated in 100-psi increments from 0 to 4000 psi. Pressure indication is sent from remote transmitters in the individual systems.

Power for the gages is furnished by the essential ac bus 26 volt instrument transformer through the A, B, L SPIKE and R SPIKE circuit breakers on the pilot's annunciator panel.

L and R Hydraulic Quantity Low Warning Lights

The L or R HYD warning light on the pilot's annunciator panel illuminates when hydraulic quantity in the respective reservoir decreases below 1.2 gallons.

A and B Hydraulic Quantity/Pressure Low Warning Lights

The A or B HYD warning light on the pilot's annunciator panel illuminates when hydraulic quantity in the respective reservoir decreases below approximately 1.2 gallons and/or the respective hydraulic pressure decreases below 2200 (±150) psi.

L AND R HYDRAULIC SYSTEMS

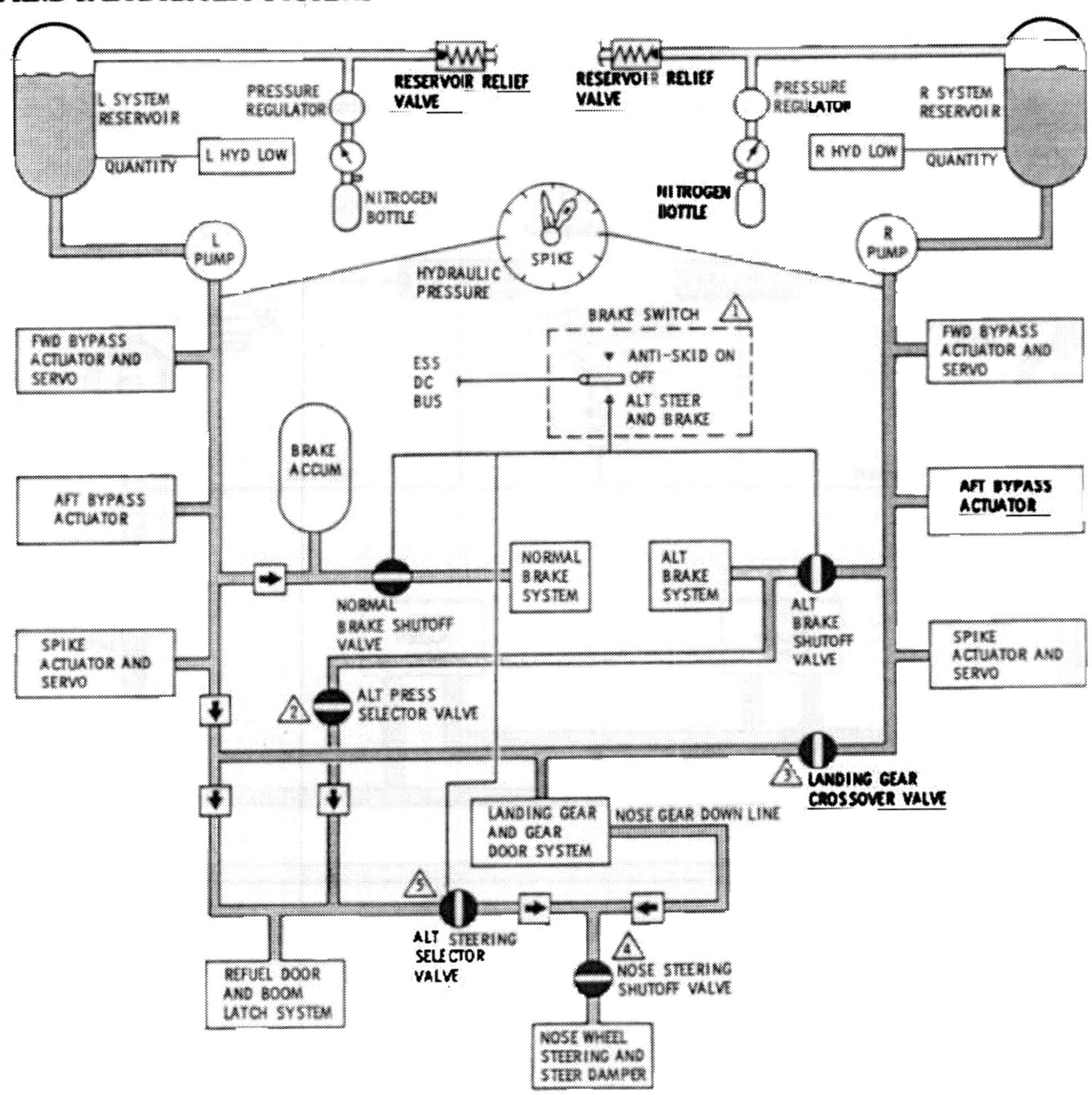

NOTE

1. OFF: "L" system powers brakes, no anti-skid protection.
   ANTI-SKID ON: Provides anti-skid protection on "Normal" brake system.
   ALT STEER AND BRAKE: Closes "Normal" and opens "Alt" brake shutoff valves, arms alternate system selector valves, and energizes "Alt. Anti-skid" system
2. With brake switch in "Alt Steer and Brake" position, valve is opened if "L" system pressure decreases below 2200 psi.
3. Crossover valve opens automatically if "L" system pressure decreases below 2200 psi, but only for gear retraction.
4. Steering controlled by CSC/NWS switch on control stick to provide nose steering on normal or alternate system pressure.
5. Valve opens when alternate steering and braking selected regardless of pressure in the L system

F203-16(a)

Figure 1-48 (Sheet 1 of 2)

A AND B HYDRAULIC SYSTEMS

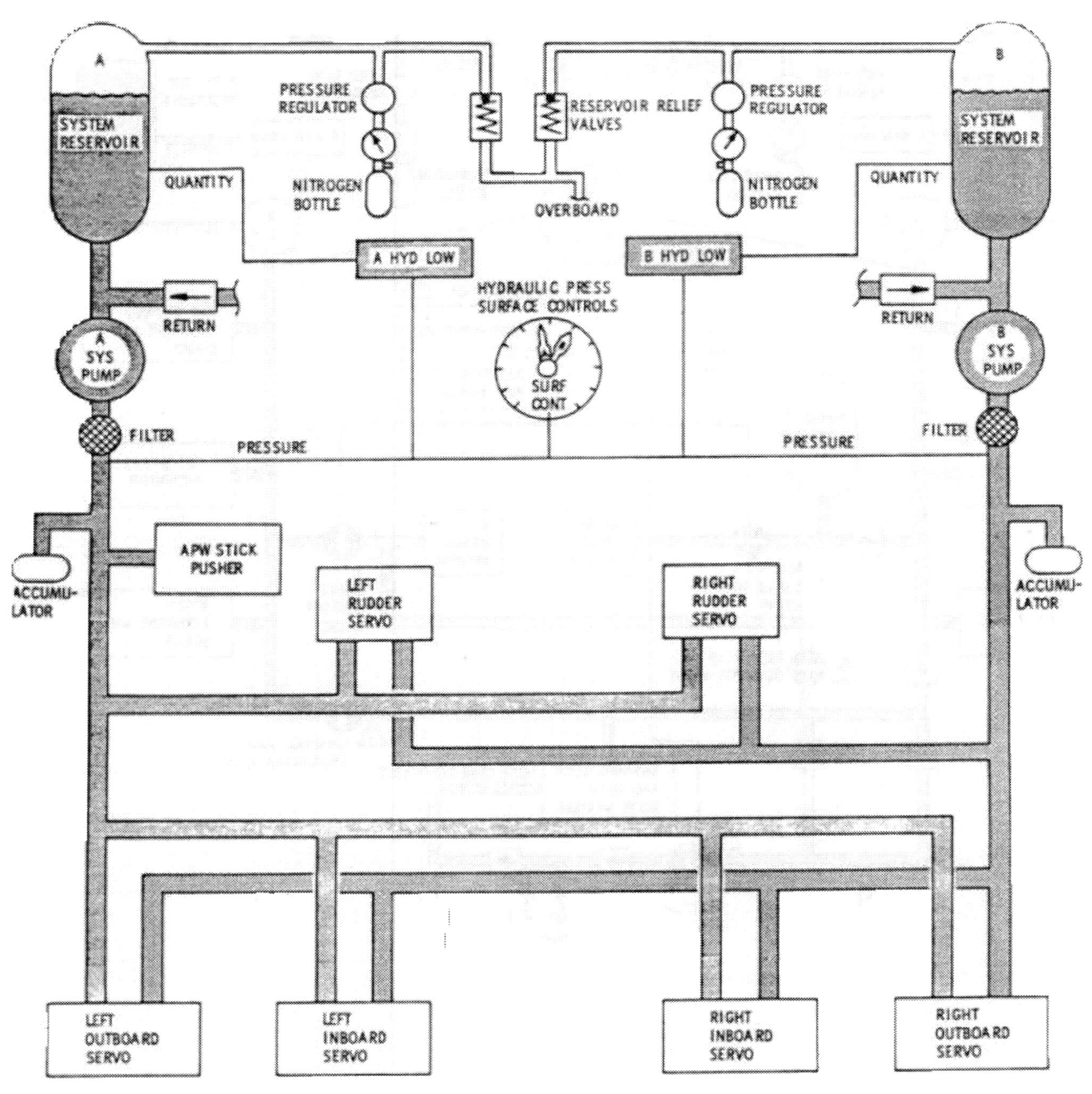

NOTE

Each hydraulic system provides actuation power to half the actuating cylinders at each servo assembly.

HYD LOW lights are illuminated by decreasing quantity with 1.2 gallons remaining in the respective reservoir, and/or by decreasing pressure at approximately 2200 psi.

F203-15(B)

Figure 1-48 (Sheet 2 of 2)

NOTE

Rapid control surface deflection while near idle rpm may result in temporary illumination of the A and/or B HYD warning light(s). The light(s) should extinguish when flow demands diminish and normal pressure is restored.

LANDING GEAR SYSTEM

The tricycle landing gear and the main wheel well inboard doors are electrically controlled and hydraulically actuated. The main gear outboard doors and the nose gear doors are linked directly to the respective gear struts. Each three-wheeled main gear retracts inboard into the fuselage and the dual-wheel nose gear retracts forward into the fuselage. The main gear is locked up by the inboard doors and the nose gear by an uplock which engages the strut. There is no hydraulic pressure on the gear when it is up and locked. Downlocks inside the actuating cylinders hold the gear in place in the extended position. Normal gear operation is by L hydraulic pressure. L system hydraulic pressure is also on the gear when in the extended position. Normal gear retraction and extension time is 12 to 16 seconds. Should L hydraulic pressure drop to 2200 psi during retraction, the power source automatically changes to the R hydraulic system. If the L system fails, R system pressure cannot be used to extend the gear and the manual gear release must be used.

A landing gear strut damper system controls gear "walking," a fore-and-aft oscillation of the main landing gear strut associated with brake application. The system is sensitive to less than 1-g change in fore-and-aft acceleration. The damping is controlled through a g-monitoring valve which automatically increases or decreases the brake pressure as required. Hydraulic pressure for the damper system is provided by the L system. The damper does not function with the brake switch in ALT STEER & BRAKE. A strut damper shutoff valve removes L system pressure from the damper valve when the landing gear is retracted.

Landing and taxi lights are on the nose gear strut. Refer to Lighting Equipment, this section.

Landing Gear Handle

A wheel-shaped landing gear handle, on the pilot's left instrument side panel, has two positions: UP and DOWN. An up-lock latch prevents the gear handle from being inadvertently placed in DOWN. An up-lock release lever which extends from the top of the gear handle, must be pushed forward to release the up-lock latch. A safety-lock solenoid prevents the gear handle from being inadvertently placed UP while the aircraft is on the ground. A manual solenoid release button is located just above the gear handle. Depressing the release button overrides the safety-lock soleoid and allows the gear handle to be moved to UP. In UP, the gear will retract if hydraulic pressure is available except that the landing gear control circuit is interlocked with the gear scissor switches to prevent retraction of the gear on the ground.

A red warning light is located in the transparent gear handle. Power for the circuit is furnished by the essential dc bus through the landing gear CONT circuit breaker on the pilot's left console.

Manual Landing Gear Release Handle

A manual GEAR RELEASE T-handle is located on the pilot's annunciator panel. If the L hydraulic system fails, the landing gear handle should be placed DOWN and the CONT circuit breaker should be pulled before pulling the GEAR RELEASE handle. If the landing gear handle cannot be placed DOWN and the landing gear CONT circuit breaker is not pulled, the landing gear will retract if there is pressure in the R hydraulic system. When the GEAR RELEASE handle is pulled, the gear extends by gravity within 90 seconds. Up to 65 pounds of force and approximately 9-1/3 inches extension of the handle is required to release the gear. The uplocks are released in the following sequence as the cable extends: nosewheel, right main gear door aft latch, right main door forward latch, left main gear door aft latch, and left main door forward latch. Gear retraction can be accomplished after emergency extension if L or R hydraulic system pressure is available.

CAUTION

The landing gear must not be retracted while the manual release handle is pulled, as damage to the system can result. Stow the handle before retracting the gear.

Landing Gear Position Lights

Three green lights on the pilot's left instrument side panel illuminate when each respective landing gear is down-and-locked. The location of each light corresponds to the gear it monitors. The lights also illuminate when the IND & LT TEST button is depressed. Power is furnished by the essential dc bus through the landing gear IND circuit breaker on the pilot's left console.

Landing Gear Warning Light and Audible Warning

The red warning light in the landing gear lever handle illuminates when:

1. Gear is cycling.
2. Gear system is not locked in the position (UP or DOWN), programmed by the landing gear handle.
3. Gear is UP and throttles are within approximately 1 inch of the IDLE stop, while below 10,000 ± 500 feet.

A pulsed-tone warning signal is produced in the pilot's and RSO's earphones when the throttles are retarded below minimum subsonic cruise setting, the landing gear is not in the down and locked position and aircraft altitude is below 10,000 (± 500) feet. The pulsed tone circuit is isolated from the gear handle light circuit so that if an emergency gear extension is necessary with the gear handle up, the tone will not occur if the gear is locked down and the throttles are retarded. The tone sounds if the IND & LT TEST button is depressed while below 10,000 ±500 feet. Power for the light and audible warning is furnished by the essential dc bus through the landing gear WARN circuit breaker on the pilot's left console.

Landing Gear Warning Cutout Button

The aural gear warning may be silenced by depressing the GEAR SIG REL button on the pilot's left instrument side panel. The circuit is reactivated when the throttles are advanced above the minimum cruise setting. Power is furnished by the essential dc bus through the landing gear WARN circuit breaker on the pilot's left console.

Landing Gear Ground Safety Pins

Removable ground safety pins are installed in the landing gear assemblies to prevent inadvertent gear retraction. Warning streamers direct attention to their removal before flight. Extra pins are provided in a container on the pilot's aft bulkhead left of the ejection seat.

NOSEWHEEL STEERING SYSTEM

The nosewheel steering system provides power steering while on the ground. It can be engaged when aircraft weight is on any gear by aligning rudder pedal position with nosewheel angle and depressing the CSC/NWS button on the control stick. A holding relay circuit keeps steering engaged when the button is released. The button must be depressed and released again to disengage steering.

A nosewheel steering engaged light is provided on the top left of the pilot's instrument panel. Illumination of the green STEER ON legend indicates nosewheel steering engagement. The light extinguishes if steering is disengaged. Steering disengages automatically when weight is not on any gear. With weight on a gear, steering disengages with loss of hydraulic pressure or when manually disengaged by the pilot.

The steering angle obtained is directly proportional to rudder pedal deflection. The nosewheel is steerable 45° either side of center. Minimum steering radius is approximately fifty-five feet. See Figure 2-3 for clearance requirements while turning.

WARNING

The landing gear side load strength is critical. Side loads during takeoff, landing, and ground operation must be kept to a minimum.

CAUTION

Do not engage nosewheel steering before nosewheel touchdown; otherwise, excessive strut and fuselage forebody loads could result from steering angles developed before nosewheel contact.

A hydraulically actuated clutch is located within the steering damper unit. The clutch engages and disengages nosewheel steering when the CSC-NWS switch is actuated.

Rudder control in-flight with the gear down is severely restricted if the clutch jams and nosewheel steering does not disengage.

WARNING

Retract the landing gear immediately to relieve restriction of rudder movement if jamming of the nosewheel steering clutch is suspected while in-flight.

NOTE

Approximately 6° of rudder would be available, through cable stretch, by applying 180 pounds of force at the rudder pedals. Rudder restriction would not be noted with the gear up.

A mechanically operated centering cam automatically centers the nosewheel when the gear retracts.

Rudder pedal movement controls a hydraulically operated nosewheel steering and shimmy damper unit by means of a cable system when steering is engaged. While on the ground with the brake switch in the ANTI-SKID ON or OFF position, hydraulic power for steering is obtained from the L system through the nose landing gear down line and the nose steering shutoff valve. If L system pressure decreases below 2200 psi, selection of the brake switch ALT STEER & BRAKE position makes R system hydraulic power available for steering. Then hydraulic power is supplied to the nose steering shutoff valve through the alternate brake shutoff valve, alternate pressure selector valve, and the alternate steering selector valve. See sheet 1 of Figure 1-48.

NOTE

With ALT STEER & BRAKE selected, the R hydraulic system cannot supply hydraulic pressure for nosewheel steering until L system pressure decreases below 2200 psi. The L system continues to supply hydraulic power while above 2200 psi.

NOTE

After the landing gear CONT circuit breaker has been opened, ALT STEER & BRAKE must be selected to open the alternate steering selector valve and obtain hydraulic power for nosewheel steering.

Electrical power for control of the nosewheel steering system is obtained from the essential dc bus through the STEER, BRK & SKID, and CONT circuit breakers. See sheet 1 of Figure 1-47 for functions lost when any of these circuit breakers is open.

WHEEL BRAKE SYSTEM

The aircraft is equipped with hydraulic operated power brakes, controlled through toe-action of the rudder pedals, and provided with artificial feel. Two interrelated brake systems are provided: a normal system using

L system hydraulic power, and an alternate system using R system hydraulic power. Selection of normal or alternate brake system is controlled by the brake switch on the pilot's left instrument side panel. Both systems use a hydraulically operated relay system to control metering of hydraulic pressure to the multiple-disc brake assemblies on each main gear wheel. Braking follows toe pressure command within one-half second.

The normal brake system has a small accumulator and a strut damper system. The accumulator is charged by L system pressure and may provide up to three brake applications if L system fails. The brake accumulator is not required to hold a charge. The probability that the accumulator will provide braking decreases as time from the loss of L hydraulic system pressure increases. Accumulator braking is not available with the brake switch in ALT STEER & BRAKE. The strut damper system dampens fore and aft oscillations of the main gear struts (associated with brake applications). Strut oscillation ("strut walk") occurs at approximately 10 cps. Strut damping is operational only on the normal brake system.

Antiskid protection is available to both the normal and alternate brake systems. The antiskid system senses wheel skid as a function of wheel rpm. Wheel rpm decreasing too rapidly causes the antiskid system to relieve brake pressure to the affected main gear. Pressure is relieved until rpm increases sufficiently to permit further brake application without skidding. If wheel rpm does not increase within 2.7 seconds after antiskid relieves brake pressure: the antiskid fail-safe circuit deactivates antiskid and illuminates the ANTI-SKID OUT annunciator caution light, and braking without antiskid protection becomes available.

The antiskid system is operational above 12 miles per hour. The brake system permits near full system pressure at the brake assemblies under extreme braking. With very heavy braking, locked wheels may occur at speeds of less than 25 knots. Momentary lockup and unlocking does not affect overall braking performance. A touchdown safety feature in the antiskid system prevents landing with the brakes applied when the brake switch is in ANTI SKID ON or ALT STEER & BRAKE.

A DRY-WET switch (on left instrument side panel) permits selection of antiskid sensitivity. The WET position increases the sensitivity of the antiskid system to wheel spindown and improves antiskid operation on wet or icy runways. The sensitivity of the antiskid system to wheel spindown in the WET mode is such that the deceleration from normal drag chute action above 90 knots will relieve brake pressure to both main gears.

BRAKE CONTROL SWITCHES AND INDICATORS

Brake Switch

The three-position brake switch is on the pilot's left instrument side panel. In OFF (center), L hydraulic pressure is available for braking without antiskid protection. In ANTI SKID ON (up), L hydraulic pressure is available for braking with antiskid protection unless (after S/B R-2695) the trigger switch is depressed. In ALT STEER & BRAKE (down), R-hydraulic pressure is available for braking immediately with antiskid unless (after S/B R-2595) the trigger switch is depressed, and nosewheel steering from R system is available when L system pressure drops below 2200 psi.

CAUTION

> If L hydraulic pressure is not available, R hydraulic system will not be available for braking or steering unless the brake switch is in ALT STEER & BRAKE.

Power is supplied to the brake switch from the essential dc bus through the BRK & SKID circuit breaker on the pilot's left console.

Antiskid Disconnect Trigger Switch

With S/B R-2695, antiskid system operation is interrupted while the trigger switch is held depressed. The hydraulic power source for brakes remains as selected by the brake switch.

WET-DRY Switch

The two-position wet-dry switch is on the pilot's left instrument side panel. In DRY (up), antiskid braking is compatible with wheel spindown characterisitics on a dry runway. The WET (down) position increases brake antiskid sensitivity to optimize braking on a wet or icy runway and reduces the probability of a blown tire and subsequent loss of antiskid braking. Electrical power is from the essential dc bus through the BRK & SKID circuit breaker.

ANTI-SKID OUT Light

The ANTI-SKID OUT caution light on the pilot's annunciator panel illuminates when the brake switch is OFF if the landing gear is down and there is weight on a gear, or if the antiskid system fail-safe circuit detects a fault.

After S/B R-2695, the ANTI-SKID OUT light also illuminates while the trigger switch is depressed if the landing gear is down and there is weight on a gear.

DRAG CHUTE SYSTEM

The drag chute system reduces landing roll and aborted takeoff rollout distance. The drag chute is stowed in an aft fuselage compartment above fuel tank 4. The drag chute attachment rides free in the compartment until locked to the aircraft during the initial stage of deployment. The drag chute and extraction system are packed in a deployment bag which contains a 42-inch vane-type pilot chute, a 10-foot extraction chute which produces aerodynamic lift, and a 40-foot ribbon-type drag chute. Normal chute deployment and jettisoning is accomplished electrically. Emergency deployment is accomplished mechanically. The drag chute handle operates both modes.

Drag Chute Handle

The T-shaped DRAG CHUTE handle is located on the upper left of the instrument panel. In the stowed (or jettison) position, the handle is horizontal, fully forward, and a red band on the shaft is not visible. Normal deployment is accomplished by pulling straight aft on the handle to the limit of its travel (approximately 1 inch). After chute deployment, jettisoning is accomplished by pushing the handle full forward. These handle positions operate switches which control the chute deploy and jettison actuator motors. Power for normal drag chute operation is provided by the essential dc bus through two DRAG CHUTE circuit breakers on the pilot's left console.

If normal deployment fails, emergency deployment is possible by rotating the handle 90 degrees counterclockwise from the normal deploy position and pulling to its aft travel limit (approximately 8 inches). The maximum pull force required is 60 pounds. When the handle is released, it is returned to the receptacle by the tension of a slack takeup spring. The drag chute cannot be jettisoned after emergency deployment because the actuator motor switches are disconnected during the manual deploy sequence.

CAUTION

If the DRAG CHUTE handle is pulled to the emergency deploy position and released immediately, damage could result to cockpit items as the handle snaps back to the receptacle.

NOTE

To avoid inadvertent emergency deployment, do not rotate the DRAG CHUTE handle during normal deployment.

Drag Chute Unsafe Light

The DRAG CHUTE UNSAFE caution light on the pilot's annunciator panel illuminates when: (1) the drag chute mechanism has been actuated to some degree either mechanically or electrically and is in an unsafe condition, or (2) power to both linear actuator dc motors is lost.

PRIMARY FLIGHT CONTROLS

The full-power irreversible flight control system consists of cockpit controls (stick and rudder pedals), four elevons, and two full-moving rudders. Two elevons are hinge-mounted to the upper trailing edge of each wing, one inboard and one outboard of the respective engine nacelle. A tetrahedral-shaped rudder is mounted to a fixed stub fin on the upper aft portion of each engine nacelle. Each rudder assembly is canted inward 15 degrees.

A servo assembly at each control surface meters dual system (A and B) hydraulic power for positioning the control surfaces. The control stick and rudder pedals are connected to the servos by cable and mechanical input systems. Feel springs in each axis provide the pilot with control feel proportional to the degree of control deflection. Artificial feel is provided since air loads are not felt by the pilot. Pilot inputs are limited to the force necessary to move the metering valve in the servos.

Control Stick

A conventional control stick operates the elevons. See Figure 1-49. Movement of the stick is approximately 9° forward and 16° aft of its neutral position and (with the control surface deflection limiters engaged) approximately 5 1/2° laterally. With the limiters disengaged, the stick can be positioned laterally approximately 8° from center at any point, and approximately 9° when not near its extreme forward or aft position. Similarly, full forward and aft stick positioning capability is reduced somewhat at the extreme "corners" of the deflection "box" when the limiters are disengaged. (This results because the elevon deflection angles are additive for combined pitch and roll commands. If surface deflection limits are reached, maximum stick pitch and roll command angles are also reached.) Full lateral movement requires approximately 10 pounds force. Approximately 25 pounds push force and 45 pounds pull force are required to reach the full forward and aft stick positions, respectively.

Three switches are located on the top and one switch on the right side of the grip. On the top left, a three-position communications switch labeled TRANS (up) and INPH (down) is springloaded to an unmarked off (center) position. The pilot's microphone is connected to the interphone system or to a selected radio transmitter when INPH or TRANS is selected, respectively. The microphone is disconnected when the center position is used unless the throttle-mounted microphone switch is pressed for radio communication or the interphone HOT MIC knob is pulled out. See Communications and Avionic Equipment, this section. A four-way (center-off) pitch and yaw axis TRIM switch operates as described under Manual Trim System, this section. A dual purpose CSC/NWS push-button switch, located to the right of the trim switch, either activates the autopilot Control Stick Command feature (if the autopilot is on) or engages/releases nosewheel steering while on the ground. Before S/B R-2674, pressing a rain removal system switch on the right side of the grip causes a quantity of rain removal fluid to be applied to the forward windshield panels if the windshield deicing switch on the left side of the pilot's annunciator panel has been set to RAIN REMOVAL ARM ON (up). After S/B R-2674, rain removal is deactivated.

NOTE

Do not operate the rain removal system unless the windshield is wet. The white fluid will stick to the glass and permanently obscure visibility if applied while the windshield is dry.

A multipurpose trigger switch is on the forward side of the grip. Operation of this switch disconnects air refueling, disengages autopilot, interrupts ac power to the pitch and yaw manual trim actuator motors (disabling the control stick trim switch and the RH RUDDER SYNCHRONIZER switch), interrupts the APW system stick pusher, and with S/B R-2695 interrupts antiskid system operation.

A stick-shaker motor is installed below the grip. It warns of a potentially high angle of

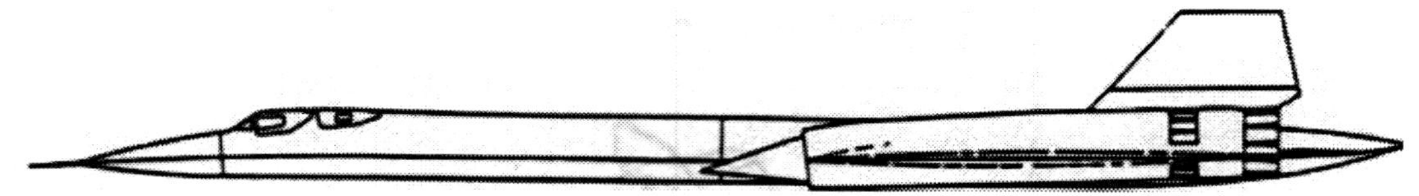

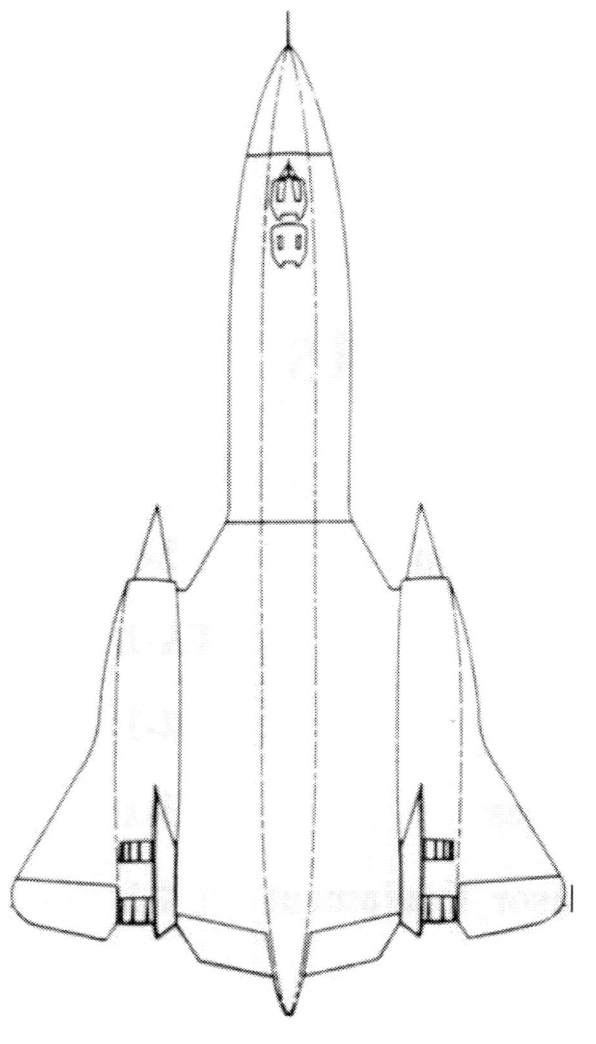

THIS PAGE INTENTIONALLY LEFT BLANK OR STILL CLASSIFIED.

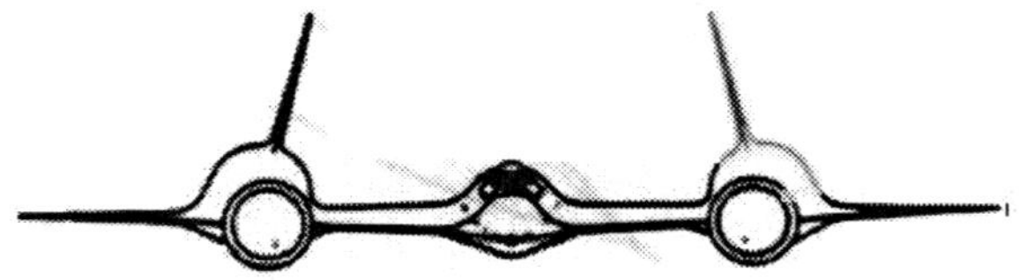

CONTROL STICK GRIP

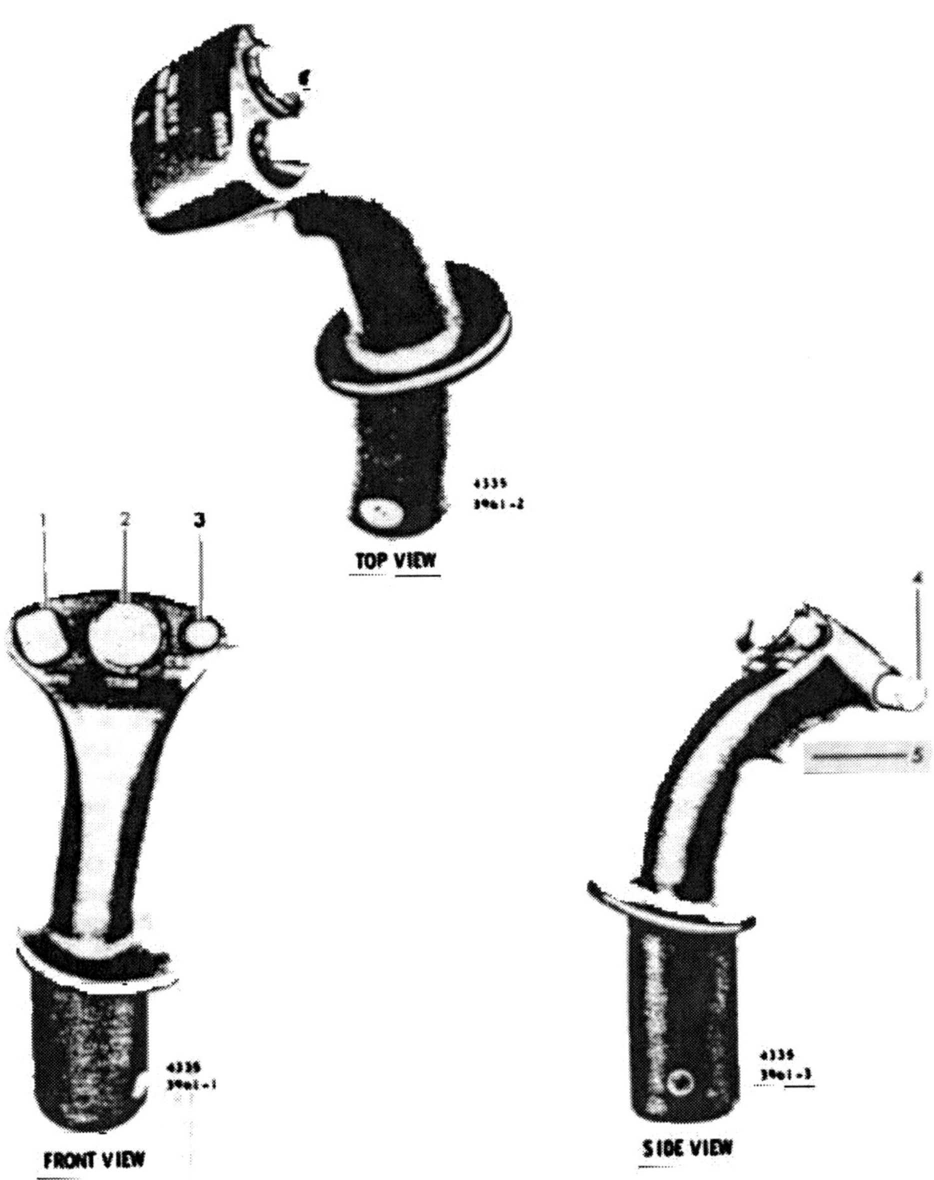

NOTE

1 Transmitter-interphone microphone selector switch
2 Pitch and yaw trim switch
3 Control stick command-nosewheel steering button
4 Rain removal switch (Deactivated with S/B R-2674)
5 Trigger switch - disconnects air refueling and autopilot; interrupts pitch and yaw trim, APW stick pusher, and antiskid (with S/B R-2695).

F203-65(a)

Figure 1-49

attack and/or pitch rate. Operation of the shaker is controlled by the APW and High Alpha Warning systems. A red SHAKER warning light, located near the apex of the pilot's glareshield, illuminates when the shaker is on. A stick-pusher mechanism is also part of the APW system. The pusher displaces the stick forward to initiate corrective action in the pitch axis. (Refer to APW and High Alpha Warning Systems, this section.) Hydraulic pressure from the A system is required for pusher operation.

ELEVON CONTROL SYSTEM

The delta wing configuration uses elevons for pitch and roll control. The elevons respond to control inputs from the control stick, pitch and roll trim systems, SAS, and the autopilot. All control inputs are applied to the inboard elevon servo assemblies, which control actuation power for elevon positioning. The outboard elevons are mechanically slaved to their respective inboard elevon.

Elevon Control Cables

The control stick is connected by mechanical linkage to pitch and roll tension regulators in the cockpit. Dual control cables run from each tension regulator to respective cable quadrants above the mixer assembly in the tail cone.

Elevon Mixer Assembly

The mixer assembly provides the mechanical geometry necessary to sum the pitch and roll inputs from their respective control cable systems. The combined inputs are then applied as a single control input to the inboard elevon servo assemblies. The pitch and roll trim actuators are an integral part of the mixer assembly and are summed along with control stick inputs. All pitch and roll trim (Mach trim and auto trim included) is applied through the mixer assembly. The pitch and roll feel springs are in the mixer. Since the trim actuators are mounted downstream of the feel springs, the control stick is not displaced by elevon trim (pitch and roll) actuations.

Inboard Elevon Servos

All pitch and roll control inputs (stick, trim, SAS, and autopilot) are applied to the inboard elevon servos. In response to these inputs, the servos meter A and B hydraulic pressures to the actuating cylinders, which position the inboard elevon surface. Each hydraulic system provides power to three actuating cylinders at each inboard servo assembly.

Input Mechanism and Spring Cartridge

The outboard elevons are slaved to their respective inboard elevon by a mechanical input which connects the inboard surface to the outboard elevon servo. Thus, any movement of the inboard surface moves the outboard servo, which positions the outboard elevon.

A spring-loaded cartridge pushrod "shotgun" is installed in the inner wing portion of the input mechanism. Cartridge spring loads maintain four rollers in a detent during normal surface control operation, permitting the cartridge to act as a solid pushrod. If excessive backloads (approximately 900 pounds) are imposed on input mechanism movement by binding or seizure outboard of the cartridge, the rollers "jump their detent" to permit operation of the inboard elevon independent of the affected outboard elevon. Without this feature, failure at the outboard servo assembly or within the input mechanism could result in structural damage and/or loss of control in pitch and roll.

When operating the controls on only one hydraulic system, half of the actuating cylinders are inoperative and sufficient back pressure may be imposed on the input mechanism to cause the spring cartridge rollers to jump the detent at lower system pressures (approximately 1500 psi).

CAUTION

When starting engines, do not move the control stick until at least 1500 psi can be maintained on the A or B hydraulic system.

SURFACE CONTROL DEFLECTION LIMITS AND RATES

MODE OF OPERATION	SURFACE AUTHORITY LIMITS (1) PITCH	SURFACE AUTHORITY LIMITS (1) ROLL (2)	SURFACE AUTHORITY LIMITS (1) YAW	SURFACE MAXIMUM RATES PITCH	SURFACE MAXIMUM RATES ROLL (2)	SURFACE MAXIMUM RATES YAW
MANUAL	10° DOWN (3) TO 24° UP (4)	24°	20° LEFT TO 20° RIGHT	32.5°/SEC	65°/SEC	37°/SEC
LIMITED MANUAL		14°	10° LEFT TO 10° RIGHT			
APW PUSHER EXTENDED (6)	10° DOWN (3) TO 10° UP (5)					
SAS	6.5° UP TO 2.5° DOWN	4°	8° LEFT TO 8° RIGHT	15°/SEC	30°/SEC	28°/SEC
AUTOPILOT	2.3° DOWN TO 2.3° UP (7)					
AUTO-TRIM MACH TRIM	5.0° DOWN TO 8.5° UP			NOMINAL 0.113° SEC		
MANUAL TRIM		9°	10° LEFT TO 10° RIGHT	NOMINAL 1.13°/SEC	NOMINAL .96°/SEC	NOMINAL .90°/SEC

NOTE

(1) Combined pitch and roll application is limited by actuating cylinder stroke extremes at 20° down to 35° up.

(2) Roll figures reflect differential roll applied.

(3) 10° down if pitch trim indicator is at or below zero. Trim above zero indication decreases down elevon authority.

(4) 24° up if pitch trim indicator is at or above zero. Trim below zero indication decreases up elevon authority.

(5) Assumes pitch trim indicator at zero. Trim above zero indication increases up elevon authority. Trim below zero indication decreases up elevon authority.

(6) Pusher deflects elevons 1.7° down from trimmed position. Up elevon movement above trimmed position requires additional force to overcome pusher.

(7) Autopilot authority limits are 2.3° above FL500, and 1.6° below FL500.

F203-128(g)

Figure 1-50

FLIGHT CONTROL SYSTEMS

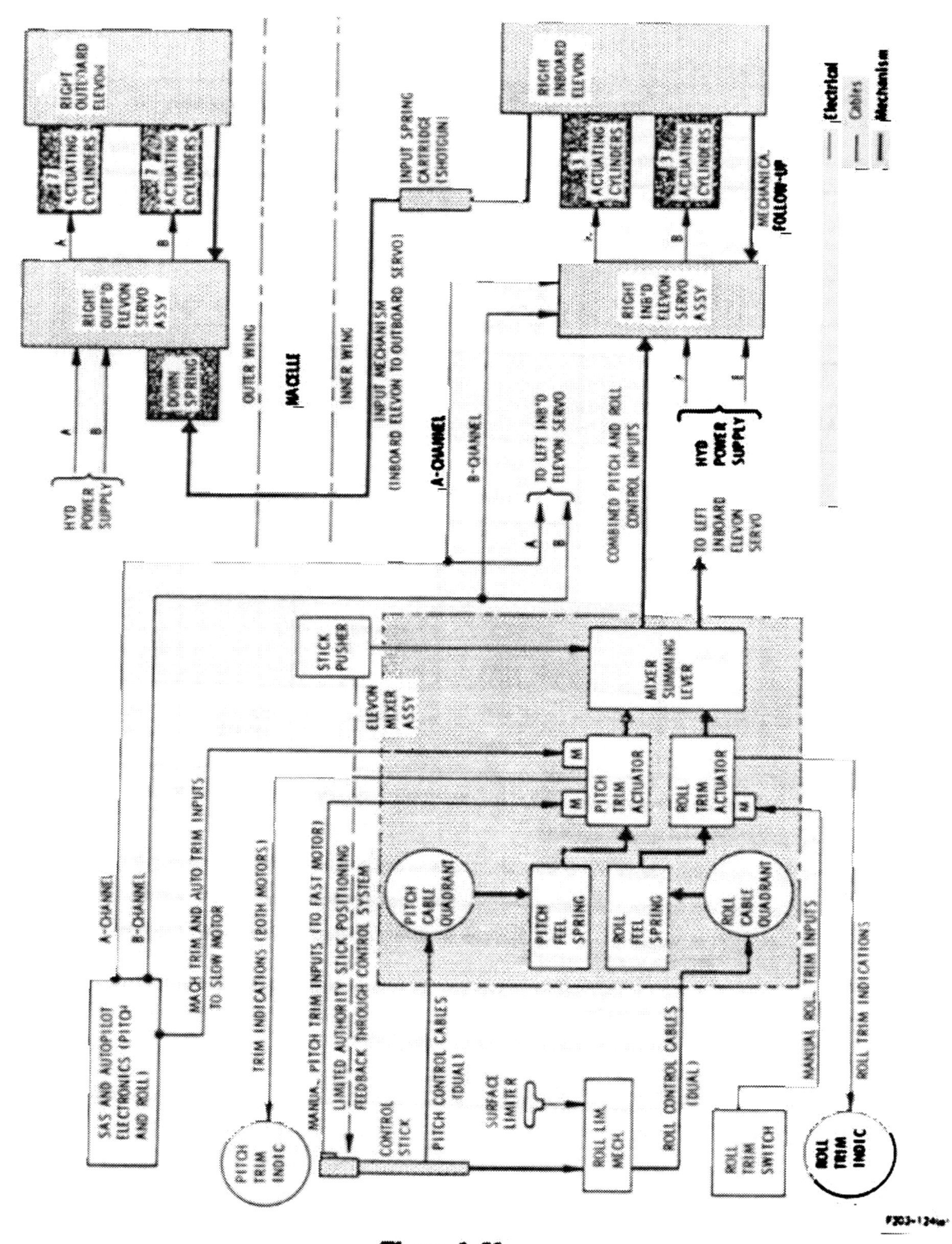

Figure 1-51

RUDDER CONTROL SYSTEM

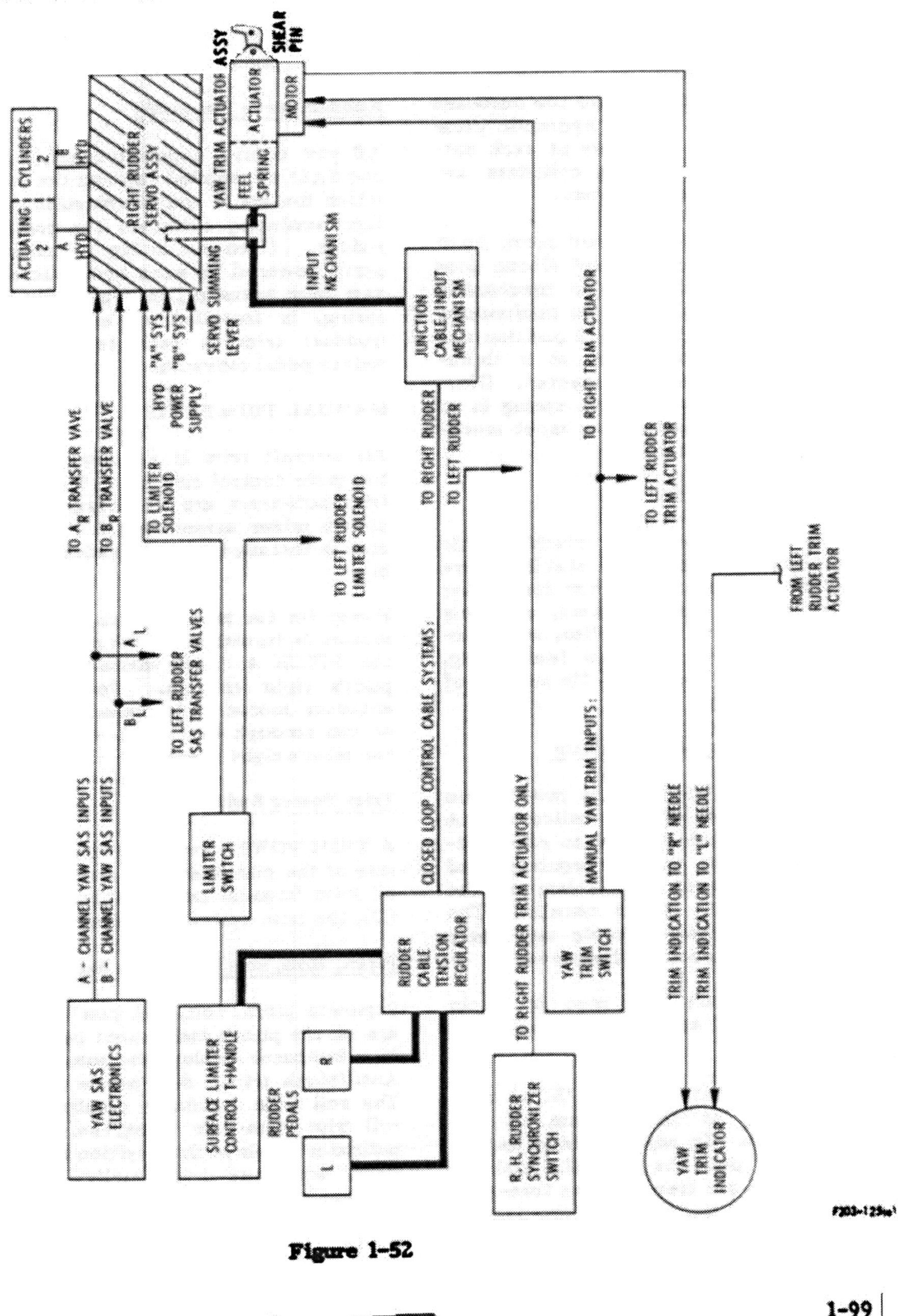

Figure 1-52

Outboard Servo and Limiter Spring

Inboard elevon movement is transmitted through the input mechanism to the outboard servo, which meters A and B hydraulic pressures to 14 actuating cylinders at each outboard elevon. Half of the cylinders are powered by each hydraulic system.

A limiter spring at the outboard servo input lever ensures that the outboard elevon does not travel full down if the input mechanism fails. Limiter spring force, in conjunction with servo bias spring loads, will position and maintain the outboard elevon at a three-degree-down position if disconnected. (Normal function of the servo bias spring is to apply a down-elevon load on the input mechanism to eliminate hysteresis.)

RUDDER CONTROL SYSTEM

The two full-moving rudders, which provide yaw (directional) control and stability, are positioned by control inputs from the rudder pedals, manual yaw trim system, and yaw SAS. Individual servo assemblies, which include a trim actuator and yaw feel spring, are installed in the fixed stub fin section of each rudder assembly.

Rudder Pedals and Input Mechanism

The rudder pedals connect to the yaw tension regulator by push-pull rods and bellcranks. A single closed-loop cable system to each rudder originates at the tension regulator and terminates at an input mechanism at the inboard side of each engine nacelle. The input mechanism transmits cable movement to the input lever at each rudder servo.

The rudder pedals are also used for main wheel braking (toe action) and nosewheel steering.

Pedal position is adjusted by the PEDAL ADJ T-handle, located at the bottom of the annunciator panel. To adjust rudder pedals, hold pedals and pull the PEDAL ADJ T-handle; the pedals are free to move fore-and-aft. Push or release the pedals to the desired position and release the T-handle to lock the pedals in place.

Rudder Servo Assembly

All yaw control inputs (pedals, manual trim, and SAS) are applied through the rudder servo which meters A and B hydraulic pressures to four actuating cylinders for positioning the rudder. (Two actuating cylinders at each servo powered by each hydraulic system.) A yaw trim actuator, incorporating a yaw feel spring, is installed at each servo. Yaw (rudder) trim is reflected in proportional rudder pedal movement.

MANUAL TRIM SYSTEM

All aircraft trim is achieved by positioning the main control surfaces. The pitch and roll trim actuators are an integral part of the elevon mixer assembly and a yaw trim actuator is installed at each rudder servo assembly.

Power for the manual pitch and yaw actuator motors is furnished by the ac hot bus through the PITCH & YAW circuit breaker on the pilot's right console. Power for the roll actuator motor is furnished by the essential ac bus through the ROLL circuit breaker on the pilot's right console.

Trim Power Switch

A TRIM POWER switch is located on the left side of the pilot's annunciator panel. In OFF, all trim (manual and auto) is inoperative. In ON, the trim system is operable.

Pitch, Roll, and Yaw Trim Indicators

Separate pitch, roll, and yaw trim indicators are on the pilot's instrument panel. The pitch trim indicator displays the sum of manual and Auto/Mach trim. SAS inputs are not shown. The roll trim indicator displays differential roll trim from 0 to 9 degrees. The yaw trim indicator displays the position of the left and right actuators individually on L and R needles. These needles are aligned (superimposed) when equal trim is applied at both rudders.

Power for the roll and yaw trim indicators is furnished by the essential ac bus 26 volt instrument transformer through the YAW and ROLL circuit breakers on the pilot's annunciator panel. Power for the pitch trim indicator is furnished by the emergency ac bus 26 volt instrument transformer through the PITCH circuit breaker on the pilot's annunciator panel.

Pitch Trim System

The pitch trim actuator may be operated by manual control, using a fast motor, or automatic control (including Mach Trim), using a slow motor. The manual (fast) trim motor trims ten times faster than the Auto/Mach (slow) trim motor. Pitch trim indication signals are provided by two position transmitters on the actuator. The slow motor transmitter also provides feedback signals to the autotrim system.

Pitch and Yaw Trim Switch

The manual pitch trim control switch is combined with the yaw trim switch on the control stick grip. Moving the switch up and down (from the spring-loaded center off position) applies nose-down and nose-up trim, respectively. Pitch trim application is limited to 5° ($-1/2^{\circ}$ $+1^{\circ}$) down and 8.5° (-1°, $+1/2^{\circ}$) up by actuator stroke limitations.

Moving the switch left and right applies left yaw and right yaw trim, respectively, to both rudders.

Manual pitch and yaw trim can be disconnected by pressing the trigger switch on the control stick grip. Mach trim is not affected.

CAUTION

To avoid runaway trim due to a sticking trim switch, assure positive switch movement to neutral after each actuation.

Roll Trim Switch

A three-position roll trim switch, located forward of the throttle quadrant, provides manual control of the roll trim actuator in the elevon mixer assembly. The self-centering switch may be toggled left and right to apply left roll and right roll trim, respectively. Roll trim application is limited (by actuator stroke length) to $\pm$ 4.5 degrees, for a maximum of 9 degrees differential roll trim.

Yaw Trim System

Yaw trim is applied by two trim actuators, one at each rudder servo assembly. The yaw function of the pitch and yaw trim switch on the control stick grip energizes both rudder trim actuators simultaneously. A right hand rudder synchronizer switch, located forward of the throttle quadrant, energizes the right rudder trim actuator only. Yaw trim is limited to approximately 10 degrees left and right by actuator stroke limits.

A shear pin is in the attach fittings of each actuator so rudder pedal inputs cannot be blocked by failure of a yaw feel spring in the trim actuator. If a seizure does occur, pedal inputs can break the shear pin and free the actuator attach point. When an actuator shear pin is broken, some yaw feel is lost and little or no yaw trim is available at the affected rudder.

NOTE

The trim actuator remains operative and the trim indicator needles will indicate normal trim operation, even though trim is applied at only one rudder. Rudder centering will be poor.

Right Hand Rudder Synchronizer Switch

Due to variations in rudder actuator motor speeds, yaw trim may not be applied equally to both rudders. A split in rudder position and yaw trim needle indications may result. A three-position, self-centering, RH RUDDER SYNCHRONIZER switch (located left of the roll trim switch) should be used to equalize left and right rudder trim positions. The switch may be toggled left and right to move the trailing edge of the right rudder left and right, respectively.

SURFACE LIMITER SYSTEM

Lateral (roll) control stick travel and rudder displacement are restricted by the surface limiter system. The system should be released (full travel mode) at speeds below Mach 0.5, and engaged (limited mode) at higher speeds. The surface limiter system is controlled by the SURF LIMIT RELEASE T-handle on the left side of the pilot's annunciator panel.

The T-handle is spring-loaded in the full forward (limited) position, and must be pulled aft to disengage the limiters and rotated 90 degrees clockwise to lock the handle aft. The T-handle is released (to engage the limiters) by rotating the handle 90 degrees counterclockwise.

A SURFACE LIMITER caution light on the pilot's annunciator panel illuminates if the limiters are in the wrong mode for the existing aircraft speed. The light is controlled by T-handle switch contacts and Mach inputs from DAFICS. With the limiters released, the light illuminates above Mach 0.5, and goes out when limiting is engaged. With limiters engaged, the light illuminates below Mach 0.5, and goes out when the full travel mode is selected.

Roll Travel Limiter System

With limiters engaged, a pin is inserted into a cam at the base of the control stick, physically limiting elevon travel to 14 degrees roll differential. Unlimited, the maximum roll differential is 24 degrees.

Rudder Travel Limiter System

Rudder travel is reduced from 20 degrees to 10 degrees left and right when the limiters are engaged. Rudder travel is limited by two different methods to ensure that no combination of rudder pedal, yaw trim, or yaw SAS can position the rudders beyond the limits. One method (mechanical), limits rudder pedal inputs by insertion of a pin between two stops on the rudder tension regulator in the forward cockpit. The second method (electro-hydraulic) limits travel of the rudder servo input lever at each servo by hydraulically extending solenoid limiter stops to restrict input lever movement.

The limiter control solenoids, located at each rudder servo, control hydraulic pressure for extension or retraction of the solenoid limiter stops. The control solenoids must be energized to permit hydraulic retraction of the stops when the limters are released. Power for solenoid control is furnished by the essential dc bus through individual RUD LIM circuit breakers on the pilot's left console. If limiter control power is lost, the solenoid stops extend to the limited position.

If power to one of the solenoid limiters is lost (loss of dc control power, relay failure, control circuit breaker trip, etc), rudder travel on the affected rudder will be limited to 10 degrees and the SURFACE LIMITER light will remain illuminated when full travel is selected. With one rudder limited and full travel selected, the non-limited rudder can be positioned well beyond the 10 degree limit by applying abnormal pressure on the rudder pedals.

DIGITAL AUTOMATIC FLIGHT AND INLET CONTROL SYSTEM (DAFICS)

The Digital Automatic Flight and Inlet Control System (DAFICS) comprises five major subsystems: stability augmentation system, autopilot/Mach trim system, automatic pitch warning and high angle of attack system, automatic/manual inlet control system, and air data system. The autopilot utilizes inputs from the ANS or the INS.

Air pressures measured independently by the A, B, and M channels of the pressure transducer assembly (PTA) are transmitted to the respective (A, B, and M) DAFICS computers, then shared between computers to determine best available measurements. DAFICS computes air data for schedules in DAFICS subsystems and cockpit displays (including the TDI).

The stability augmentation system provides automatic stability augmentation in the pitch, roll and yaw axis. The autopilot provides automatic flight control in the pitch and roll axes, and the Mach trim system provides speed stability augmentation in the pitch axis. The automatic pitch warning system provides a control stick shaker and stick pusher when approaching flight limits in the pitch axis. The inlet control system provides automatic and manual control of inlet air flow. (Refer to Figures 1-53 and 1-54.)

Computer Reset Switches

Individual A, B, and M CMPTR RESET switches on the pilot's annunciator panel allow manual restart of a computer. Holding a reset switch in the up position stops the respective DAFICS computer. Releasing the reset switch to the spring-loaded down position initiates restart. Deliberate computer(s) shutdown inflight is not authorized. The switches are powered by their respective A, B, and M CMPTR dc circuit breakers on the pilot's left console.

COMPUTER BUILT IN TEST

Computer inflight BIT performs a series of tests throughout flight to insure A, B, and M computer health and if any test fails, the computer will automatically shut itself down and turn on the appropriate CMPTR OUT light on the pilot's annunciator panel.

DAFICS PREFLIGHT BUILT IN TEST

The DAFICS preflight built in test (BIT) is normally accomplished before and after flight. The DAFICS PREFLIGHT BIT switch is located on the pilot's right console.

If the following requirements are not satisfied, the DAFICS PREFLIGHT BIT switch will not engage.

1. Hydraulic pressure - A system.
2. CMPTR OUT lights (3) not illuminated.
3. CSC/NWS switch - Released.
4. APW switch - PUSHER/SHAKER.
5. SPIKES & FWD BYPASS doors -AUTO.
6. RESTART switches - Off.
7. Throttle Restart switch - Off.
8. SAS channel engage switches - ON.
9. AUTOPILOT PITCH & ROLL switches - ON.
10. KEAS HOLD switch - ON.
11. HEADING HOLD switch - ON.

If B, L, or R hydraulic pressure is not available, the DAFICS PREFLIGHT BIT will fail.

During the DAFICS PREFLIGHT BIT, the APW pusher operates momentarily, the SHAKER warning light flashes momentarily, the A, B and M CMPTR OUT annunciator lights flash several times, the SAS SENSOR and SERVO lights flash momentarily, control surfaces move slightly, and the spikes move. The test cycle can be terminated manually by stopping any DAFICS computer. The BIT TEST light illuminates steady green while the test is running.

The test takes about one minute to complete. If the test is successful, the BIT TEST light flashes green at completion of the test. If any malfunctions are detected, the test continues, but at completion of the test the BIT TEST light extinguishes and the BIT FAIL light illuminates steady red.

When the DAFICS preflight BIT check is finished, indications are:

1. BIT FAIL light - Extinguished.
2. BIT TEST light - Flashing green.

or

1. BIT FAIL light - Steady red.
2. BIT TEST LIGHT - Extinguished.
3. MASTER CAUTION light - On.
4. SAS OUT annunciator panel light - Flashing.
5. Autopilot pitch and roll switches - Off.
6. HEADING HOLD switch - Off.
7. KEAS HOLD switch - Off.
8. AUTOPILOT OFF light - On.
9. OFF flags on both TDI's.
10. CIP barber pole at zero.
11. Spikes full forward.
12. DAFICS PRELIGHT BIT switch - OFF, automatically.

When the BIT terminates, if a steady red BIT TEST light, any SENSOR light, any SERVO light, or any CMPTR OUT light illuminates, notify maintenance.

The DAFICS remains in the test mode. Pressing one of the six SENSOR/SERVO recycle switches resets DAFICS to the flight mode. When DAFICS is reset to the flight mode, indications are:

1. A, B, and M CMPTR OUT lights - Flash, momentarily.
2. BIT TEST light - Off.
3. CIP Barber pole - Normal position.
4. Both TDI's will initiate resynchronization and run up to 55,000 ft, Mach 2.0, and 300 KEAS. AOA will indicate 10°. AOA will return to 0° in approximately 1 minute and 15 seconds and TDI indications will return to normal in approximately 2 minutes and 15 seconds after resetting DAFICS to the flight mode.

WARNING

- Failure to recycle a SENSOR/SERVO recycle switch will cause the DAFICS system to remain in the ground test mode. The SAS is non-functional while in the ground test mode. DAFICS will not operate normally until the system is reset.
- Do not attempt to activate the DAFICS PREFLIGHT BIT switch during flight. The DAFICS PREFLIGHT BIT check is inhibited unless there is weight on wheels, Mach number is less than 0.09, or KEAS is less than 101.

NOTE

Once DAFICS is reset to the flight mode, a flashing red BIT FAIL light indicates loss of SAS analytical redundancy.

STABILITY AUGMENTATION SYSTEM (SAS)

The stability augmentation system (SAS), a combination of electronic and hydraulic equipment, is an integral part of the basic aircraft control system. The system is normally engaged in all flight conditions, although it can be disengaged manually. Each axis of SAS (pitch, roll and yaw) is provided with two SAS channels. The SAS detects aircraft attitude changes and initiates control surface deflections to counteract the changes. Normally, the DAFICS A computer runs the A channel in pitch and yaw and the DAFICS B computer runs the B channel in pitch and yaw. The DAFICS M computer can take over for A computer or B computer or both in the pitch and yaw axis should A and/or B computer fail. The M computer drives through servo amplifiers in the A and B computer to provide surface control. The roll SAS is configured so that either A or B computer is capable of driving both roll servo channels. Sensor and servo monitors provide detection and automatic disengaging capability for faults.

During normal flight conditions, the aircraft experiences many small changes in attitude due to air loads or control inputs. These attitude changes are sensed by pitch, yaw, and roll sensors in each axis (three rate gyros in the pitch axis, three rate gyros plus three lateral accelerometers in the yaw axis, and two rate gyros in the roll axis). Analytical redundancy derived by the DAFICS computers from attitude displacements provides added redundancy for pitch, yaw and roll rate gyros, but not for the lateral accelerometers. The attitude changes detected by the sensors are sent to the DAFICS computers which electrically command the transfer valve positions of the SAS servos. The transfer valve converts the electrical signal into a proportional hydraulic flow into the SAS servo actuators. The SAS

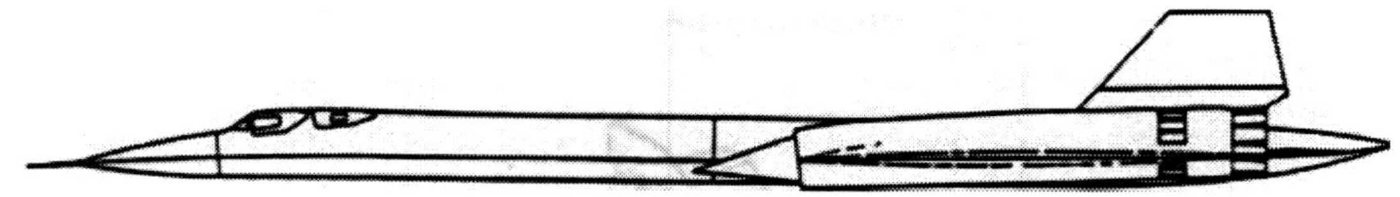

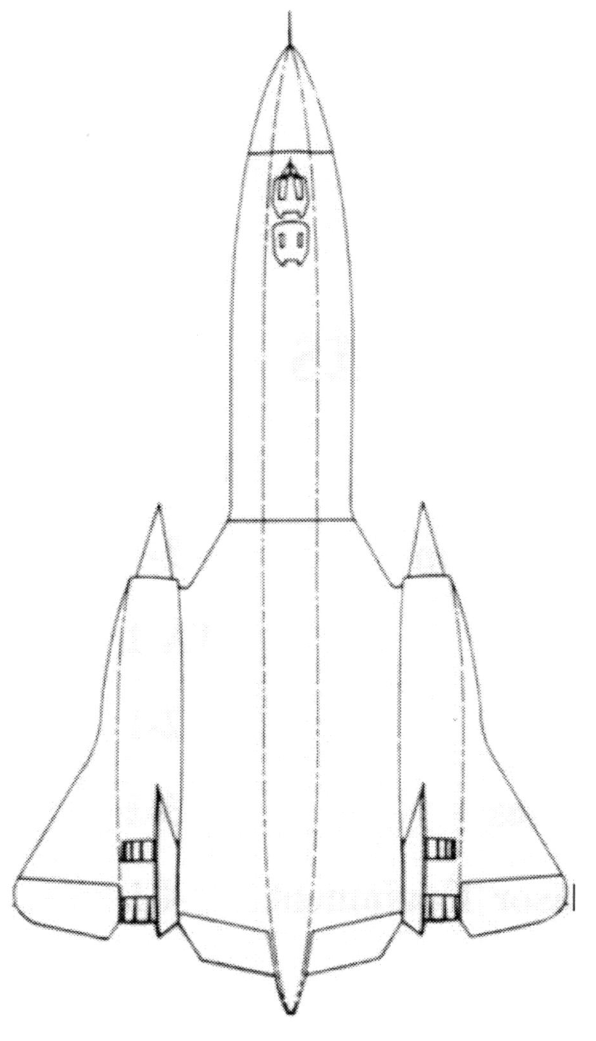

THIS PAGE INTENTIONALLY LEFT BLANK OR STILL CLASSIFIED.

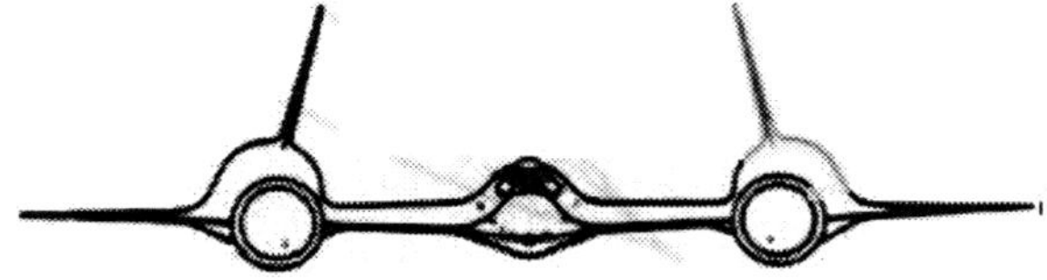

COMPUTER INPUTS

COMPUTER INPUTS	DAFICS									
	SAS			AUTOPILOT & MACH TRIM	AIR DATA	APW & HIGH q WARNING	AIR INLET SYSTEM	COMPUTERS		
	PITCH	YAW	ROLL					A	B	M
A CMPTR RESET Switch								X		
B CMPTR RESET Switch									X	
M CMPTR RESET Switch										X
P_t	X	X	X	X	X	X	X	X	X	X
P_s	X	X	X	X	X	X	X	X	X	X
ΔP_α						X	X	X	X	X
ΔP_β							X			X
Duct Pressure Ratio Transducers (Left and Right)							X	X	X	
INLETS: AUTO SPIKE Switch Positions							X	X	X	X
AUTO FWD BYPASS Switch Positions							X	X	X	X
Manual Spikes							X			X
Manual Forward Bypass Doors							X			X
Manual BIAS Schedule Controls							X			X
G-BIAS (Ng)							X	X	X	
Total Temperature (From Left Inlet) ①								X	X	X
ELEVON SERVOS (2 Left and 2 Right)	X		X					X	X	X
RUDDER SERVOS (2 Left and 2 Right)		X						X	X	X
SURFACE RELEASE T-handle Position					X			X	X	
SAS: A Pitch Rate Sensor	X			X		X		X	X	
B Pitch Rate Sensor	X			X		X			X	X
M Pitch Rate Sensor	X			X		X		X		X
A Yaw Rate Sensor		X						X	X	
B Yaw Rate Sensor		X							X	X
M Yaw Rate Sensor		X						X		X
A Roll Rate Sensor			X	X				X	X	
B Roll Rate Sensor			X	X				X	X	
A Lateral Accelerometer Sensor		X						X	X	
B Lateral Accelerometer Sensor		X							X	X
M Lateral Accelerometer Sensor		X						X		X
A & B Pitch Channel Switches	X							X	X	X
A & B Yaw Channel Switches		X						X	X	X
A & B Roll Channel Switches			X					X	X	
A PITCH SENSOR Switch/Light	X							X	X	
B PITCH SENSOR Switch/Light	X								X	X
M PITCH SENSOR Switch/Light	X							X		X
A YAW SENSOR Switch/Light		X						X	X	
B YAW SENSOR Switch/Light		X							X	X
M YAW SENSOR Switch/Light		X						X		X
A PITCH SERVO Switch/Light								X		X
B PITCH SERVO Switch/Light									X	X
A YAW SERVO Switch/Light								X		X
B YAW SERVO Switch/Light									X	X
A ROLL SERVO Switch/Light								X	X	
B ROLL SERVO Switch/Light								X	X	
DAFICS PREFLIGHT BIT Switch	X	X	X	X	X	X	X	X	X	X
AUTOPILOT: All AUTOPILOT Panel Switches & Trim Wheels				X				X	X	
CSC Switch				X				X	X	
Trigger Switch (A/P Pusher Disconnect Switch)				X		X		X	X	
ATT REF SELECT Switch	X		X	X		X		X	X	X
INS: Pitch	X	X	X	X				X	X	X
Roll	X	X	X	X				X	X	X
Heading	X	X	X	X				X	X	X
Attitude Ready				X				X	X	
Heading Valid				X				X	X	
ANS: Pitch	X		X	X		X		X	X	X
Roll	X		X	X		X		X	X	X
Heading	X		X	X		X		X	X	X
Steering Commands				X				X	X	
NAV READY				X				X	X	
Hydro Pressure	X	X	X	X		X	X	X	X	X
Weight-On-Wheels (WOW)				X		X		X	X	X
Gear Up Locks						X		X	X	X
Gear Down Locks							X			X

F305-274(a)

① Used in ANS and Map Projector

Figure 1-53

COMPUTER OUTPUTS

	COMPUTER		
	A	B	M
TDI & TDI OFF FLAG (Pilot's and RSO's)	X	X	
INLETS:			
AUTO Spikes	X	X	
AUTO Forward Bypass Doors	X	X	
L & R UNST Lights	X	X	
ELEVON SERVOS (2 Left and 2 Right)	X	X	X
RUDDER SERVOS (2 Left and 2 Right)	X	X	X
SURFACE LIMITER Annunciator Panel Light	X	X	
SAS:			
A PITCH SENSOR Caution Light	X	X	
B PITCH SENSOR Caution Light		X	X
M PITCH SENSOR Caution Light	X		X
A YAW SENSOR Caution Light	X	X	
B YAW SENSOR Caution Light		X	X
M YAW SENSOR Caution Light	X		X
A or B or BOTH ROLL SENSOR Caution Light	X	X	
A PITCH SERVO Caution Light	X		X
B PITCH SERVO Caution Light		X	X
A YAW SERVO Caution Light	X		X
B YAW SERVO Caution Light		X	X
A ROLL SERVO Caution Light	X	X	
B ROLL SERVO Caution Light	X	X	
SAS OUT Annunciator Panel Light	X	X	X
DAFICS PREFLIGHT BIT Switch	X	X	X
BIT TEST/FAIL Lights	X	X	X
MACH TRIM	X	X	X
DAFICS:			
A CMPTR OUT Annunciator Panel Light	X		
B CMPTR OUT Annunciator Panel Light		X	
M CMPTR OUT Annunciator Panel Light			X
2 PTA CHANNELS OUT Annunciator Light	X	X	X
AUTOPILOT:			
Auto Trim	X	X	
Flight Director Steering Indicator	X	X	
Pitch Trim Indice	X	X	
Roll Trim Indice	X	X	
APW & HIGH α WARNING:			
Pusher	X	X	
Shaker	X	X	
SHAKER Warning Light	X	X	
APW Annunciator Panel Light	X	X	
ANGLE OF ATTACK Indicator	X	X	
ADI (APW Boundary Only)	X	X	

Figure 1-54 (Sheet 1 of 2)

COMPUTER OUTPUTS

	COMPUTER		
	A	B	M
KEAS Warning:	X	X	
KEAS Warning Light	X	X	
Low KEAS Aural Warning			X
DMRS (DAFICS related portions)			X
CIP (Barber Pole Only)			X
Altitude Reporting (IFF Mode C)			X
Altitude and TAS to ANS:			X
Altitude from ANS to V/H System - NAV Switch Position			
Altitude from ANS to CAPRE SLR for RADAR ALTITUDE			
TAS from ANS to Navigation Control & Display Panel			
TAS to Pilot's & RSO's Map Projector - Automatic Map Rate			X
Heading to PVD			X

Figure 1-54 (Sheet 2 of 2)

servos position the flight control surfaces to compensate for the original sensed rate of attitude change. The three pitch and the three yaw gyros are mounted in the No. 2 fuel tank; the two roll gyros are in the R bay; the lateral accelerometers are in the nose wheel well.

SAS Control

SAS engagement is controlled by an engage solenoid on each transfer valve. The solenoids must be energized to permit hydraulic flow into the transfer valve and engage the SAS. With the solenoid deenergized, hydraulic flow is shut off and the SAS is disengaged as SAS electronic inputs cannot position control surfaces when hydraulic flow to the transfer valves is shut off. The SAS is normally engaged (solenoids energized) by operating the SAS control switches to ON. The SAS may be disengaged manually by moving the control switches to OFF, or automatically by DAFICS.

SAS Redundancy

Basic SAS system redundancy is provided by the presence of two channels in each axis. Either channel alone is capable of providing satisfactory damping in the respective axis. SAS reliability is further assured by the fact that a single electrical or hydraulic failure cannot result in loss of both channels in any axis.

Separate sources of hydraulic power are provided for each channel; A hydraulic system for A channel SAS, and B hydraulic system for B channel SAS. Complete loss of either hydraulic system will not adversely affect operation of the remaining SAS channel, providing the pitch, roll and yaw engage switches of the failed SAS channel are OFF.

Similar protection through redundancy is provided in the SAS electrical supply and distribution systems. The AC and DC power supplies are from the two highest priority buses in the aircraft, the emergency AC and the essential DC buses. Both of these buses may be energized by aircraft battery power in the event of multiple failures within the electrical systems.

DC control power into the computers is through the A, B, and M CMPTR circuit breakers on the pilot's left console. Loss of A or B computer power does not eliminate roll SAS. Loss of power to any two computers does not eliminate pitch and yaw SAS. AC power distribution to the DAFICS and hence SAS is from the emergency AC bus, through nine circuit breakers on the right console in the aft cockpit (A COMPUTER Aφ, Bφ, Cφ, B COMPUTER Aφ, Bφ, Cφ, M CMPTR Aφ, Bφ, Cφ). Each computer has isolated power. Each computer can have a single ac circuit breaker fail and continue to operate normally.

SAS servo engage power is separated from computer dc power to improve redundancy. Six circuit breakers on the pilot's left console (PITCH A, B; ROLL A, B; YAW A, B) provide independent engage power for each servo.

Pitch SAS

The SAS pitch axis control system consists of three pitch sensors (A, B, and M), the three DAFICS computers (A, B, and M) and two servo channels (A and B). Normally, the A computer drives the A servo channel the B computer drives the B servo channel. In case of A and/or B computer failures, the M computer can drive the A servo channel or the B servo channel or both. To obtain triple redundancy for pitch rate sensors, the DAFICS computers employ an analytical pitch rate derived from the pitch, roll and heading displacements of the platform (ANS or INS) selected on the pilot's ATT REF SELECT switch. This analytical redundancy pitch rate is used to isolate gyro faults after a second pitch rate gyro has failed, so that the remaining good gyro can be selected. The sensor selection process is shown on Figure 1-55.

The computers generate control inputs to the pitch SAS transfer valves in response to vehicle pitch rates. The transfer valves convert electronic metering signals into hydraulic flow to the SAS servos which are positioned to cause elevon deflections to counteract gyro-sensed attitude changes.

PITCH SAS BLOCK DIAGRAM

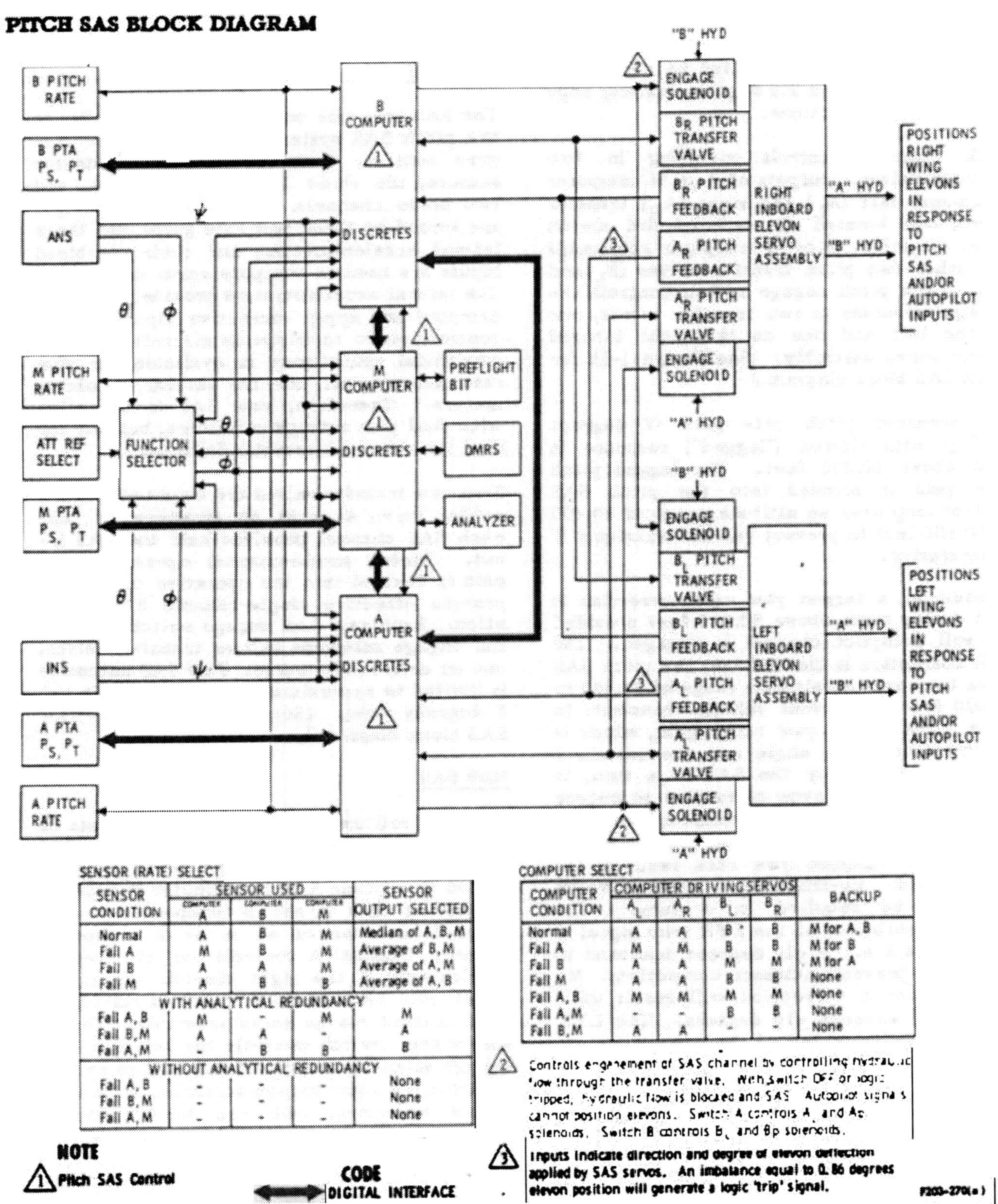

SENSOR (RATE) SELECT

SENSOR CONDITION	SENSOR USED COMPUTER A	COMPUTER B	COMPUTER M	SENSOR OUTPUT SELECTED
Normal	A	B	M	Median of A, B, M
Fail A	M	B	M	Average of B, M
Fail B	A	A	M	Average of A, M
Fail M	A	B	B	Average of A, B
WITH ANALYTICAL REDUNDANCY				
Fail A, B	M	-	M	M
Fail B, M	A	A	-	A
Fail A, M	-	B	B	B
WITHOUT ANALYTICAL REDUNDANCY				
Fail A, B	-	-	-	None
Fail B, M	-	-	-	None
Fail A, M	-	-	-	None

COMPUTER SELECT

COMPUTER CONDITION	COMPUTER DRIVING SERVOS A_L	A_R	B_L	B_R	BACKUP
Normal	A	A	B	B	M for A, B
Fail A	M	M	B	B	M for B
Fail B	A	A	M	M	M for A
Fail M	A	A	B	B	None
Fail A, B	M	M	M	M	None
Fail A, M	-	-	B	B	None
Fail B, M	A	A	-	-	None

2. Controls engagement of SAS channel by controlling hydraulic flow through the transfer valve. With switch OFF or logic tripped, hydraulic flow is blocked and SAS Autopilot signals cannot position elevons. Switch A controls A_L and A_R solenoids. Switch B controls B_L and B_R solenoids.

3. Inputs indicate direction and degree of elevon deflection applied by SAS servos. An imbalance equal to 0.86 degrees elevon position will generate a logic 'trip' signal.

NOTE

1. Pitch SAS Control

CODE

⟷ DIGITAL INTERFACE

F203-270(a)

Figure 1-55

Pitch SAS authority is limited to 6.5 degrees trailing edge up and 2.5 degrees trailing edge down elevon deflections.

Each channel controls metering in two transfer valves. Outputs of A or M computer commands left (A_L) and right (A_R) transfer valve, one located on each inboard elevon servo assembly. B or M computer commands the other two pitch transfer valves (B_L and B_R). Each pitch engage switch controls the engage solenoids in two transfer valves, one on the left and one on the right inboard elevon servo assembly. (See Figure 1-55 for Pitch SAS block diagram.)

An increased pitch rate gain (7 degrees $\delta e/^{o}q$) with slowed ("lagged") response is used above 50,000 feet. The lagged pitch rate gain is blended into the pitch SAS control loop over an altitude range of 49,400 to 50,600 feet to prevent vehicle transients if maneuvering.

In addition, a lagged yaw rate correction is used during turns above 50,000 feet provided the roll autopilot channel is engaged. The LYR correction is blended into the pitch SAS servo loop over an altitude range of 49,400 to 50,600 feet to prevent vehicle transients in turns. The lagged yaw rate signal, which is derived from bank angle and the sustained yaw rate sensed by the SAS in a turn, is applied as an up elevon correction to oppose the down elevon command resulting from the sustained pitch rate sensed by the pitch SAS in a turn. Lagged yaw rate reduces the amount of up-trimming which would otherwise be required in a turn. The maximum authority of the LYR trim signal is 2.3 degrees δ e. A $\pm$10 degrees deadband in bank angle prevents nuisance corrections. No LYR up elevon correction will result until bank angle exceeds $\pm$10 degrees. The LYR feature is switched out of the control system when the roll autopilot is disengaged. No abrupt trim change requirement occurs during the transition. If disabled while turning, the LYR term fades out gradually to zero in approximately 15 seconds.

Yaw SAS

The SAS yaw axis control system is similar to the pitch SAS system, with three yaw rate gyro sensors, three lateral accelerometer sensors, the three DAFICS computers, and two servo channels. Yaw attitude changes are sensed by three yaw rate gyros and three lateral accelerometers, and their combined inputs are used to compute servo commands. The lateral accelerometers provide long term damping and apply corrective inputs to the control system to minimize aircraft sideslips. Analytical redundancy is available for yaw rate sensors but not for lateral accelerometers. Therefore, yaw SAS is operative with dual yaw rate gyro failures, but not for dual lateral accelerometer failures.

Two yaw transfer valves are mounted on each rudder servo in such an arrangement that each SAS channel provides half the SAS input. During single-channel operation, the gain is doubled into the operative channel to provide effective single-channel SAS operation. Each yaw SAS engage switch controls the engage solenoids in two transfer valves, one on each rudder servo. Yaw SAS authority is limited to approximately 8 degrees left and 8 degrees right. (See Figure 1-56 for Yaw SAS block diagram.)

Roll SAS

The SAS roll axis control system consists of two roll sensors (A and B), two DAFICS computers (A and B), and two servo channels (A and B). Either A or B computer is capable of driving both A and B independent servo channels in case of an A or B computer failure. Channel A controls the left elevons and channel B the right elevons through a single roll SAS transfer valve mounted on each inboard elevon servo assembly. The A roll engage switch controls the left roll SAS engage solenoid, and the B roll engage switch controls the right engage solenoid. With both servos operating, roll SAS is limited to approximately 4 degrees differential elevon.

When only one roll channel is engaged, automatic logic protection is unavailable. Since only one elevon is moving in response to roll

YAW SAS BLOCK DIAGRAM

B YAW RATE
B LAT ACCEL
B PTA P_S, P_T
ANS
M YAW RATE
M LAT ACCEL
M PTA P_S, P_T
INS
A YAW RATE
A LAT ACCEL
A PTA P_S, P_T
FUNCTION SELECTOR
B COMPUTER
M COMPUTER
A COMPUTER
DISCRETES
PREFLIGHT BIT
DMRS
ANALYZER
"B" HYD
ENGAGE SOLENOID
B_R YAW TRANSFER VALVE
B_R YAW FEEDBACK
A_R YAW FEEDBACK
A_R YAW TRANSFER VALVE
"A" HYD
RIGHT RUDDER SERVO ASSEMBLY
"A" HYD
"B" HYD
ACTUATING CYLINDERS POSITION RIGHT RUDDER IN RESPONSE TO YAW SAS, TRIM OR PEDAL INPUTS
CODE
DIGITAL INTERFACE
B_L YAW TRANSFER VALVE
B_L YAW FEEDBACK
A_L YAW FEEDBACK
A_L YAW TRANSFER VALVE
LEFT RUDDER SERVO ASSEMBLY
ACTUATING CYLINDERS POSITION LEFT RUDDER IN RESPONSE TO YAW SAS, TRIM AND PEDAL INPUTS

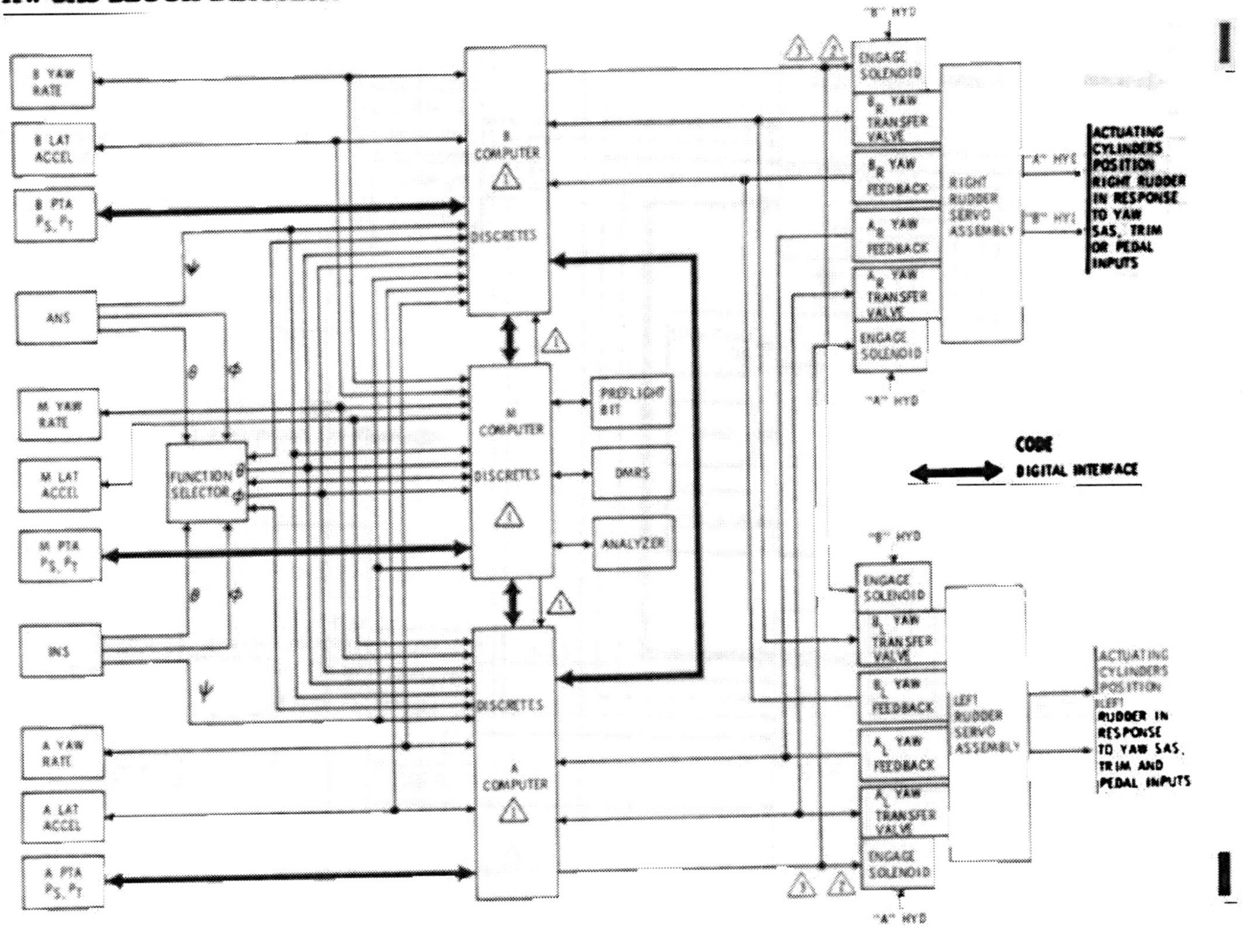

SENSOR (RATE) SELECT

SENSOR CONDITION	SENSOR USED A	SENSOR USED B	SENSOR USED M	SENSOR OUTPUT SELECTED
Normal	A	B	M	Median of A, B, M
Fail A	M	B	M	Average of B, M
Fail B	A	A	M	Average of A, M
Fail M	A	B	B	Average of A, B
RATE GYRO WITH ANALYTICAL REDUNDANCY				
Fail A, B	M	-	M	M
Fail B, M	A	A	-	A
Fail A, M	-	B	B	B
RATE GYROS WITHOUT ANALYTICAL REDUNDANCY OR LATERAL ACCELEROMETER				
Fail A, B	-	-	-	None
Fail B, M	-	-	-	None
Fail A, M	-	-	-	None

COMPUTER SELECT

COMPUTER CONDITION	COMPUTER DRIVING SERVOS A_L	A_R	B_L	B_R	BACKUP
NORMAL	A	A	B	B	M for A, B
FAIL A	M	M	B	B	M for B
FAIL B	A	A	M	M	M for A
FAIL M	A	A	B	B	None
FAIL A, B	M	M	M	M	None
FAIL A, M	-	-	B	B	None
FAIL B, M	A	A	-	-	None

NOTE

1. YAW SAS Control
2. Logic outputs are generated by a "difference" between compared inputs, which disengage the faulty channel and energize SAS lamps to indicate whether fault was in SAS sensors or SAS servo. Gain is doubled in the active channel to maintain effective Yaw SAS.

F303-273(a)

ROLL SAS BLOCK DIAGRAM

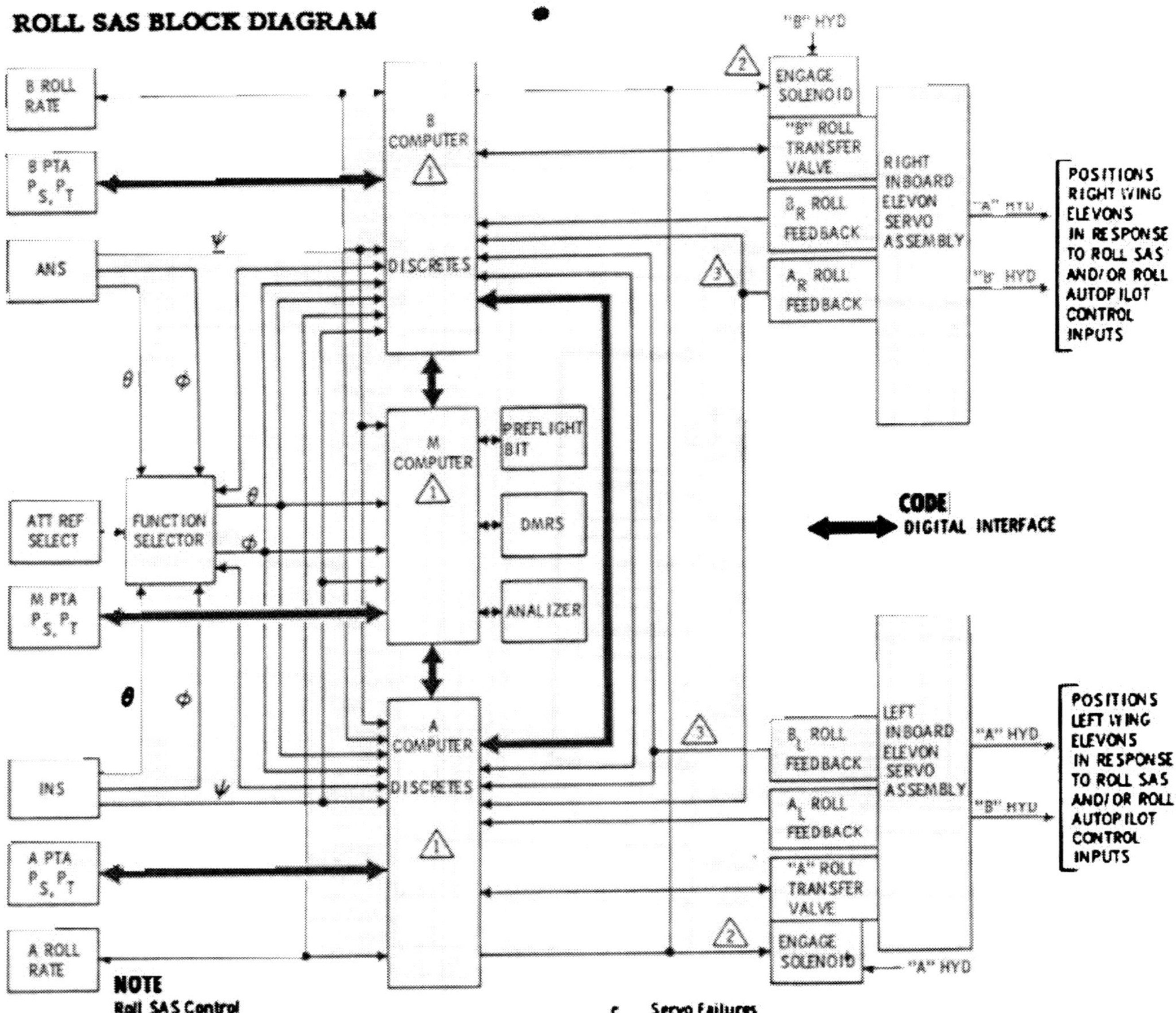

NOTE

Roll SAS Control

a. Both switches OFF to ON engages both SAS channels and "ARMS" roll logic.

b. Sensor Failures
- First sensor failure - Normal roll SAS capability.
- Second sensor failure - No roll SAS capability.

⚠1 **SENSOR (RATE) SELECT**

SENSOR CONDITION	SENSOR USED		SENSOR
	COMPUTER A	COMPUTER B	OUTPUT SELECTED
NORMAL	A	B	AVERAGE OF A, B
WITH ANALYTICAL REDUNDANCY			
Fail A	B	B	B
Fail B	A	A	A
Fail A, B	-	-	None
WITHOUT ANALYTICAL REDUNDANCY			
Fail A	-	-	None
Fail B	-	-	None

⚠2 Controls engagement of SAS channel by controlling hydraulic flow through transfer valve. With switch OFF or logic "TRIPPED", hydraulic flow is blocked and SAS/Autopilot control signals cannot position elevons.

c. Servo Failures
- Servo logic "trip" disengages both channels.
- For a hard failure, single channel operation can be selected by disengaging both A and B engage switches and re-engaging channel opposite from failed servo light.
- Single channel operation overrides all servo logic protection.
- For momentary failure, recycle either engage switch to resume normal operation.

COMPUTER SELECT

COMPUTER CONDITION	COMPUTER DRIVING SERVOS		BACKUP
	LEFT	RIGHT	
Normal	A	B	Other Computer
Fail A	B	B	None
Fail B	A	A	None

⚠3 Inputs indicate direction and degree of elevon deflection applied by SAS servos. An imbalance equal to 0.6 degrees elevon position will generate a logic "TRIP" signal.

F203-27 (e)

Figure 1-57

SAS, differential elevon available is reduced from 4 to 2 degrees with some coupling into the pitch and yaw axes.

Analytical redundancy in the DAFICS computers derives a third roll rate to use for voting after a single rate gyro has failed. This capability provides dual sensor redundancy. (See Figure 1-57 for Roll SAS block diagram.)

SAS Controls and Indicators

Manual engage, disengage, and recycle control of SAS is provided by controls on the function selector panel on the pilot's right console. Indicator lights (recycle light/-switches) on the function selector panel display status of the system sensors and servos. When not in the ground test mode, the red BIT FAIL light flashes if analytical redundancy is not available. The SAS OUT light on the annunciator panel and the master CAUTION light illuminate for SAS failures but not for analytical redundancy failure. Refer to Figure 1-58, SAS Controls and Indications Table, for details of switch functions.

SAS LOGIC

DAFICS logic monitors voltages within all three axes of SAS so that malfunctions within the SAS can be detected at an early stage. The system is extremely sensitive and voltage tolerances are small, so even minor errors are detected to provide maximum protection. The logic system will disengage the affected servo when a difference in voltage is detected which is equivalent to approximately 2 degrees surface deflection error in pitch and yaw, and 0.6 degree in roll.

Pitch SAS

Sensors

Computer sensor select capability selects median between A, B and M gyro outputs and uses that signal for A, B or M computer SAS control. If tracking requirements are exceeded (first failure) the gyro that is out of tolerance is voted out and the appropriate PITCH SENSOR light on the function selector and the SAS OUT light on the annunciator panel are turned ON. The remaining two gyro outputs are now averaged and that signal is used for A, B or M computer SAS control. If the remaining two gyros exceed the tracking requirement (second failure) both gyros are compared with analytical redundancy derived pitch rate to pick out the failed gyro (appropriate PITCH SENSOR light on function selector and SAS OUT light on annunciator panel are turned ON). The remaining good gyro is now used for A, B or M computer SAS control. If the remaining gyro now exceeds tracking limit (third failure) all pitch SAS channels are disengaged (A, B and M PITCH SENSOR lights on function selector and SAS OUT light on annunciator panel are turned on).

If analytical redundancy is not available (flashing BIT FAIL light), when second gyro failure occurs, both pitch SAS channels are disengaged (A, B, and M PITCH SENSOR lights on function selector and SAS OUT lights on annunciator panel are turned on).

Servos

Computer servo monitoring compares left servo LVDT's with right servo LVDT's. If tracking requirement is exceeded, both LVDT's are compared with a servo model to isolate failure (appropriate PITCH SERVO light on function selector and SAS OUT light on annunciator panel are turned on) and that servo channel is disengaged.

Pitch servo LVDT primary and secondary monitors determine if impedence is within acceptable limits. If limits are exceeded, appropriate PITCH SERVO light on function selector and SAS OUT light on annunciator panel are turned on and the failed servo channel is disengaged.

Yaw SAS Logic

Sensors

Yaw sensor failure can be either a rate gyro or lateral accelerometer. Yaw rate gyro failures are indicated by a steady A, B, or M YAW SENSOR light. Lateral accelerometer failures are indicated by a flashing A, B, or M YAW SENSOR light. If the same yaw rate gyro and lateral accelerometer fail (A, B or M sensor), that YAW SENSOR light will be on steady. Computer sensor select capability for yaw rate gyros is the same as for pitch rate gyros, therefore, yaw SAS is available for single or dual gyro failures (if analytical redundancy is operative). Analytical redundancy is not provided for lateral accelerometers. Computer sensor select capability for lateral accelerometers is the same as for pitch rate gyros when analytical redundancy is not available, therefore, yaw SAS is not operative with dual lateral accelerometer failures.

Servos

Computer servo monitoring is the same as pitch SAS.

Yaw servo LVDT primary and secondary monitors are the same as for pitch SAS.

Roll SAS Logic

Sensors

Computer sensor select capability averages the A and B gyro outputs and uses that signal for A and B computer SAS control. If A and B gyro outputs exceed tracking requirement (first failure) both gyros are compared with the analytical redundancy derived roll rate signal. The remaining gyro that is within tolerance is selected and used for A and B computer SAS control (the ROLL SENSOR light on function selector and SAS OUT light on annunciator panel are turned on).

If analytical redundancy is not available (flashing BIT FAIL light), when first gyro failure occurs both roll SAS channels are disengaged (roll sensor light on function selector and SAS OUT light on annunciator panel are turned on). With analytical redundancy a second failure disengages roll SAS (ROLL SENSOR light on function selector and SAS OUT light on annunciator panel remain ON).

Servos

Computer servo monitoring compares left servo LVDT's with right servo LVDT's. If tracking requirement is exceeded both LVDT's are compared with a servo model to isolate failure and appropriate ROLL SERVO light on the function selector and SAS OUT light on annunciator panel are turned on and Roll SAS disengages. To select single channel Roll SAS engage the channel opposite from failed ROLL SERVO light on function selector.

Roll servo LVDT primary and secondary monitors determine if resistance is within acceptable limits. If limits are exceeded appropriate ROLL SERVO light on function selector and SAS OUT light on annunciator panel are turned on and Roll SAS disengages.

A subsequent servo failure in the reengaged good channel will neither disengage the servo nor illuminate the SERVO light in that channel.

AUTOPILOT

DAFICS A and B computers control the pitch and roll autopilot. Both the pitch and roll channels can be engaged, or either one may be operated independently. The autopilot pitch channel may be operated in the basic attitude hold mode, or with one of two other special features (Mach hold or KEAS hold). An automatic pitch trim system is energized when the pitch channel of the autopilot is engaged. A Mach trim system, not part of the autopilot, is enabled when the pitch channel of the autopilot is off. The roll channel may be operated in the basic attitude hold mode, or with one of two other features (heading hold or automatic navigation).

The autopilot pitch channel cannot be engaged unless at least one pitch SAS channel is engaged, and the autopilot roll channel cannot be engaged unless at least one roll SAS channel and one yaw SAS channel is engaged. Elevon control responses due to SAS and/or autopilot inputs are not reflected in control stick movement as they are applied

SAS CONTROLS AND INDICATIONS

CONTROL OR INDICATOR	POSITION OR INDICATION	FUNCTION
CHANNEL ENGAGE SWITCHES. One ON-OFF toggle switch for channel (A & B) of each axis. (On function selector.)	ON (A or B)	Energizes A or B channel of respective axis by energizing the engage solenoids of associated transfer valves.
	Off (A or B)	Disengages respective channel by de-energizing associated transfer valve engage solenoid. SERVO legend(s) corresponding to disengaged channel(s) illuminate.
SENSOR/SERVO Light-Switches, SENSOR half. One SENSOR light for each sensor (A, B & M) of Pitch SAS and Yaw SAS. (On function selector.)	All SENSOR lights Off.	Indicates normal pitch and yaw SAS conditions with all sensor inputs tracking within required limits.
	A, B or M SENSOR light On steady.	In pitch SAS, indicates failure and exclusion of sensor input. SAS remains engaged and operative. In yaw SAS, indicates exclusion of sensor input due to rate gyro failure. SAS remains engaged and operative.
	Push-button Recycling	Pushing any of the three push-buttons for that SAS axis recycles SAS logic to re-input the excluded sensor. SEE NOTE.
	A, B or M YAW SENSOR light Flashing. △1	Indicates exclusion of yaw sensor input due to lateral accelerometer failure. SAS remains engaged and operative.
	Push-button Recycling	Pushing any of the three push-buttons for the yaw axis recycles SAS logic to re-input the excluded sensor. SEE NOTE.
	A & M or B & M or A & B SENSOR lights On steady. △2	In pitch SAS, indicates failure and exclusion of two sensor inputs (analytical redundancy operating). SAS remains engaged and operative. In yaw SAS, indicates exclusion of two sensor inputs due to rate gyro failures (analytical redundancy operating). SAS remains engaged and operative.
	Push-button Recycling	Pushing any of the three push-buttons for that SAS axis recycles SAS logic to re-input one sensor. Pulse aircraft both directions in that axis, then press any of the three push-buttons again to re-input the other sensor. SEE NOTE.

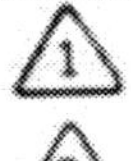

If both the yaw rate gyro and the lateral accelerometer fail in the same yaw sensor, that YAW SENSOR light will be On steady.

△2 **Assumes at least two DAFICS computers operating.**

Figure 1-58 (Sheet 1 of 4)

SAS CONTROLS AND INDICATIONS (Cont.)

CONTROL OR INDICATOR	POSITION OR INDICATION	FUNCTION
SENSOR/SERVO (Cont.)	A, B & M SENSOR lights On steady.	In Pitch SAS, indicates failure of all three sensors, or of two sensors and analytical redundancy. Both pitch servos disengage. In yaw SAS, indicates failure of all three sensors due to rate gyro malfunctions or of two rate gyros and analytical redundancy. Both yaw servos disengage.
	Push-button Recycling	Pushing any of the three push-buttons for that SAS axis recycles SAS logic to re-input one sensor. Pulse aircraft both directions in that axis, then press any of the three push-buttons again to re-input the next sensor. Repeat for last sensor. SEE NOTE.
	A, B & M YAW SENSOR lights Flashing	Indicates failure of at least two yaw sensors due to lateral accelerometer malfunctions. Both servos disengage.
	Push-button Recycling	Recycling has no effect.
Roll SENSOR Light. (On function selector between ROLL engage switches.)	Light Off	Indicates normal roll SAS condition with both sensor inputs tracking within required limits.
	Light On	Indicates failure and exclusion of one sensor, both sensors, or one sensor and analytical redundancy. Two failures cause both roll servos to disengage. With failure and exclusion of one sensor, roll SAS remains engaged and operative.
	Engage Switch Recycling	Recycling either SAS ROLL engage switch recycles SAS logic to re-input one sensor. If light remains on, roll aircraft both left and right, then recycle either engage switch to re-input the other sensor. SEE NOTE.

Figure 1-58 (Sheet 2 of 4)

SAS CONTROLS AND INDICATIONS (Cont.)

CONTROL OR INDICATOR	POSITION OR INDICATION	FUNCTION
SENSOR/SERVO Light-Switches, SERVO half. One SERVO light for each servo (A & B) of each SAS axis. (On function selector.)	All SERVO lights Off.	Normal condition with SAS engaged. (Engage switches ON and logic recycled.)
	A or B SERVO light ON. Pitch or Yaw SAS. 3	Indicates channel disengagement due to engage switch Off or servo logic trip in that channel (failure in transfer valve/servo portion of channel). SAS is still operational but servo redundancy is lost.
	Push-button Recycling	For pitch or yaw SAS servo logic trip, pressing any of the three push-buttons for that SAS axis recycles logic and re-engages the tripped channel. SEE NOTE.
	A or B SERVO light On. Roll SAS.	Indicates channel engage switch Off or servo logic trip in that channel. Both roll servos disengage.
	Engage Switch Recycling	Recycling either ROLL engage switch recycles logic and re-engages the tripped channel. SEE NOTE. To override if failure remains, disengage both switches and re-engage good servo only (to provide proper control gain). Subsequent failure in good servo will not disengage that channel or illuminate that servo light.
	A & B SERVO Lights On. Pitch or Yaw SAS.	Indicates both channel engage switches Off or servo logic trip in both channels. Both servos in that axis disengage.
	Push-button Recycling	For pitch or yaw SAS logic trip, pressing any one of the three push-buttons for that SAS axis will re-engage one servo. Press any of the three push-buttons again to re-engage the other servo. SEE NOTE.
	A & B SERVO Lights On. Roll SAS.	Indicates both channel engage switches Off or servo logic trip in both channels in infrequent occasions when system comparison of servos against model does not yield a positive identification of the servo which has failed. Both roll servos disengage.
	Engage Switch Recycling.	Recycling either ROLL engage switch re-engages both channels. SEE NOTE. To override if failure remains, disengage both switches and re-engage good servo only (to provide proper control gain). Subsequent failure in good servo will not disengage that channel or illuminate that servo light.

3 Unless (A & M) or (B & M) computers failed.

SAS CONTROLS AND INDICATIONS (Cont.)

CONTROL OR INDICATOR	POSITION OR INDICATION	FUNCTION
DAFICS Preflight BIT FAIL light (red).	Flashing	Indicates analytical redundancy is not available.
SAS LITE TEST switch. Press-to-test switch on function selector panel.	PRESS	Energizes all lights (13) on function selector panel and DAFICS PREFLIGHT BIT TEST/FAIL lights.
SAS OUT light on annunciator panel.	OFF	Normal condition with all SAS channels engaged.
	ON	Indicates failures and/or disengagement of a SAS sensor(s) or servo(s) or DAFICS in PREFLIGHT BIT mode.

NOTE: SAS logic trips may be due to transient, passive, or active failures.

A transient trip is usually due to temporary and excessive fluctuations of electrical or hydraulic supplies into SAS, and the tripped channel will re-engage and remain engaged when recycled (unless fluctuations continue or recur).

A passive trip is usually caused by a failure such as loss of ac or dc electrical power, electrical shorts/opens, or hydro/mechanical malfunction of the SAS transfer valves or servos. A channel with a passive failure will usually not engage upon recycling.

An active trip occurs when signals of some magnitude are generated within the affected channel. When recycled, the channel will re-engage and remain engaged until aircraft attitude changes initiate signals in the affected axis. Exercising the flight controls in that axis so as to generate strong signals in the suspect channel will permit an active check of channel integrity. (If channel remains engaged, the trip may be considered a transient type.)

Figure 1-58 (Sheet 4 of 4)

through series servos, common to both. However, pitch autotrim applications from the autopilot or Mach trim are displayed on the pitch trim indicator, as autotrim is applied through the pitch trim actuator in the mixer assembly.

Essential dc power is provided through the A/P MACH TR A and A/P MACH TR B circuit breakers on the pilot's left console. Emergency ac power is provided through the RSO's AP MACH TR circuit breaker.

NOTE

Autopilot redundancy is lost when one of the two A/P MACH TR dc circuit breakers is opened. To disable the autopilot, both A/P MACH TR A & B circuit breakers must be opened or the AP MACH TR ac circuit breaker in the aft cockpit must be opened.

Autopilot Reference Signals

The autopilot receives attitude, heading, and navigational inputs from the Astroinertial Navigation System. The Inertial Navigation

System, provides only attitude and heading inputs. The autopilot uses the attitude reference (ANS or INS) selected by the pilot's ATT REF SELECT switch. If the rate of change of pitch angle of the selected attitude reference is excessive compared to SAS pitch rate, the pitch autopilot disengages and the AUTOPILOT OFF light on the pilot's annunciator panel illuminates. DAFICS schedules autopilot signal strength (gain) relative to altitude and airspeed, and provides Mach and KEAS references.

Pre-engage Synchronize Mode

When not engaged, the autopilot operates in the pre-engage synchronize mode, which aligns autopilot signals with the existing aircraft attitude to minimize control transients when the autopilot is engaged.

Autopilot Alignment (Trim) Indicators

When the autopilot is engaged, the alignment indicators reflect the direction and magnitude of autopilot control inputs. Needles will normally be close to center, and a continuous deflection in the same direction indicates an out-of-trim condition.

NOTE

> The pilot should be prepared for control transients when engaging or disengaging the autopilot regardless of alignment indications, as out-of-synchronization conditions may exist which are not registered on the alignment indicators.

Attitude Hold Mode

Each autopilot channel is initially engaged in the attitude-hold mode. After initial engagement, secondary modes may be engaged as desired: KEAS hold or Mach hold in pitch channel, and heading hold or Auto-Nav in roll channel.

In attitude hold, the autopilot controls elevon deflections to maintain attitude, using change in attitude as an error signal. Manual control inputs from stick or trim will cause only momentary departures from the engaged attitude, as the autopilot will counteract such inputs to regain the reference attitude.

The reference attitude may be changed by use of the respective channel trim wheel or the CSC switch without disengaging the autopilot. The pitch and roll trim wheels permit vernier adjustments by inserting corrections at approximately 1 degree per 15 degrees wheel rotation in pitch and 1 degree per 8° wheel rotation in roll. Depressing the CSC push-button disengages the autopilot secondary modes and permits control stick changes to aircraft attitude without override or disengagement of the autopilot. The attitude existing when the CSC pushbutton is released becomes the new autopilot reference attitude.

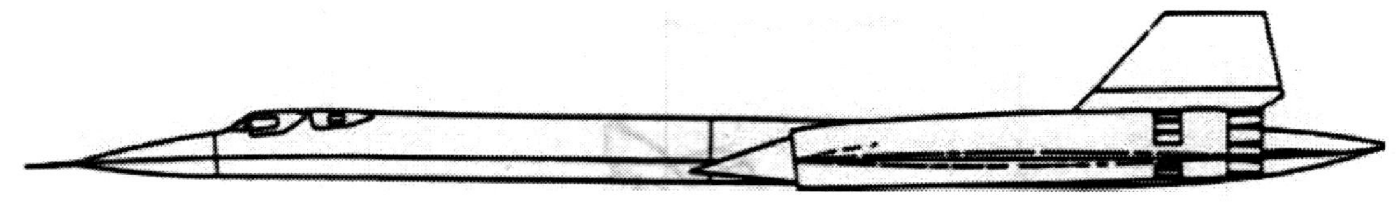

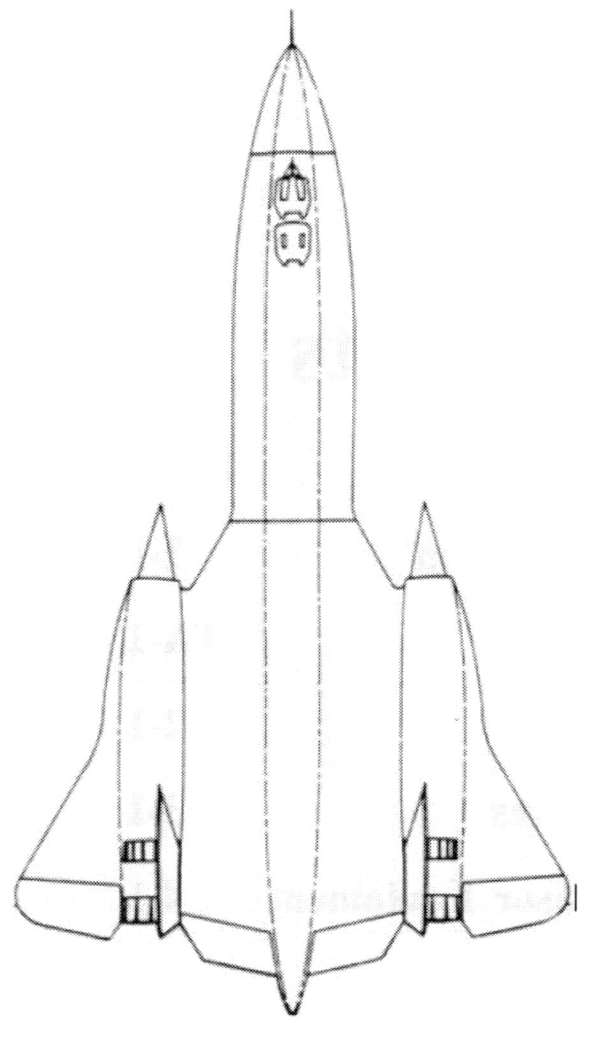

THIS PAGE INTENTIONALLY LEFT BLANK OR STILL CLASSIFIED.

Mach Hold and KEAS Hold

Mach hold or KEAS hold may be engaged after the pitch autopilot is engaged in attitude hold. When either secondary mode is engaged, the attitude-hold mode is rendered inoperative, and the autopilot controls pitch attitude to maintain the Mach or KEAS existing when the respective mode was engaged.

The DAFICS computers provide Mach and KEAS signals to both the autopilot and the triple display indicator, so if TDI indications are in error, do not engage Mach hold or KEAS hold. With the pitch autopilot engaged, stabilize the aircraft for a few seconds at the desired Mach or KEAS before engaging either mode. When switching between Mach hold and KEAS hold, a slight time delay is provided to permit autopilot synchronization to the new reference, and the pilot should monitor respective indications to ensure that the autopilot has engaged in the desired mode.

NOTE

> Manual control inputs from stick or trim will cause attitude transients while the autopilot is engaged in Mach hold or KEAS hold. If autopilot authority is exceeded while in Mach hold or KEAS hold, the autopilot will not maintain the engaged value. Use of CSC will disengage all autopilot secondary modes. If Mach hold or KEAS hold is engaged, autopilot pitch trim wheel inputs will disengage the autopilot pitch channel.

KEAS Bleed

The KEAS hold and Mach hold modes automatically control the autopilot pitch channel so that a speed schedule from 500 KEAS and 2.1 Mach to 380 KEAS and 3.3 Mach will not be exceeded. The bleed line is a linear function of 10 knots per 0.1 Mach number. (See Figure 1-59.)

When KEAS hold is engaged during climb, pitch attitude is controlled to maintain engage equivalent airspeed until the KEAS bleed schedule is intercepted. When KEAS hold is used for a normal 450 KEAS climb, bleed begins when the bleed line is intercepted at 2.6 Mach, and will continue to 380 KEAS and Mach 3.3. If a 430 KEAS climb speed was selected, bleed-off would begin at Mach 2.8.

If KEAS hold is disengaged after climb and reengaged at a speed below bleed line, the engage KEAS will be held constant for subsequent cruise or descent. However, if KEAS hold is not disengaged after climb, and Mach is reduced, the autopilot will follow the bleed line until the initial engage KEAS is reached, and then maintain that KEAS. If KEAS hold is engaged while airspeed is above the bleed line, the autopilot will slow the aircraft to the bleed-line schedule by nose-up attitude changes, and will follow the bleed schedule during subsequent cruise, climb, and descent.

If Mach hold is engaged, pitch attitude is controlled to maintain engage Mach unless the KEAS bleed line is intercepted. During climb at constant Mach, KEAS decrease. During descent at constant Mach, KEAS increase; however if the KEAS bleed line is intercepted the autopilot follows the KEAS bleed line. If Mach hold is engaged at Mach 3 and power is reduced, the autopilot holds Mach 3 by descending until equivalent airspeed reaches 410 KEAS and then descends following the KEAS bleed schedule (KEAS increase as Mach decreases). If power is then added, the autopilot follows the KEAS bleed schedule back to Mach 3, and then increases pitch to maintain Mach 3.

KEAS Warning Light and Aural Warning

A red KEAS warning light is installed on the upper left side of the pilot's instrument panel. Illumination of the KEAS light warns the pilot that an abnormally high or low airspeed condition has been reached and that appropriate corrective action must be taken.

At high airspeed, the light flashes if 470 KEAS is exceeded when below Mach 2.6. Above Mach 2.6, the light flashes 20 KEAS above the KEAS bleed schedule.

The light will also flash when airspeed is below 300 KEAS and above Mach 1.3.

The light will also flash and a steady tone will sound when airspeed is below approximately 250 KEAS. The low airspeed warning is deactivated below 0.5 Mach to prevent nuisance warnings during takeoff and landing. The pilot should increase airspeed immediately when the low airspeed KEAS warning is activated. If airspeed decreases after a low airspeed warning, the airplane could enter a flight region where recovery from a gust or high nose-up attitude might not be possible.

A small hysteresis band can be expected when passing through 250 KEAS or 0.5 Mach number, as follows:

a. Decreasing airspeed - Light on at 245 to 255 KEAS while above 0.5 Mach number.

b. Increasing airspeed - Light off at 250 to 260 KEAS.

c. Decreasing Mach number - Light off at 0.50 to 0.55 Mach number.

Refer to Figure 1-59, Autopilot KEAS Hold and KEAS Warning Light Schedules. Operation of the light is controlled by the DAFICS computers as shown by Figures 1-62, and 1-64.

Intensity of the light will be full bright if the console lights rheostat is OFF, and it will be dim in all other rheostat positions. Operation of the light can be tested by depressing the IND & LT TEST switch. Power for the light is supplied from the essential dc bus through the WARN 2 circuit breaker on the pilot's left console.

AUTOPILOT KEAS BLEED AND KEAS WARNING LIGHT SCHEDULES

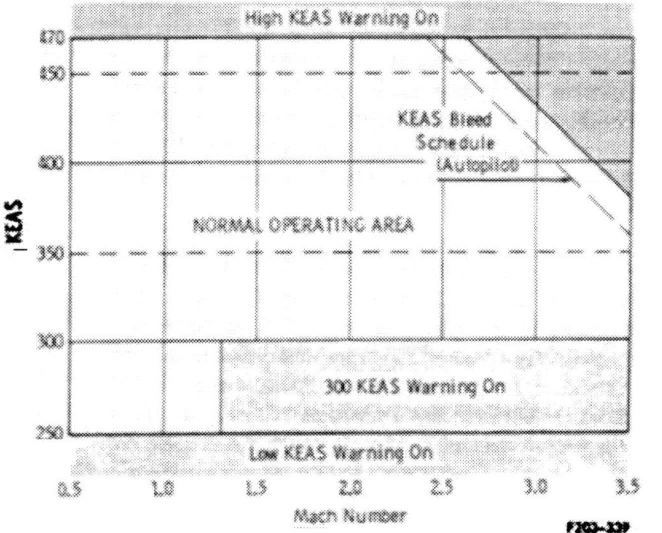

Figure 1-59

Heading Hold and Automatic Navigation

Heading hold or automatic navigation (AUTO NAV) may be engaged after the roll autopilot is engaged in attitude hold. When either secondary mode is engaged, roll attitude hold is overridden and the autopilot roll trim wheel is inoperative. These secondary modes cannot be engaged together. Use of CSC will disengage either secondary mode.

HEADING HOLD may be engaged at bank angles up to 50 degrees. The aircraft heading at engagement becomes the reference heading and the autopilot will roll the aircraft to capture and maintain the reference heading. The reference heading cannot be changed while heading hold is engaged.

AUTO NAV may be engaged only when the forward cockpit ATT REF SELECT switch is

in ANS and the ANS-ready signal exists. With AUTO NAV engaged, the roll autopilot responds to bank inputs from the ANS. Bank angles in the auto-nav mode are limited to 45° by the ANS system.

Autopilot Controls and Indicators

Except for the control stick command (CSC) button and autopilot disengage trigger switch on the control stick grip, all autopilot controls are on the function selector panel. See Figure 1-60.

A/P OFF Light and Switch

A square A/P OFF switch/light is installed on the pilot's right console. The switch operates in conjunction with the AUTO PILOT OFF light on the annunciator panel: when an autopilot pitch and/or roll channel is disengaged, the AUTO PILOT OFF annunciator light illuminates; pressing the A/P OFF switch will cause the OFF portion of the switch to glow amber and also will extinguish the AUTO PILOT OFF annunciator light. The A/P OFF light remains on either until both channels of the autopilot have been engaged or until a pitch or roll channel is subsequently disengaged. Cycling of the A/P OFF switch has no effect after it illuminates. If one autopilot axis (pitch or roll) is still engaged, subsequent disengagement of that channel causes the AUTO PILOT OFF annunciator light to illuminate again.

Autopilot Operational Restrictions

Pitch and Roll Channels

Autopilot authority is so limited that malfunctions within the autopilot cannot cause damaging or uncontrollable aircraft attitude changes.

1. Roll channel authority is limited by roll SAS authority limitations (4 degrees roll differential with both roll SAS channels engaged, and 2 degrees differential with one roll SAS channel engaged).
2. Pitch channel authority is limited to 1.6 degrees up and down elevon below 50,000 feet pressure altitude, and 2.3 degrees at high altitudes.

The autopilot will disengage if bank angle exceeds 50 degrees.

Do not use the pitch autopilot with bank angles exceeding 45°.

The autopilot will disengage when changing attitude reference using the pilot's ATT REF SELECT switch.

The autopilot will disengage if the ready signal from the selected attitude reference source (ANS or INS) is interrupted.

A rate frequency monitor disengages the respective channel if a failure occurs. A condition of high turbulence could cause the frequency monitor to disengage a normally functioning autopilot, but repeated disengagements indicate failure within the autopilot.

NOTE

If the APW system stick warning switch is in PUSHER/SHAKER, the autopilot pitch channel will disengage when the pusher boundary is reached.

Pitch Channel

Do not engage Mach hold or KEAS hold if TDI indications are in error.

WARNING

Monitor flight instruments during autopilot operation. Assure that speed does not exceed the normal operating envelope.

AUTOPILOT CONTROLS AND INDICATIONS

CONTROL OR INDICATOR	POSITION OR INDICATION	FUNCTION OF CONTROL OR INDICATOR
ALIGNMENT INDICATORS One floating-indices type indicator for each autopilot channel on function selector panel.	Respective channel not engaged.	Indicator needle centers.
	Channel engaged	Needle fluctuates with autopilot control inputs but should remain near center. Continuous deflections in same direction may indicate out-of-trim condition.
A/P ENGAGE SWITCHES One ON-OFF switch for each channel on function selector panel, solenoid held ON.	OFF	Respective autopilot channel is disengaged. Secondary modes cannot be engaged (Mach Hold, KEAS Hold, Auto-Nav, Heading Hold).
	ON	Respective channel is engaged in Attitude Hold mode. Secondary modes may be engaged.
TRIM WHEELS One serrated trim wheel for each channel on function selector panel. Top third of wheel is visible.	Channel not engaged	Trim wheels are inoperative.
	Channel engaged in Attitude Hold	Trim wheel permits vernier changes to reference attitude without disengaging autopilot. Roll trim wheel is inoperative in Heading Hold and Auto-Nav modes. If Mach hold or KEAS hold is engaged, pitch trim wheel inputs will disengage the autopilot pitch channel and illuminate the AUTOPILOT OFF annunciator light.
MACH HOLD/KEAS HOLD SWITCH A 3-position switch on function selector panel. Solenoid held in MACH HOLD or KEAS HOLD, springloaded OFF.	OFF	Mach Hold and Keas Hold disengaged.
	MACH HOLD	Pitch A/P channel is engaged in Mach Hold mode if the pitch channel engage switch is ON. A/P controls pitch attitude to maintain engage Mach or follow KEAS-bleed line. Switch will trip OFF if CSC is used.
	KEAS HOLD	Pitch A/P channel is engaged in KEAS Hold mode if pitch channel engage switch is ON. A/P controls pitch attitude to maintain engage KEAS or follow KEAS-bleed line. Switch will trip OFF if CSC is used.
HEADING HOLD SWITCH ON-OFF switch on function selector panel, solenoid held ON.	OFF	Heading Hold mode is disengaged.
	ON	Roll A/P channel is engaged in Heading Hold if roll channel engage switch is ON. If engaged in a turn, A/P will roll aircraft to capture the heading existing at time of engagement. Switch will trip OFF if Auto-Nav is selected or CSC is used.

[illegible] 1 of 2)

AUTOPILOT CONTROLS AND INDIC

CONTROL OR INDICATOR	POSITION OR INDICATION	FUNCTION OF CONTROL OR INDICATOR
AUTO NAV SWITCH ON-OFF switch on function selector panel, solenoid held ON.	OFF	Auto-Nav mode is disengaged.
	ON	Roll A/P channel is engaged in auto-nav mode if roll channel engage switch is ON and attitude reference selector is in ANS position. Switch will trip OFF if heading hold is selected or CSC is used.
ATTITUDE REF. SELECT SWITCH. A 2-position (ANS-INS) switch on pilot's instrument panel.	ANS	Permits autopilot use of attitude, heading, and navigational inputs from the ANS.
	INS	Permits autopilot use of attitude and heading inputs from the INS. Auto-nav mode is inoperative and cannot be engaged with INS selected.
	ANS to INS INS to ANS	Both Pitch and Roll A/P channels are disengaged, and must be manually re-engaged.
CSC SWITCH Pushbutton switch on control stick grip.	Released	No affect.
	Depressed	Disengages all secondary modes of Pitch and Roll A/P (Mach hold, KEAS hold, Auto-Nav and Heading hold switches trip OFF). Permits aircraft attitude changes without override or disengagement of A/P Attitude hold modes. Pitch and Roll channels of A/P accept attitude existing when CSC is released as the new reference attitude for Attitude-Hold modes. (Secondary modes must be manually re-engaged.)
A/P DISCONNECT SWITCH Multi-function trigger switch on control stick grip.	Released	No affect.
	Pressed	Disengages entire autopilot immediately. All modes must be manually reengaged. Also disconnects air refueling system and disables pitch and yaw trim switch and APW system stick pusher while depressed.
AUTOPILOT OFF Annunciator Panel light	Illuminated	Illuminated when pitch and/or roll autopilot disengages.
A/P OFF Rt instr side panel	Illuminated	When depressed, extinguishes the AUTOPILOT OFF lite on annunciator panel.
AUTO NAV OUT RSO Annunciator Panel	Illuminated	Illunimates when auto-nav is not being used.

Figure 1-60 (Sheet 2 of 2)

AUTOPILOT AND MACH TRIM BLOCK DIAGRAM

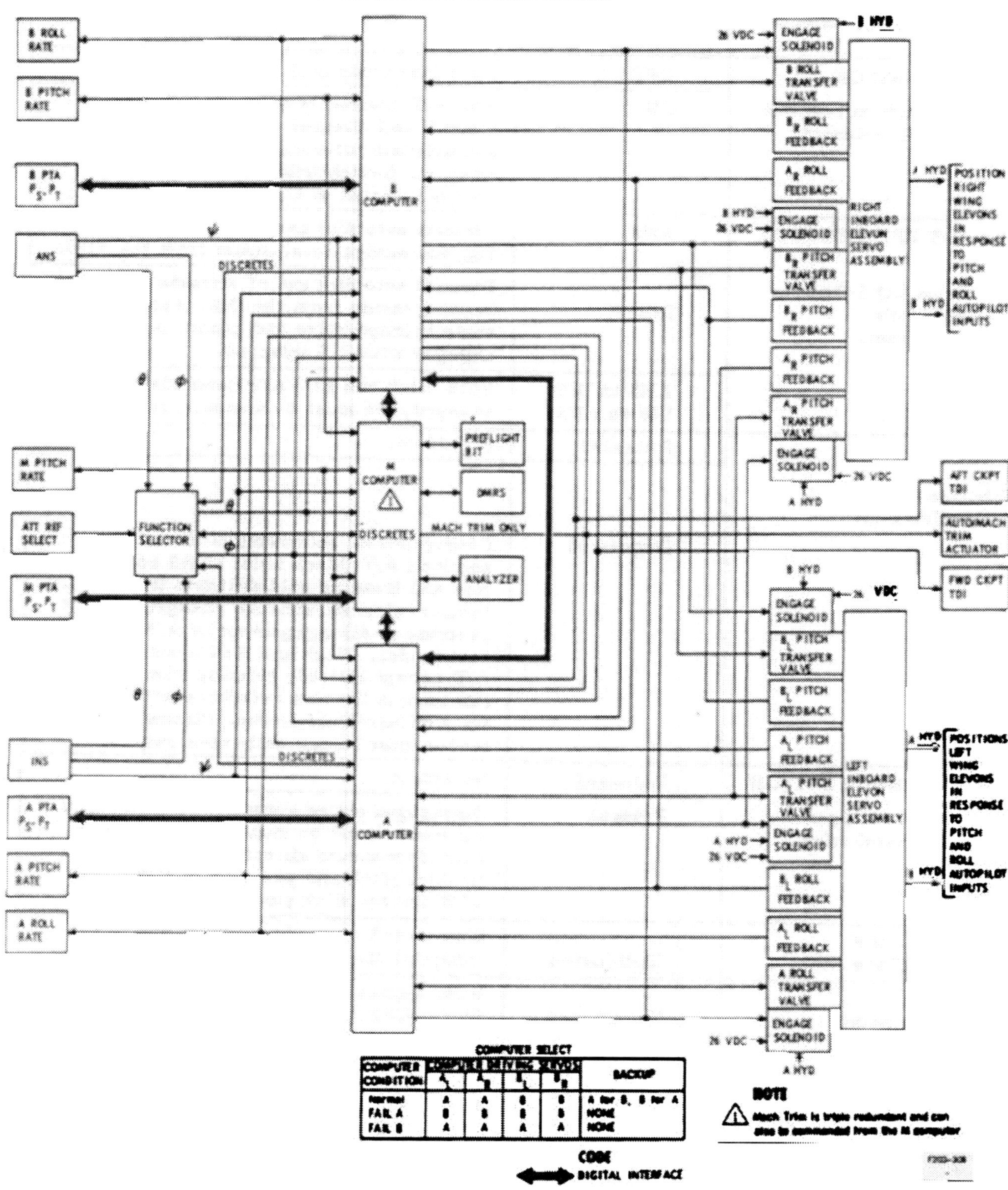

COMPUTER CONDITION	COMPUTER DRIVING SERVOS A_L	A_R	B_L	B_R	BACKUP
Normal	A	A	B	B	A for B, B for A
FAIL A	B	B	B	B	NONE
FAIL B	A	A	A	A	NONE

Figure 1-61

Pitch autopilot rate capability may be exceeded when elevon deflection requirements exceed autopilot authority plus autotrim rates. For example, this condition could appear as a loss of pitch trim wheel effectiveness during high pitch rate transonic acceleration maneuvers and supersonic turning decelerations.

Roll Channel

AUTO NAV commands a maximum bank angle of 45°.

The auto-nav mode can be disengaged by using either the AUTO NAV switch, the HEADING HOLD switch, or the CSC button. There should not be any transient roll attitude changes unless the alignment indices show an out-of-synchronization condition.

AUTO NAV will not engage unless the ATT REF SELECT switch is in ANS.

HEADING HOLD cannot be engaged if INS is the selected attitude reference and the INS is operating in the ATTITUDE mode.

Autopilot Operation

To engage autopilot in attitude hold:

1. Engage SAS
2. Manually trim the aircraft.
3. Pitch and/or roll engage switches - ON.

For minor changes to attitude, use pitch or roll trim wheels.

For major changes to engaged attitude:

4. CSC button - Hold depressed until aircraft is stabilized in new attitude.
5. CSC button - Release.

   Autopilot will reengage in existing pitch and roll attitude.

To engage Mach hold or KEAS hold:

6. With the pitch autopilot engaged, stabilize the aircraft for a few seconds at the desired Mach or KEAS before engaging either mode.
7. Mach hold/KEAS hold switch - As desired.

To engage Heading hold:

8. Heading hold switch - ON.

To engage Auto-Nav:

9. Attitude reference select switch - ANS.

   If the switch is in INS, the autopilot will disengage when the switch is moved to ANS.

10. Auto-nav switch - ON.

    Turns should be anticipated while in AUTO NAV and manual control stick inputs applied, if necessary, to avoid excessive roll or bank angles.

To disengage autopilot:

11. Alignment indicators - Monitor.

    Continuous deflections of indicator needle in same direction indicates that a transient will probably occur on disengagement.

To disengage all but attitude-hold modes:

12. Depress CSC button.

To disengage entire autopilot:

13. Depress control stick trigger switch.

To disengage respective channels:

14. Move corresponding pitch and/or roll axis engage switches to OFF.

To extinguish the annunciator panel AUTO PILOT OFF light:

15. Depress A/P OFF light.

MACH TRIM SYSTEM

Speed stability augmentation is provided by the Mach trim system while accelerating or decelerating in the 0.2 to 1.5 Mach range. Mach trim is scheduled by the DAFICS computers to restore conventional stick forces and trim requirements by applications of pitch trim through the slow-speed motor of the pitch trim actuator. No control switch is provided, although Mach trim is enabled with pitch autopilot OFF, and is disabled with pitch autopilot ON. All manual and auto trim (including Mach trim) can be disabled by moving the TRIM POWER switch on the annunciator panel to OFF.

Power is supplied to the Mach trim system by the essential dc bus through the A/P MACH TR A, A/P MACH TR B, and CMPTR M circuit breakers on the pilot's left console and the emergency ac bus through the A/P MACH TR circuit breaker in the aft cockpit.

NOTE

> Mach trim redundancy is reduced when one of the two A/P MACH TR dc circuit breakers is opened and redundancy is lost when both A/P MACH TR circuit breakers are opened. To disable the Mach trim system, both A and B channel circuit breakers and CMPTR M circuit breaker must be opened <u>or</u> the AP MACH TR ac circuit breaker in the aft cockpit must be opened.

AUTOMATIC PITCH WARNING AND HIGH ANGLE OF ATTACK WARNING SYSTEMS

An Automatic Pitch Warning (APW) system provides control stick pusher and stick shaker features. APW stick shaker operation warns that a potentially unsafe condition of angle of attack plus pitch rate has been reached. For those situations where both warnings are provided, the boundary for stick pusher operation is slightly more extreme than the boundary for the stick shaker. In addition to the APW System, a separate High Angle of Attack (High Alpha) warning system can also operate the stick shaker, but not the pusher.

The stick shaker can operate throughout the flight regime without restriction due to Mach number or stick or gear position. APW system inputs to the stick shaker can be disabled by setting the APW switch to OFF. The High Alpha shaker warning system starts the shaker anytime its Mach number vs angle of attack schedule is exceeded regardless of the APW switch position. Proximity to the APW stick shaker boundary is displayed by the glide slope indicator needle on the Attitude Director Indicator when the DISPLAY MODE SEL switch is not in ILS or ILS APPROACH. This display is independent of APW switch position. The SHAKER warning light near the pilot's glareshield illuminates when the stick shaker operates. An APW caution light on the pilot's annunciator panel illuminates if the APW switch is turned off, if the APW fails self check, or if the 2 PTA CHAN OUT light illuminates (the 2 PTA CHAN OUT light remains on until reset by maintenance, but the APW light may subsequently extinguish).

The stick pusher can only operate while subsonic with two of the three landing gear doors locked in the up position. Two of the three DAFICS computers must be operating. The pusher warning can be disabled manually by the APW switch, by depressing the control stick trigger switch, or by positioning the control stick more than 2.5 degrees (1.5 inches) forward of neutral.

The APW and High Alpha Warning systems are required because there is no buffet or other natural warning of approach to unstable angle of attack conditions. The pitch boundary indication on the ADI can provide the pilot with a display of angle of attack plus pitch rate relative to the APW stick shaker boundary. If the boundary is broached, activation of the stick shaker warns the pilot that he has reached a potentially dangerous angle of attack or that pitch rate is such that a potentially dangerous condition will be reached unless immediate action is taken. Refer to Figure 1-63 for the

shaker boundary schedules. The shaker warning is a 30 cps vibration of the control stick.

The stick pusher starts control stick correction in pitch. An abrupt push force of approximately 30 pounds is applied when the pusher boundary has been reached until the stick is 2.5° (1.5") forward of neutral. The push force can be overcome by the pilot, if desired. The pusher function can be disabled by the control stick trigger switch or by moving the APW switch to OFF or SHAKER ONLY.

The boundaries for operation of the APW stick shaker and pusher warnings are functions of angle of attack and pitch rate summation vs Mach number. The warning schedule reflects the changing susceptibility to pitch-up at subsonic and supersonic speeds. The boundary for operation of the High Alpha system shaker warning is fixed at angles of attack of 14° for low speed and 8° for high speed operation, with the switchover at Mach 1.4 or 1.55 (decelerating or accelerating, respectively). Observance of the APW and High Alpha system warning boundaries provides adequate margin to avoid pitch-up at all speeds.

WARNING

Avoidance of the stick shaker or pusher/shaker warning boundaries does not, by itself, assure that load factor or angle of attack limits will be observed.

Reduce angle of attack and adjust attitude nose-down if a high angle of attack warning occurs or if an alpha limit is approached.

WARNING

When subsonic, if an APW system or high angle of attack warning occurs, or if angle of attack and airspeed are not within limits, make angle of attack and speed corrections before adjusting the throttles. These actions alone may clear engine stall conditions, and are mandatory to avoid pitch-up, if at high angle of attack and/or low airspeed.

Essential dc power for the stick shaker is obtained either through the APW or STALL WARN circuit breaker on the pilot's left console. The power routing depends on which system activates the shaker. This APW circuit breaker also supplies power for the stick pusher solenoid which controls operating power from the "A" hydraulic system. The A and/or B computer must be operative for either the APW or High Alpha Warning system to operate. The High Alpha Warning system also requires essential dc power from the BRK & SKID circuit breaker on the pilot's left console. This power causes the main gear scissors switches and relays to close when aircraft weight is removed from the main gear. With the right main gear relay open, no alpha signals are received by the High Alpha Warning system.

APW AND HIGH ALPHA WARNING SYSTEM CONTROLS AND INDICATORS

There is no single control of all APW and High Alpha Warning system functions. The ADI pitch boundary indication is controlled by the Display Mode Select switch. APW control stick warning is controlled by the APW PUSHER/SHAKER switch. The APW stick shaker boundary changes and the stick pusher is cut out when two of the three landing gear doors are not locked in the up position. Each gear-up signal is routed to one DAFICS computer so if two DAFICS computers are inoperative the APW reverts to gear-not-up logic. The angle of attack input to the APW is inhibited with weight on the left main gear. The High Alpha Warning System shaker is inhibited with weight on the right main gear. The stick pusher can be cut out manually by the APW switch, the control stick trigger switch and by stick positioning. APW system status is indicated by the ADI glide slope needle, SHAKER warning light, stick shaker and pusher, APW caution light, and DAFICS A and B CMPTR OUT caution lights. No controls are provided for the High Alpha Warning system. The High Alpha Warning system control stick shaker function is always operative while in flight.

APW Pusher/Shaker Switch

A three-position APW pusher/shaker switch is provided on the left side of the pilot's instrument panel. The APW system stick shaker and stick pusher are inoperative when OFF (center) is selected; however, the High Alpha Warning system remains operative and the pitch boundary indication to the ADI continues to operate. See Figure 1-66.

The APW stick shaker is operative when the SHAKER ONLY (down) position is selected. The APW stick shaker and stick pusher are operative when the PUSHER/SHAKER (up) position is set. The APW caution light illuminates when the APW switch is OFF. APW angle of attack and pitch rate schedules are shown by Figure 1-63, Sheet 1 and 2.

With a normally operating APW system, do not position the APW pusher/shaker switch to OFF.

Shaker Warning Light

A red SHAKER warning light is located near the apex of the glareshield in the pilot's cockpit. It illuminates whenever dc power is supplied to operate the stick shaker motor by either the High Alpha Warning or APW system. The light can be tested by pressing the pilot's IND & LT TEST push-button switch. Essential dc power for the light is obtained through the APW or STALL WARN circuit breaker, whichever system is controlling the stick shaker motor.

APW Caution Light

An amber APW caution light on the pilot's annunciator panel illuminates when the APW system stick warning switch is OFF. With PUSHER/SHAKER or SHAKER ONLY selected, the light illuminates when both A and B computers are operative and the APW outputs from the A computer do not match the B computer. The APW caution light also illuminates (although it may subsequently extinguish) when the 2 PTA CHAN OUT annunciator light illuminates. If both A and B CMPTR OUT annunciator lights illuminate, both the APW system and the High Alpha Warning System are inoperative, but the APW caution light will not illuminate.

NOTE

The APW stick shaker and pusher are disabled when the APW annunciator light is on.

APW AND HIGH ALPHA WARNING SYSTEM OPERATION

APW computations are performed by the A and B DAFICS computers. Both computers use the same triple redundant digital input values for angle of attack and pitch rate as shown by Figures 1-62, and 1-64. If both the A and B computers are operating, both computers activate the shaker and pusher when the appropriate boundaries for the existing Mach number are exceeded. Pusher warning can not occur unless the APW shaker warning is also on. If the digital APW outputs from the A computer do not exactly match the digital APW outputs from the B computer, the APW annunciator light illuminates and both the pusher and shaker are deactivated. If either the A or B CMPTR OUT light illuminates, the remaining computer continues all APW functions including pusher, shaker, and ADI pitch boundary indication, but redundancy is lost.

APW AND HIGH α WARNING SYSTEM

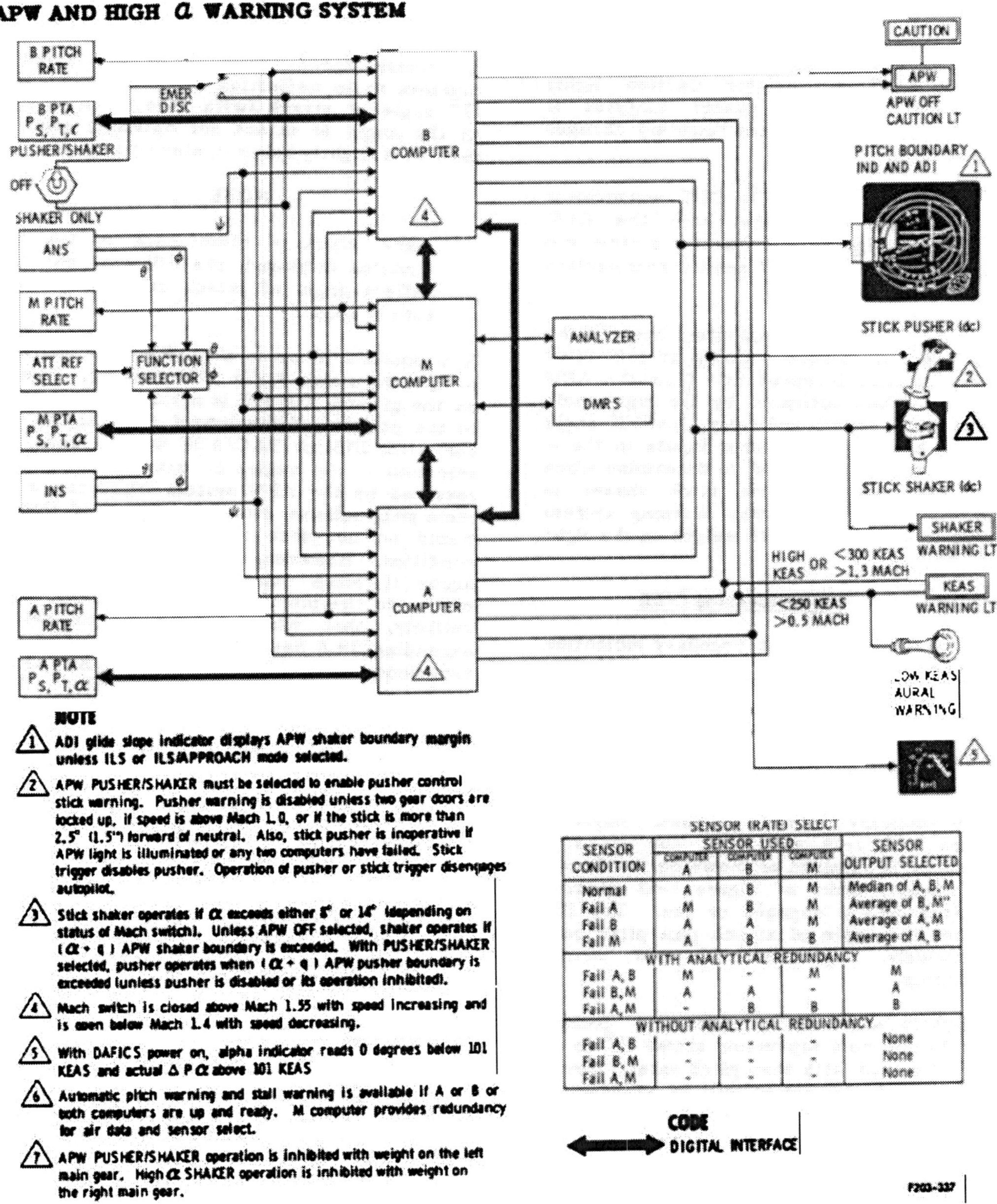

NOTE

1. ADI glide slope indicator displays APW shaker boundary margin unless ILS or ILS/APPROACH mode selected.
2. APW PUSHER/SHAKER must be selected to enable pusher control stick warning. Pusher warning is disabled unless two gear doors are locked up, if speed is above Mach 1.0, or if the stick is more than 2.5° (1.5") forward of neutral. Also, stick pusher is inoperative if APW light is illuminated or any two computers have failed. Stick trigger disables pusher. Operation of pusher or stick trigger disengages autopilot.
3. Stick shaker operates if α exceeds either 8° or 14° (depending on status of Mach switch). Unless APW OFF selected, shaker operates if (α + q) APW shaker boundary is exceeded. With PUSHER/SHAKER selected, pusher operates when (α + q) APW pusher boundary is exceeded (unless pusher is disabled or its operation inhibited).
4. Mach switch is closed above Mach 1.55 with speed increasing and is open below Mach 1.4 with speed decreasing.
5. With DAFICS power on, alpha indicator reads 0 degrees below 101 KEAS and actual $\Delta P \alpha$ above 101 KEAS
6. Automatic pitch warning and stall warning is available if A or B or both computers are up and ready. M computer provides redundancy for air data and sensor select.
7. APW PUSHER/SHAKER operation is inhibited with weight on the left main gear. High α SHAKER operation is inhibited with weight on the right main gear.

SENSOR (RATE) SELECT

SENSOR CONDITION	SENSOR USED COMPUTER A	COMPUTER B	COMPUTER M	SENSOR OUTPUT SELECTED
Normal	A	B	M	Median of A, B, M
Fail A	M	B	M	Average of B, M"
Fail B	A	A	M	Average of A, M
Fail M	A	B	B	Average of A, B
WITH ANALYTICAL REDUNDANCY				
Fail A, B	M	-	M	M
Fail B, M	A	A	-	A
Fail A, M	-	B	B	B
WITHOUT ANALYTICAL REDUNDANCY				
Fail A, B	-	-	-	None
Fail B, M	-	-	-	None
Fail A, M	-	-	-	None

CODE

DIGITAL INTERFACE

F203-337

Figure 1-62

If both (A and M) CMPTR OUT or (B and M) CMPTR OUT annunciator caution lights illuminate, the APW pusher function is deactivated and the shaker boundary changes to gear down values.

If the A and B CMPTR OUT annunciator caution lights illuminate, both the APW system and High Alpha Warning system are inoperative, but the APW annunciator caution light will not illuminate.

The DAFICS A and B computers operate the high angle of attack warning system stick shaker function independently from the APW System. When activated by the right main gear scissors switch and relay, DAFICS angle of attack and Mach number inputs to the A and B computers are used to determine when the High Alpha system stick shaker is reached. The High Alpha Warning system shaker is inoperative with weight on the right main gear.

APW Pitch Boundary Indication (PBI)

While in flight, the pitch boundary indication (PBI) on the Attitude Director Indicator shows proximity to the APW stick shaker boundary (unless ILS or ILS APPROACH Display Mode is selected). A three-quarters scale deflection of the glide slope displacement indicator (second dot from the top of the scale) indicates that the APW shaker boundary has been reached. Refer to Figures 1-62, 1-64, and 1-66. The PBI pitch rate signal is obtained as shown on the Sensor (rate) Select chart of Figure 1-63 whether the pitch SAS is engaged or not. The PBI displays the angle of attack plus pitch rate continuously, regardless of APW switch positioning.

A slightly-below-center indication corresponds to optimum supersonic cruise (5° to 6° angle of attack with zero pitch rate). Three-quarters scale deflection will be approached at supersonic maximum cruise altitude and during intermediate altitude turns.

At cruise speeds near Mach 0.9 three-quarters scale deflection represents 11° to 12° angle of attack (with zero pitch rate). At the angle of attack for optimum cruise the PBI is slightly below center scale.

NOTE

Since alpha + pitch rate is the function displayed, the PBI does not reflect angle of attack unless pitch rate is zero.

It is possible to cause the APW stick shaker to operate momentarily while taxiing. When on the ground, the PBI is normally deflected to the position of the lowest dot marking if the ANS, INS, or TAC/ADF display mode is selected. No angle of attack signal is received by the APW system while the left main gear scissors switch is open, and there would be no pitch rate while in a static condition. However, needle fluctuations will occur if pitch rate gyros sense fuselage motion in the pitch axis. It is possible, but unlikely, that while taxiing, pitch rates exceeding 14.6 degrees per second could be experienced which would cause the APW stick shaker to operate momentarily. The high angle of attack shaker warning can not occur, as the right main gear scissors switch relay opens this warning circuit while on the ground.

NOTE

Positioning of the PBI at the bottom index while in one "g" flight could indicate no angle of attack input to the APW system. This could result from failure of the left main gear scissors switch.

High Alpha Stick Shaker Boundaries

In flight, the High Alpha Warning system powers the stick shaker when its angle of attack boundary is reached —regardless of pitch rate and the pitch boundary indication. The angle of attack settings for this system are 14° when below Mach 1.4 (or when below Mach 1.55 if accelerating) and 8° when above

(Note: Page 1-130A/(1-130B Blank) deleted)

APW AND HIGH α WARNING SYSTEM (With S/B R-2559)

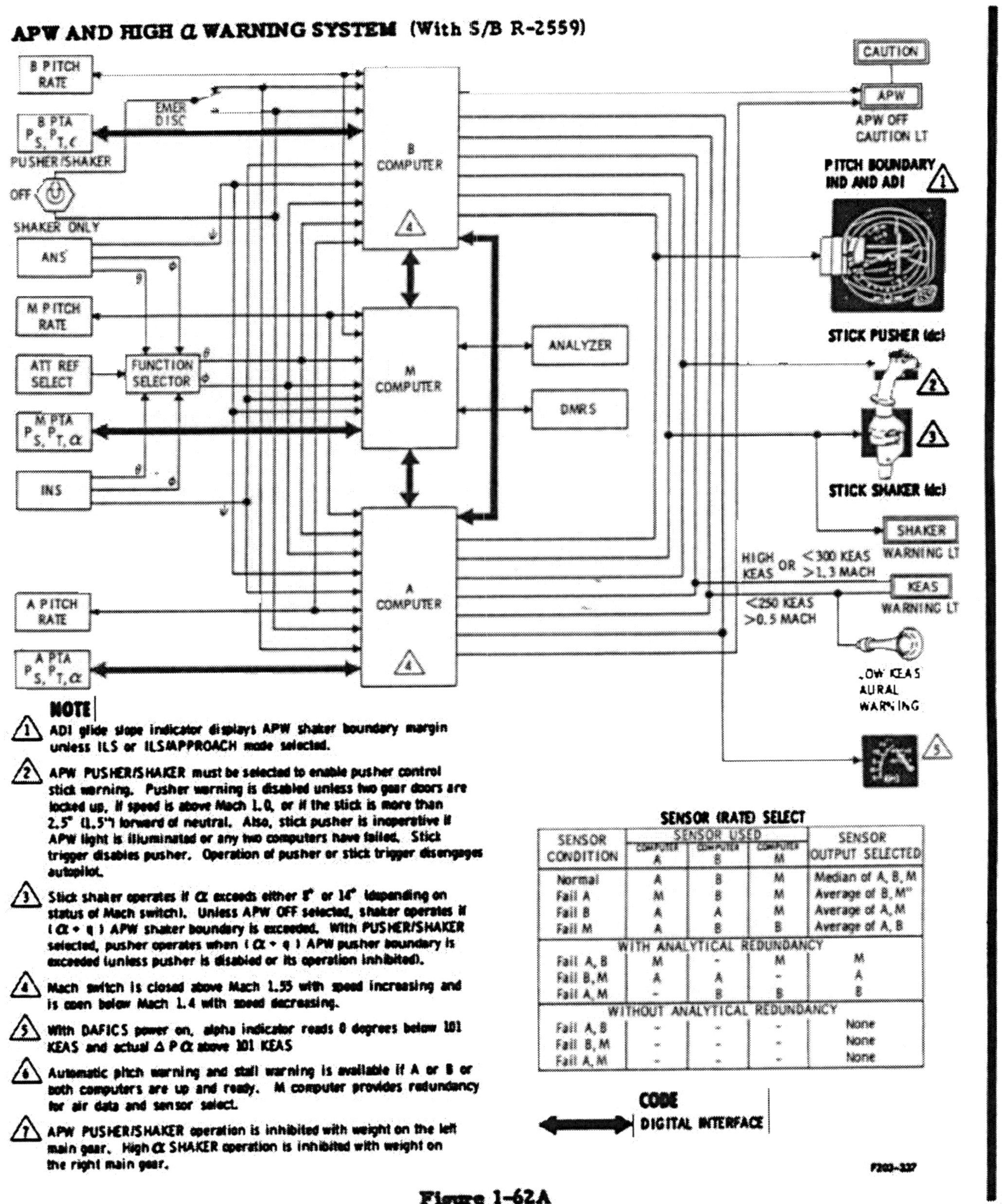

NOTE

1. ADI glide slope indicator displays APW shaker boundary margin unless ILS or ILS/APPROACH mode selected.
2. APW PUSHER/SHAKER must be selected to enable pusher control stick warning. Pusher warning is disabled unless two gear doors are locked up, if speed is above Mach 1.0, or if the stick is more than 2.5° (1.5") forward of neutral. Also, stick pusher is inoperative if APW light is illuminated or any two computers have failed. Stick trigger disables pusher. Operation of pusher or stick trigger disengages autopilot.
3. Stick shaker operates if α exceeds either 8° or 14° (depending on status of Mach switch). Unless APW OFF selected, shaker operates if (α + q) APW shaker boundary is exceeded. With PUSHER/SHAKER selected, pusher operates when (α + q) APW pusher boundary is exceeded (unless pusher is disabled or its operation inhibited).
4. Mach switch is closed above Mach 1.55 with speed increasing and is open below Mach 1.4 with speed decreasing.
5. With DAFICS power on, alpha indicator reads 0 degrees below 101 KEAS and actual Δ P α above 101 KEAS
6. Automatic pitch warning and stall warning is available if A or B or both computers are up and ready. M computer provides redundancy for air data and sensor select.
7. APW PUSHER/SHAKER operation is inhibited with weight on the left main gear. High α SHAKER operation is inhibited with weight on the right main gear.

SENSOR (RATE) SELECT

SENSOR CONDITION	SENSOR USED COMPUTER A	COMPUTER B	COMPUTER M	SENSOR OUTPUT SELECTED
Normal	A	B	M	Median of A, B, M
Fail A	M	B	M	Average of B, M**
Fail B	A	A	M	Average of A, M
Fail M	A	B	B	Average of A, B
WITH ANALYTICAL REDUNDANCY				
Fail A, B	M	-	M	M
Fail B, M	A	A	-	A
Fail A, M	-	B	B	B
WITHOUT ANALYTICAL REDUNDANCY				
Fail A, B	-	-	-	None
Fail B, M	-	-	-	None
Fail A, M	-	-	-	None

CODE

DIGITAL INTERFACE

F203-337

Figure 1-62A

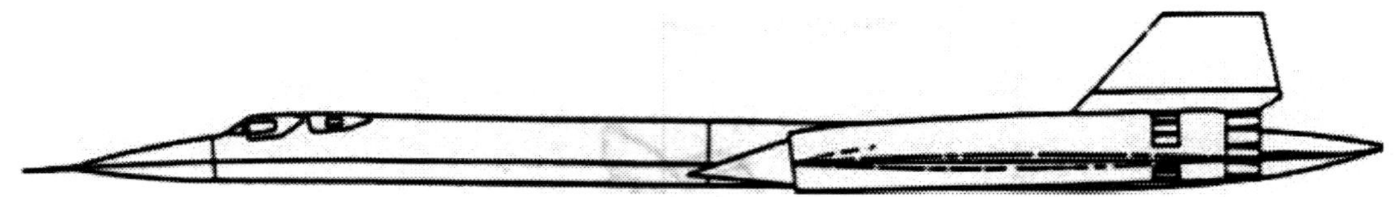

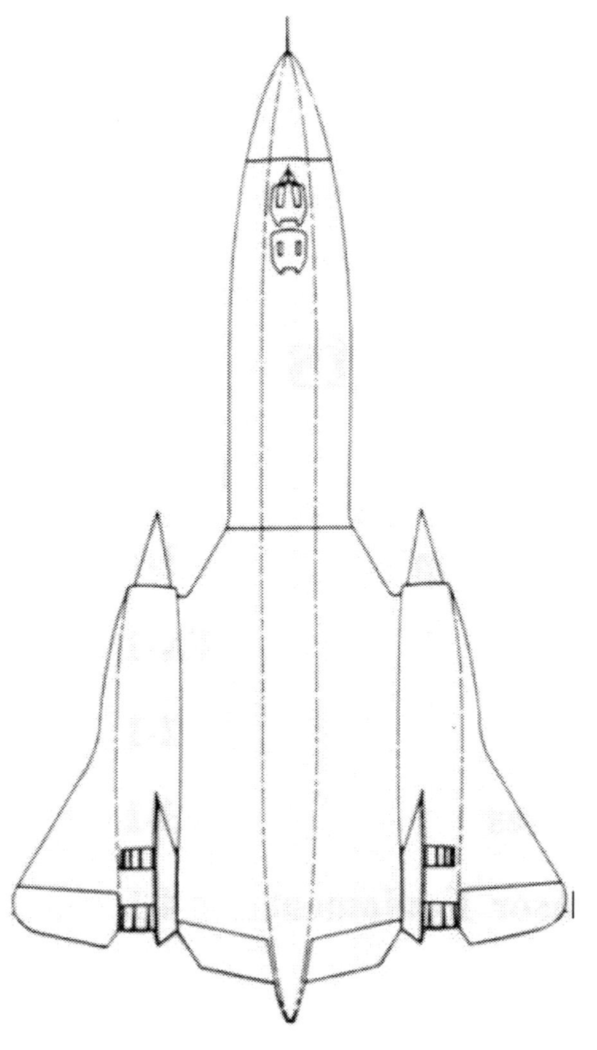

THIS PAGE INTENTIONALLY LEFT BLANK OR STILL CLASSIFIED.

STICK WARNING SCHEDULES

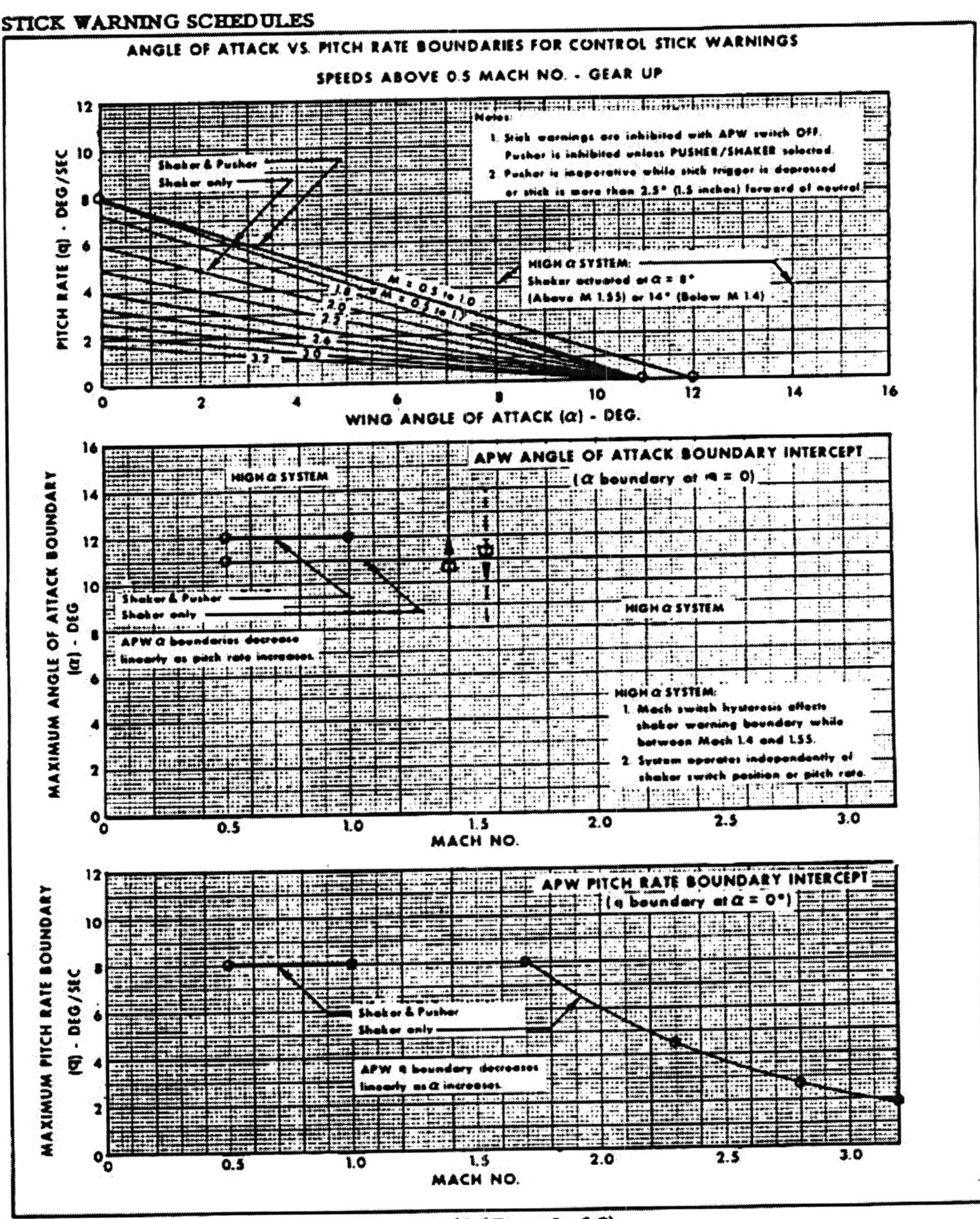

Figure 1-63 (Sheet 1 of 2)

STICK WARNING SCHEDULES

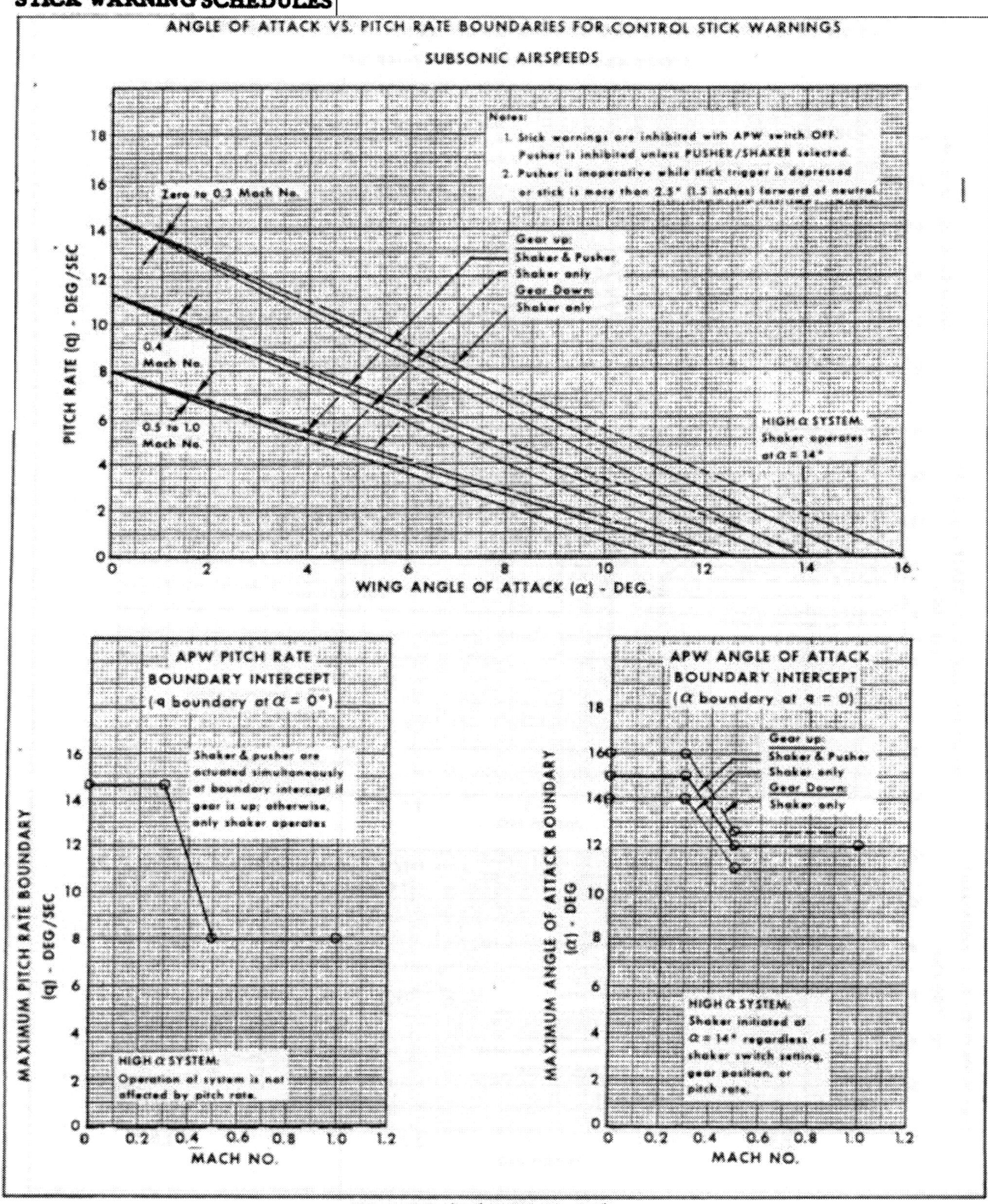

Figure 1-63 (Sheet 2 of 2)

Mach 1.55 (or when above Mach 1.4 when decelerating). See Figure 1-63.

APW Control Stick Warning Schedules

Refer to Figure 1-63, Sheet 1 and 2. Note that Sheet 1 only shows control stick warning schedules for gear up operation at speeds above Mach 0.5. These schedules apply to the major portion of flight. Schedules are included for lower speeds, typical for conditions before landing or after takeoff.

Notice that the APW alpha + pitch rate (q) boundary schedules, shown for the stick shaker and stick pusher, represent 100% of the APW boundary condition when the opposite factor (pitch rate or angle of attack) equals zero. For example: at Mach 3.2, an APW system angle of attack signal of 11° or a pitch rate signal of 1.8 degrees per second will cause the stick shaker to operate when the opposite pitch rate or alpha factor is zero. (See Note). Similarly, a 5.5° angle of attack obtained simultaneously with a 0.8 degree per second pitch rate will cause the shaker to operate. This is illustrated by the Mach 3.2 line on the angle of attack vs pitch rate boundary chart. Each of the 5.5° and 0.8°/sec angle of attack and pitch rate factors is 50% of its boundary intercept for Mach 3.2 speed. Any similar combination for the same speed would start the stick shaker.

NOTE

> Since the high angle of attack warning system setting is 8° in the high speed region, any angle of attack of 8° or more would start the stick shaker.

The stick pusher boundary intercept for gear up operation is 1° above the shaker boundary at zero pitch rate, as shown by Figure 1-63, Sheet 2 of 2. The difference decreases to zero at the pitch rate boundary intercept (alpha = 0°). The pusher is inhibited unless two gear are up, to eliminate pusher activation close to the ground immediately after takeoff or before landing. The shaker warning boundary is automatically shifted to a higher alpha schedule when two gear are not up, to minimize nuisance operation.

PITOT-STATIC SYSTEMS

A dual pitot-static and alpha/beta system supplies the total and static pressure necessary to operate the basic flight instruments and DAFICS. The pitot-static and alpha/beta probe is mounted on the nose of the aircraft. Pressure entering the tip of the probe is divided internally to produce two separate total pressure sources. Eight static ports on the probe furnish two separate static sources. One pitot and one static source are referred to as system 1 and the other two sources are referred to as system 2. System 1 supplies pressure to the pilot's altimeter, vertical speed indicator, and Mach-airspeed indicator. System 2 supplies pressure to the 3 channels of the pressure transducer assembly (PTA) which supplies digitized pressure information to the DAFICS computers. An alpha/beta probe provides angle of attack and sideslip pressure signals to the PTA. (See Figure 1-64.)

Heating elements for the probes are controlled by the pitot heat switch. Although there is no time limit for operation of the heater system while on the ground, the life span of the pitot heater element is lowered if pitot heat is left on unnecessarily.

Pitot Heat Switch and Indicator Light

A two-position PITOT HEAT switch is located on the pilot's annunciator panel. In ON, the essential ac bus powers heating elements of the pitot-static and alpha-beta probes through the PITOT HTR circuit breaker on the pilot's right console.

The circuit also incorporates a PITOT HEAT annunciator caution light, and two altitude pressure switches. The light illuminates when the pitot heat switch is not in the correct position for the aircraft altitude. With the pitot heat switch ON, the light is on when above 63,000 ± 4000 feet. With the pitot heat switch OFF, the light illuminates below 40,000 ± 2000 feet, and during climb extinguishes between 40,000 and 50,000 feet.

PITOT-STATIC, PTA, INLET, SAS, AUTOPILOT, & APW INTERFACE

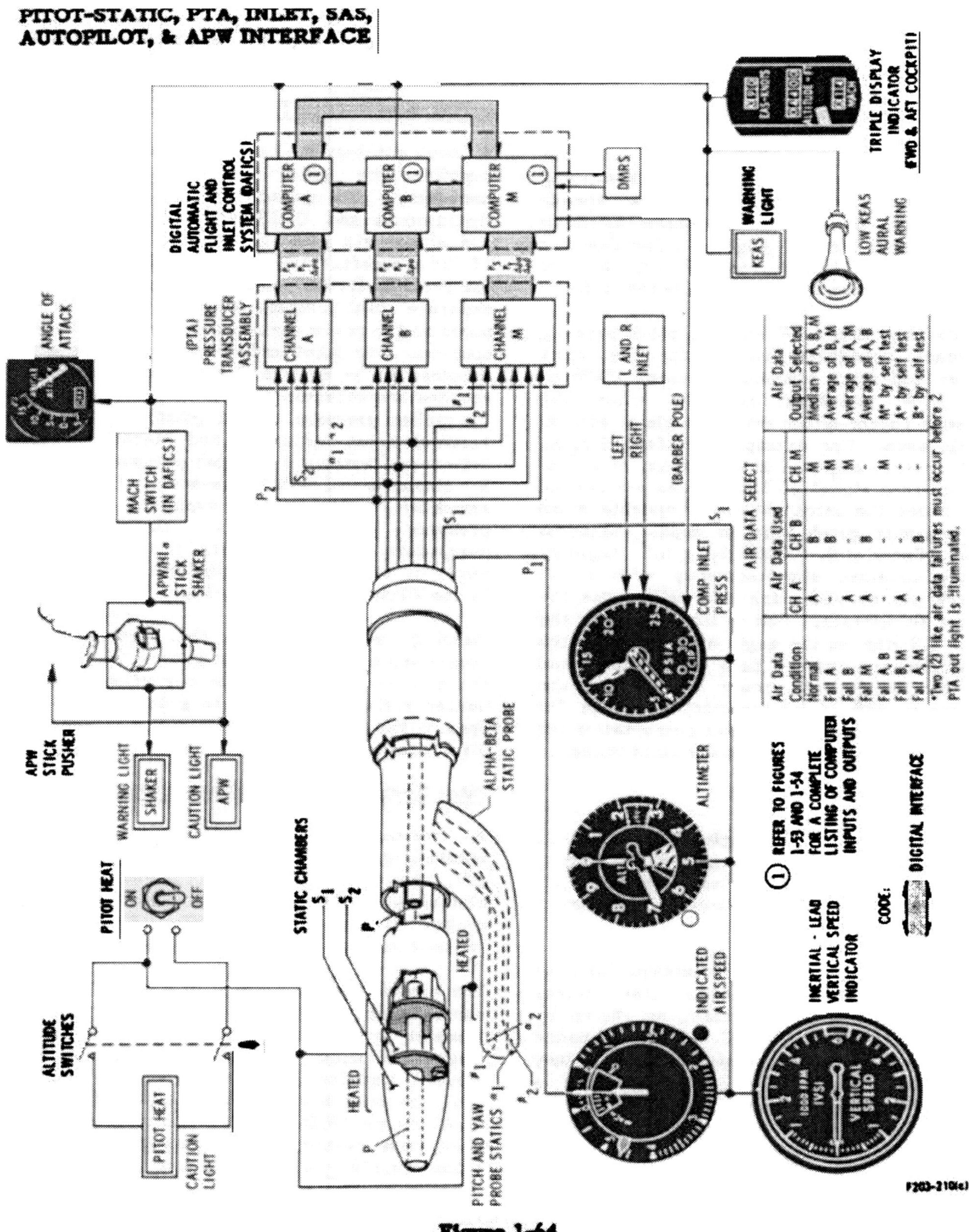

AIR DATA SELECT

Air Data Condition	Air Data Used CH A	CH B	CH M	Air Data Output Selected
Normal	A	B	M	Median of A, B, M
Fail A	-	B	M	Average of B, M
Fail B	A	-	M	Average of A, M
Fail M	A	B	-	Average of A, B
Fail A, B	-	-	M	M* by self test
Fail B, M	A	-	-	A* by self test
Fail A, M	-	B	-	B* by self test

*Two (2) like air data failures must occur before 2 PTA out light is illuminated.

Figure 1-64

PRESSURE TRANSDUCER ASSEMBLY (PTA)

The PTA changes pressures (P_s, P_T, $\Delta P\alpha$, $\Delta P\beta$) to electrical signals for DAFICS. The PTA has three electrically independent channels (A, B, M) which produce digitized P_s, P_T, and $\Delta P\alpha$ data. In addition the M channel transduces $\Delta P\beta$ data. The PTA is located in the left forward chine bay, aft of the nose bulkhead and is cooled by cockpit exit air.

CAUTION

> Loss of cockpit pressurization or cockpit air off at supersonic cruise will cause overheat and subsequent failure of the PTAs if a descent is not begun within 15 minutes.

Each PTA Channel (A, B, and M) only sends data to its corresponding (A, B, and M) DAFICS computer. The DAFICS computers share PTA data and select the best air data for DAFICS computations. (See Air Data Select chart, Figure 1-64) All operating DAFICS computers use the same selected PTA data. If a DAFICS computer is inoperative, the corresponding PTA channel is also inoperative.

The A, B, and M channels of the PTA are powered by the emergency ac bus through the three A, B, and M COMPUTER circuit breakers for the respective computer located in the aft cockpit. With two circuit breakers opened to a computer, the corresponding PTA channel is disabled.

FLIGHT AND NAVIGATION INSTRUMENTS

TRIPLE DISPLAY INDICATOR (TDI)

A TDI on the instrument panel in each cockpit provides digital displays of airspeed (KEAS), pressure altitude, and Mach number as computed by DAFICS.

The altitude indication range of each TDI is 0 to 99,950 feet. At 100,000 feet, the first digit is dropped, indicating 09,950 feet at 109,950 feet (the maximum limit of the DAFICS signal to the instruments).

The TDI Mach number indication range is 0 to 3.99. The minimum indication at static conditions normally varies from 0.11 to 0.2 Mach number and the maximum limit of the DAFICS signal to the instruments is Mach 3.5.

The TDI knots equivalent airspeed (KEAS) indication range is 0 to 599 KEAS. The minimum indication normally varies from 25 to 110 KEAS. The maximum limit of the DAFICS signal to the instruments is 560 KEAS.

OFF flags appear on the faces of the TDI instruments and TDI values are unusable if the DAFICS A and B computers fail. OFF flags appear on both TDI instruments if the emergency ac bus AP MACH TR circuit breaker in the aft cockpit is open, but indications remain valid. The OFF flags will not necessarily appear in conjunction with a 2 PTA CHAN OUT annunciator caution light.

If the OFF flags should appear with the 2 PTA CHAN OUT caution, it indicates that none of the available PTA channels is passing self check. Refer to 2 PTA Channels Out Procedures, Section III.

Power for both instruments is from the emergency ac bus through the TDI circuit breakers. The forward cockpit TDI circuit breaker provides power to the forward cockpit TDI; the aft cockpit TDI circuit breaker provides power to the aft cockpit TDI, which in turn provides TAS signals to the Pilot's & RSO's Map Projector - Automatic Map Rate.

AIRSPEED - MACH METER

A combination airspeed and Mach meter, operating directly from pitot-static pressure, is installed on the pilot's instrument panel. Airspeed and Mach values shown are indicated values as opposed to equivalent airspeed and true Mach displayed on the TDI. A limit airspeed needle varies with altitude to show the KIAS limit corresponding to a preset KEAS vs altitude schedule. See Airspeed - Mach Meter, Section V.

NOTE

At low altitudes and airspeeds the high end of the Mach scale will show in the window.

ALTIMETER

A pressure altimeter is installed on the pilot's instrument panel. The instrument has 1000-foot and 100-foot pointers, and a 10,000 foot pointer (with a triangular marker at its extremity) which extends to the edge of the dial. The center of the instrument has a cutout through which yellow and black warning stripes appear when aircraft altitude is less than 16,000 feet. The barometric pressure scale appears in a cutout at the right side and barometric pressure is set by a knob located at the lower left side of the instrument.

NOTE

The static altimeter and the IVSI exhibit appreciably more lag while supersonic than subsonic.

Altimeter Correction Card

An altimeter altitude correction card, located on the pilot's right sunshade, shows the indicated corresponding pressure altitudes to fly for a corresponding true pressure altitude. The difference between the columns is altimeter instrument error only.

INERTIAL-LEAD VERTICAL SPEED INDICATOR (IVSI)

An IVSI, installed on the pilot's instrument panel, shows the rate of change of altitude in feet per minute up to 6000 feet per minute. See Figure 1-64. It is operated by static pressure changes sensed by the system 1 static ports. It operates like conventional vertical velocity instruments except that an internal mechanism, sensitive to load factors between 0.7 and 1.4 g's, momentarily alters static pressure within the IVSI under changing load factors. The pressure change causes a corresponding change in rate-of-climb or rate-of-descent indication when the load factor is increasing or decreasing, respectively. The indication change will usually precede aircraft altitude changes. Changing static pressure conditions, such as during climb or descent, cause the vertical speed indications to continue registering in the appropriate direction as the effect of the load factor pulse diminishes. The instrument appears to be "lag free" when subsonic. At high altitudes, reduced air density decreases instrument response and accuracy. Indications may be less than 1/3 of actual conditions. Response to acceleration is retained, but to a lesser degree.

Below approximately 40,000 feet, IVSI indications respond quickly to pitch control. In a level turn the IVSI will show a small rate of climb during roll-in and rate of descent during roll-out due to load factor changes. Expect a change of approximately 50 fpm at cruise.

ANGLE OF ATTACK (AOA) INDICATOR

An angle of attack indicator is mounted on the pilot's instrument panel. The dial displays positive and negative angles of attack in one-degree increments through a range from -5° to $+22^\circ$.

The instrument responds to synchro signals from DAFICS which uses differential pressures sensed by the alpha ports of the alpha/beta probe ($\Delta P \alpha$) related to total pressure (P_T). See Figure 1-64. A and/or B computer must be operating to provide angle of attack.

An OFF flag is displayed on the face of the indicator if power to the instrument is interrupted.

The AOA indicator reads 0° on the ground and reads aircraft actual AOA at 100 KEAS or higher. In level flight, pitch angle is 1.2° greater than the wing angle of attack because of the negative angle of incidence of the wing. See Angle of Attack, Section VI.

ATTITUDE-DIRECTOR INDICATOR (ADI)

An ADI, located on the pilot's instrument panel, combines an attitude indicator, turn-and-slip indicator, and a flight director with ILS glide slope presentation. Pitch and roll signals from the ANS or INS systems are routed to the instrument through the ATT REF SELECT switch. The attitude sphere allows presentation of pitch and roll through 360 degrees. The sphere moves behind a small aircraft symbol fixed at the center of the instrument face. A trim knob allows manual positioning of the sphere in pitch.

The sphere is marked with a horizon line, small dots for 5-degree increments, short lines for 10-degree increments, and numeral markers for each 30-degree increment; large dots indicate the poles. Indices are provided for bank angles of 10, 20, 30, 45, 60 and 90 degrees.

The turn-and-slip indicator is mounted at the bottom of the ADI. A deflection of one-needle-width indicates a 4-minute, 360-degree, standard rate turn. The rate-of-turn transmitter receives power from the essential dc bus through the TURN GYRO circuit breaker on the pilot's left console.

Bank and pitch steering command bars are superimposed on the ADI. The bank steering bar shows bank required to position the aircraft on a desired heading or course and will center when (1) on course with wings level, (2) flight attitude is correct for return to course, or (3) bank is correct for rollout on-course. When in AUTO NAV, the maximum bank angle command is 45°. The maximum bank command is 35° in INS, TACAN ADF and ILS modes. The maximum bank and pitch commands are 15° and 10°, respectively, in ILS APPROACH mode.

The pitch steering bar indicates pitch attitude change required to intercept the ILS glide slope. The bar centers when (1) on glide slope with proper pitch angle, (2) pitch angle is correct for return to glide slope, or (3) pitch angle is correct for leveling-out on glide slope.

The steering bars indicate corrective action required and not direction or displacement from a desired course or glide slope. The vertical (bank) and horizontal (pitch) steering bars will be centered when power is off. The pitch steering bar is stowed at the bottom of the instrument when the DISPLAY MODE SEL switch is in INS, or TACAN ADF. The bank steering bar receives command signals in all display modes.

The pitch steering bar indicates rate of climb or descent when the DISPLAY MODE SEL switch is in ANS. See Figure 1-66. The bar indicates zero vertical velocity when it is aligned with the small airplane symbol. Full scale deflection of the bar to the top or bottom dot on the glide slope displacement scale represents a vertical velocity of 3484 fpm or more. This provides a scale sensitivity of approximately 1000 fpm for each quarter-inch displacement. The minimum vertical speed which can be sensed by the ANS and displayed is approximately 55 fpm. The displacement of the steering bar is opposite in direction to the vertical velocity indicator, to indicate the direction of pitch attitude change required to offset the climb or descent. The pitch steering bar vertical velocity indication can be used to maintain precise control of altitude at high Mach.

A warning flag comes into view at the top of the ADI if the instrument receives unreliable steering information. If INS attitude reference is selected and the INS is in the attitude mode, the flag indicates that heading and INS steering information are not reliable. If ANS attitude reference is selected and the DISPLAY MODE SEL switch is in ANS, the flag indicates the ANS is not ready and attitude, true heading, and steering information are not reliable. The glide slope pointer at the left center of the instrument shows aircraft position above or below the ILS glide slope when ILS or ILS APPROACH display mode is selected and valid ILS signals are received. The glide slope warning flag appears if there is insufficient ILS signal strength. The glide slope pointer also is the pitch boundary indicator when ILS or ILS APPROACH display mode is not selected. Refer to Figure 1-66 and to the APW and High Angle of Attack

Warning Systems description, this section. A power failure warning flag appears in the lower left of the instrument when ac power is interrupted, ATT REF SELECT switch is in INS and the INS ready signal is not present, or the ATT REF SELECT switch is in ANS and the nav-ready signal is not present.

Display Mode Select Switch

The DISPLAY MODE SEL switch on the pilot's instrument panel has five positions: ANS, INS, TACAN ADF, ILS, and ILS APPROACH. The switch controls display inputs to components of the HSI and ADI. See Figure 1-66.

Attitude Reference Select Switch

The ATT REF SELECT switch on the pilot's instrument panel has two positions: ANS (up) and INS (down). It selects which system supplies pitch and roll signals to the attitude director indicator, and pitch, roll, and heading signals to the autopilot. Autopilot AUTO NAV is only operative when the ATT REF SELECT switch is in ANS. Changing the ATT REF SELECT switch disengages the autopilot. Autopilot HEADING HOLD will not engage if INS attitude reference is selected and the INS is in attitude mode.

WARNING

INS reference should be selected in both cockpits if ANS nav-not-ready warnings (pilot's ANS REF and RSO's ANS FAIL caution lights) appear. Otherwise, unreliable attitude information may be displayed.

PERIPHERAL VISION DISPLAY (PVD)

The PVD is a twilight and night attitude orientation device that projects on the pilot's instrument panel, a laser-generated, thinly focused red line parallel to the horizon. A mask in the PVD projector prevents display of the line on the ADI. Like an attitude indicator, the PVD line remains fixed in inertial space. The PVD is not intended to be consciously included in the pilot's instrument crosscheck. Instead, the laser line is perceived indirectly through peripheral vision and subconsiously supports spatial orientation just as a visible outside horizon supports orientation.

The PVD derives pitch and roll attitude from the attitude source (ANS or INS) selected by the pilot's ATT REF SELECT switch. Roll movement of the line corresponds with changes in roll attitude and has no limit. Pitch movement of the line corresponds with changes in pitch attitude. Pitch movement of the line stops when a display boundary at the edge of the instrument panel is reached. The pitch angle at which the line will reach a boundary depends on the position of the pitch adjust switch and the pitch scale switch. The line flashes at 4 Hz to warn of a potentially hazardous flight condition if pitch angle exceeds 35 degrees nose up or 15 degrees nose down.

Twelve regularly spaced variations in line intensity create subtle segments in the line that remain fixed in heading relative to inertial space. Movement of the segments across the instrument panel allows a sense of heading change during turns. Because rate of turn decreases at high speed, the effect is most noticable at low speed. DAFICS M computer provides rate of turn inputs to the PVD processor.

The PVD processor, located in the R bay, continually performs built-in-test. If an internal processing fault is detected, the laser line will not be generated. If DAFICS rate of turn input is lost or a rate of turn fault is detected, the line segments will not move (zero turn rate), but pitch and roll will still be displayed. To inhibit the PVD when inputs from the selected attitude reference are unreliable, the laser line is not generated if a DAFICS analytical redundancy (ANR) failure occurs (flashing DAFICS Preflight BIT FAIL light).

Power for the PVD is provided through three PVD circuit breakers on the pilot's right console: 115 volt emergency ac, 26 volt emergency ac and 28 volt essential dc. The laser line will not be displayed if any circuit breaker is open.

PVD Projector

The PVD projector is located on the lower right side of the forward canopy.

Laser Safety

The location of the PVD projector makes it difficult for a pilot in a pressure suit to inadvertently expose an eye to direct laser energy. Laser beam reflections are not hazardous. The PVD should be OFF when the canopy is open to prevent ground support personnel from inadvertently looking directly into the laser beam.

WARNING

> Do not look directly into the laser beam.

PVD Control Panel

The PVD control panel is located on the pilot's right console. See Figure 1-65.

Roll Adjust Switch

The ROLL adjust switch allows ±5 degrees of display adjustment in roll. Normally the switch should be aligned with the horizontal index.

Pitch Adjust Switch

The PITCH adjust switch allows vertical adjustment of the reference point on the instrument panel for level flight. When the switch is aligned with the index mark, the laser line will be between the ADI and the HSI when the aircraft is in the nominal attitude for supersonic cruise (6 degrees pitch angle).

Intensity Switch

The OFF detent, at the counterclockwise position of the intensity switch, deenergizes the PVD. Intensity of the laser line increases with clockwise rotation of the switch. The full clockwise position is labeled BRT. The PVD may be manually reset by rotating the switch to OFF and back on or by depressing the white pushbutton projecting above the center of the switch.

Pitch Scale Switch

The pitch SCALE switch allows the pilot to select the ratio of laser line movement in pitch compared to movement of the real horizon. In the normal (NORM) position, the angular movement of the laser line, when observed by the pilot, is the same as the angular movement of the outside horizon. In the 2, 1/2, 1/4 and 1/8 positions, the line moves 2 times, 1/2, 1/4 and 1/8 as much as the angular movement of the outside horizon, respectively.

The position of the laser line changes when the SCALE switch is moved, unless the aircraft is in the nominal attitude for supersonic cruise.

PVD CONTROL PANEL

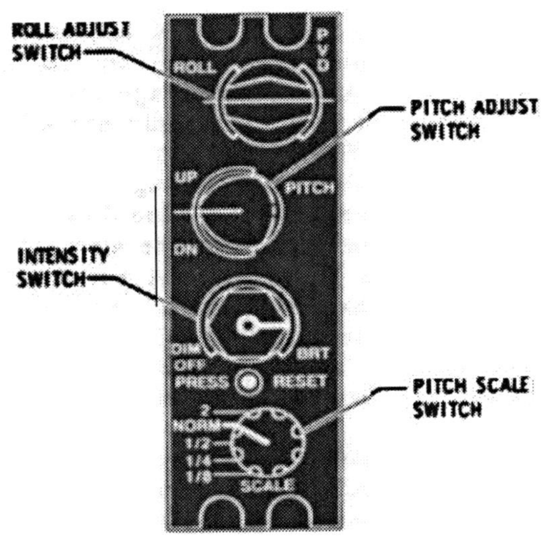

Figure 1-65

2-INCH STANDBY ATTITUDE INDICATOR (Without S/B R-2466)

A self-contained standby attitude indicator is located at the top of the pilot's instrument panel. A cylinder rotates to indicate angles through 80° climb, 90° dive, and 360° roll. The cylinder is inscribed with a horizon and 5° graduations for pitch angle. Bank angle is marked on the instrument case in 10° increments up to 30° and then 30° increments up to 90°. See Figure 1-12.

A cutout disables the self-erection feature when pitch or roll angle exceeds 12 ± 1/2° to prevent the indicator from erecting to a false vertical. This puts the indicator into a free drift mode when above this angle.

The indicator has a built-in 7° pitch bias. With this bias, the gyro horizon will center relative to the case index marks when the pitch attitude is 7° nose-up (5.8° angle of attack). When the aircraft is at a 0° pitch attitude on the ground, the gyro will erect to a 7° nose-down indication.

A combination pitch trim and caging knob is provided on the lower right corner of the indicator. The knob can be turned to position the miniature airplane from 5° up to 5° down. The knob is also used to cage the gyro by pulling the knob out to its fully extended position. The OFF flag appears when the caging knob is pulled. Releasing the knob allows the gyro to erect to 7° nose down and 0° roll within five minutes if the aircraft is level. Pulling the knob out and turning it clockwise locks the cylinder in the caged position. The cylinder should be unlocked during normal use. The nominal erection rate is 2.5° per minute to the apparent vertical.

The indicator receives emergency ac power through the STBY ATT circuit breaker on the pilot's right console. An OFF flag appears if power is interrupted. The indicator displays useful pitch and roll information for at least nine minutes after power loss.

Standby Attitude Indicator Characteristics

The standby attitude indicator has a maximum free drift rate of 0.9 degrees per minute. In level flight, any tendency to drift is continuously corrected to the apparent vertical at a rate of two to three degrees per minute. An acceleration sensor disables the erection system at bank angles greater than twelve degrees to prevent bank errors during turns. In this free gyro mode, a random combination of errors due to gyro drift, earth's rate and earth's profile can accumulate at a maximum of 1.6 degrees per minute.

In wings-level flight, the normal aircraft acceleration and deceleration does not disable the erection system and the indicator will erect to the apparent vertical. During climbout and descent, pitch errors up to approximately four and eight degrees, respectively, can be expected. If a turn is initiated immediately after climbout or descent, the free gyro errors may add to the acceleration induced error. For example, the indicator error could reach four degrees in roll after a normal supersonic descent and a 90° subsonic turn (with bank angle greater than 12°).

If the bank or pitch error does not exceed approximately twelve degrees, the indicator will automatically correct the above errors at the normal erection rate when the aircraft is in level flight and at constant velocity. However, the indicator can be aligned more rapidly by pulling the caging knob. This control should be limited to level flight conditions, as the gyro is erected to the case reference.

NOTE

If a bank or pitch error of more than twelve degrees is accumulated, the cut-out feature prevents automatic correction when level. Manual caging will be necessary.

3-INCH STANDBY ATTITUDE INDICATOR (With S/B R-2466)

A self-contained standby attitude indicator is located at the top of the pilot's instrument panel. A cylinder rotates to indicate angles of 85° climb and dive, and 360° roll. The cylinder is inscribed with an artificial horizon and 5° graduations for pitch angle. Bank angle is marked on the instrument case in 10° increments up to 30° and then 30° increments up to 90°. See Figure 1-12.

The indicator has a built-in 7° pitch bias. With this bias, the gyro horizon will center relative to the case index marks when the pitch attitude is 7° nose-up (5.8° angle of attack). When the aircraft is at a 0° pitch attitude on the ground, the gyro will erect to a 7° nose-down indication.

A combination pitch trim and caging knob is provided on the lower right corner of the indicator. The knob can be turned to position the miniature airplane from 5° up to 10° down. When the caging knob is pulled to the fully extended position, the OFF flag appears and the gyro horizon is caged to the case level-flight index. Pulling the knob out and turning it clockwise locks the cylinder in the caged position. The cylinder should be unlocked during normal use. The nominal erection rate is 2.5° per minute to the local vertical.

The indicator receives essential dc power through the STBY ATT circuit breaker on the pilot's right console. An OFF flag appears if power is interrupted. The indicator displays pitch and roll information accurate to within ±5° for at least nine minutes after power loss.

Attitude Indicator Operating Characteristics

The attitude indicator is maintained erect to the apparent vertical at a rate of two to three degrees per minute when the gyro spin axis displacement is less than 7° in pitch and roll (level flight). The gyro erects to gravity at a reduced rate of 0.75°/minute during turns and fore and aft velocity changes that exceed the 7° erection criteria.

The indicator will erect to the apparent vertical. During climbout and descent, pitch errors of approximately four and eight degrees, respectively, can be expected.

HORIZONTAL SITUATION INDICATOR (HSI)

The HSI, located on the pilot's instrument panel, integrates information from the TACAN, ANS, INS, and the ILS receiver. See Figure 1-66. Power for the HSI is furnished by the essential ac bus.

Range Indicator

The range indicator on the upper left side of the HSI displays distance in nautical miles. A shutter covers the numerals when a distance signal is not present. The maximum range readout is 1999 nautical miles. A "K" shutter covers the first digit of the range readout window when range is less than 1000 miles. When the shutter opens (over 999 NM range) a fixed numeral 1 is displayed. The shutter remains closed unless the DISPLAY MODE SEL switch is in ANS and the BEARING SELECT switch is in NORMAL.

With the DISPLAY MODE SEL switch in ANS and the BEARING SELECT switch in NORMAL, the range readout is to the ANS destination point (DP) or to the ANS-computed turn point, depending on the mode (DP or TURN, respectively) selected on the ANS control panel. Refer to Navigation Control and Display Panel DP/TURN switch, Section IV. With the DISPLAY MODE SEL switch in INS and the BEARING SELECT switch in NORMAL, the range readout is to the INS DP. With the BEARING SELECT switch in TAC/ADF, the range readout is to the selected TACAN station. With the DISPLAY MODE SEL switch not in ANS or INS, range readout is to the selected TACAN station regardless of BEARING SELECT switch position.

Course Selector Window

The COURSE window displays ANS command course when the DISPLAY MODE SEL switch is in ANS. In all other modes, it displays the course that is manually selected with the COURSE SET knob.

Bearing Select Switch

A BEARING SELECT switch, located below the DISPLAY MODE SEL switch on the pilot's instrument panel, has two positions, NORMAL (down) and TAC/ADF (up). This switch is operated with the DISPLAY MODE SEL switch to select inputs to the HSI bearing pointer and HSI range indicator. See Figure 1-66.

Rotary Compass Card

The rotating azimuth ring is read at a stationary lubber line at the 12 o'clock position.

The card displays ANS true heading with the DISPLAY MODE SEL switch in ANS; in any other position, the display is INS computed magnetic heading. The RSO can provide true heading by manually selecting "0" mag var on the INS control panel. When the INS is operating in ATT mode (either by RSO selection or INS computer failure), the heading is set by the RSO using the heading slew knob.

Heading Marker

The heading marker is a rectangular marker located just outside of the rotating compass card. The marker is manually set by the HEADING SET knob except when the DISPLAY MODE SEL switch is in ANS, then, the heading marker displays INS heading.

Bearing Pointer

The bearing pointer is a heavy arrowhead located outside of the rotating compass card. Operation of the bearing pointer is affected by three switches: DISPLAY MODE SEL, BEARING SELECT and the UHF radio function selector.

With the DISPLAY MODE SEL switch in ANS, the bearing pointer provides true bearing to a selected TACAN station or UHF transmitter regardless of the BEARING SELECT switch position. With the DISPLAY MODE SEL switch in INS, the bearing pointer is referenced to INS heading and with the BEARING SELECT switch in NORMAL, the bearing pointer displays bearing to the INS DP; with the BEARING SELECT switch in TAC/ADF, it displays bearing to a selected TACAN station or UHF transmitter. With the DISPLAY MODE SEL switch in TACAN ADF, the bearing is INS magnetic to a selected TACAN station or UHF transmitter, regardless of the BEARING SELECT switch position. With the DISPLAY MODE SEL switch in ILS or ILS APPROACH, the bearing pointer is servoed to the lubber line if the BEARING SELECT switch is in NORMAL; if in TAC/ADF, it will display INS magnetic bearing to a selected TACAN station or UHF transmitter.

Selecting the ADF position of the function selector switch on the UHF radio overrides TACAN information and provides UHF/ADF bearing.

Course Arrow and Course Deviation Bar

The course arrow and course deviation bar are on the face of the rotating compass card. The course arrow displays a commanded or selected course and the course deviation bar displays displacement from the commanded or selected course. When the DISPLAY MODE SEL switch is in ANS, the course arrow and course window display desired ground track as computed by the ANS. The course deviation bar indicates left or right location of the desired course. Full deflection of the course deviation bar is proportional to 1 nm off-course. With the DISPLAY MODE SEL switch in INS, the course arrow is manually set to the desired INS course and the course deviation bar indicates deviation from the INS course. With the DISPLAY MODE SEL switch in TACAN ADF, the course may be selected manually with the COURSE SET knob and course deviation bar signals are received from the TACAN. With the DISPLAY MODE

NAVIGATION INSTRUMENTS - Forward Cockpit

ATTITUDE DIRECTOR INDICATOR

* With ANS, INS, or TACAN/ADF display mode selected.
  1. Indicator position if APW shaker boundary reached.
  2. Pitch boundary or glide slope status indicator.
  3. Indicator position for Angle of Attack and pitch rate • Q.

♦ 45° Index marks provided on ADI.

HORIZONTAL SITUATION INDICATOR

NOTE

▲ When UHF/ADF is operating it has priority on bearing pointer

The attitude reference select switch determines which system, ANS or INS, supplies pitch and roll to the attitude director indicator, pitch, roll, and heading to the autopilot. (The autopilot receives steering signals only in the ANS position).

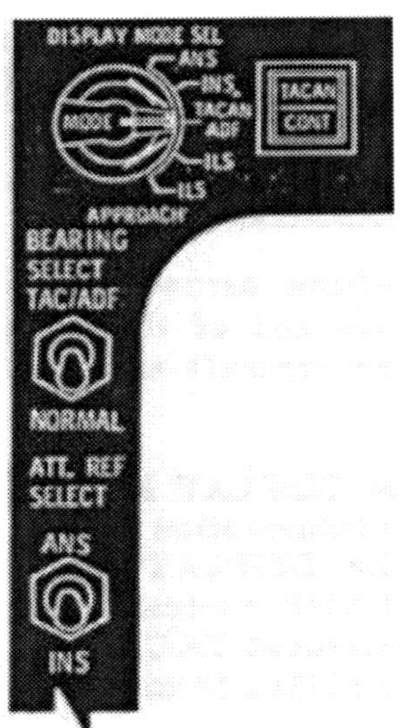

⚠1 If the INS is operating in the attitude mode and the DISPLAY MODE SEL switch is in any position other than ANS, the course warning flag comes into view.

		DISPLAY MODE SELECTOR SWITCH									
		ANS		INS		TACAN/ADF		ILS		ILS/APCH	
		BEARING SEL. SW.		BEARING SEL. SW.		BEARING SEL. SW.		BEARING SEL. SW		BEARING SEL. SW.	
INDICATOR	INDICATOR FUNCTION	NORMAL	TAC/ADF	NORMAL	TAC/ADF	NORMAL	TAC/ADF	NORMAL	TAC/ADF	NORMAL	TAC/ADF
HORIZONTAL SITUATION INDICATOR	HEADING MARKER	INS Heading		Manually Set							
	BEARING POINTER	To Selected TACAN or ADF ▲		INS DP Bearing	TO SELECT TACAN OR ADF ▲	To Selected TACAN or ADF ▲		SERVOED TO LUBBER LINE	TO SELECT TACAN OR ADF ▲	SERVOED TO LUBBER LINE	TO SELECT TACAN OR ADF ▲
	COURSE ARROW	Command Course		Manually Set INS Course		Manually Set TACAN Course		Manually Set Inbound Localizer Course			
	COURSE DEVIATION	Left or Right of Command Course		Deviation from INS Course		Deviation From TACAN Course		Deviation From Localizer Course			
	TO-FROM INDICATOR ▲	Out of View		To Selected INS DP		To Selected Tacan Station		Out of View			
	RANGE INDICATOR	Dist. to DP/TURN	Dist to TACAN Station	Distance to INS DP	Distance to TACAN Station						
	K SHUTTER	Dist. to DP/TURN	Masked								
	AIRCRAFT HEADING	True		INS Heading (Magnetic or Alternate Mode Selected Heading)							
ATTITUDE DIRECTOR INDICATOR	BANK STEERING BAR	Command Course		INS Selected course		TACAN Course		Localizer Course			
	PITCH STEERING BAR	Vertical Velocity		Out of View						Pitch Direction for Glide Slope Beam	
	PITCH BOUNDARY OR G/S INDICATOR	Out of View / Angle of Attack and Pitch Rate Status Relative to APW Boundary						Position Relative to Glide Slope Beam			
⚠1	COURSE WARNING FLAG	← Out of View → ANS Valid		INS Heading Valid		TACAN Valid		Localizer Valid			
	GLIDE SLOPE WARNING FLAG	← Out of View →						Glide Slope Valid			
	POWER WARNING FLAG	Out of View - Attitude Reference Valid									

F203-277(b)

Figure 1-66

SEL switch in ILS or ILS APPROACH, the course arrow should be manually set to the localizer course with the COURSE SET knob; course deviation bar signals are from the ILS.

To-From Indicator

The to-from arrow is located radially between one end of the course arrow and the miniature aircraft at the center of the instrument.

With the DISPLAY MODE SEL switch in INS, to-from information is relative to an INS DP. With the DISPLAY MODE SEL switch in TACAN ADF, to-from information is relative to the selected TACAN station. The to-from arrow is hidden from view in all other display modes.

BEARING, DISTANCE, HEADING INDICATOR

The bearing, distance, heading indicator (BDHI) on the RSO's instrument panel contains a rotating compass card, range window, and two pointers.

The compass card displays ANS or INS heading, depending on the position of the BDHI SEL-HEADING switch. The No. 1 pointer displays ADF, INS DP, or TACAN bearing, depending on the position of the BDHI SEL-NO. 1 NDL switch. The No. 2 pointer always displays ANS command course.

Range Readout Window

The range readout window behind the shutter is a three-digit counter which displays slant range to a TACAN station in nautical miles. The maximum range readout is 999 nautical miles. When TACAN information is unreliable, a shutter covers the range readout.

BDHI Heading Select Switch

The two-position BDHI heading select switch on the RSO's instrument panel is labeled HEADING. With the switch in INS (up), the compass card displays INS heading. With the switch in ANS (down), the compass card displays ANS heading.

Heading Slew Control

The HEADING SLEW control, on the RSO's instrument panel, is used to set the INS heading when the INS is in the ATT (attitude) mode.

BDHI SEL NO. 1 Pointer Select Switch

The BDHI SEL NO. 1 pointer select switch on the RSO's instrument panel has three positions: ADF (up), INS (center) and TACAN (down). The No. 1 pointer displays an ADF, INS DP, or TACAN bearing, as selected.

ATTITUDE INDICATOR-RSO

An attitude indicator on the RSO's instrument panel receives pitch and roll signals from either the ANS or INS. Power for the indicator is furnished by the emergency ac bus through the ATT IND circuit breaker on the RSO's right console.

Attitude Indicator Selector Switch

The ATT IND switch on the RSO's instrument panel has two positions: INS (up) and ANS (down). The switch selects the attitude reference source for the attitude indicator and for ready signals to the power-off flag.

ACCELEROMETER

A mechanical accelerometer on the pilot's instrument panel has one indicating pointer and two recording pointers (one each for loads greater or less than 1-G). The recording pointers remain at the maximum travel positions reached by the indicating pointer, thus providing a record of maximum positive and negative G-loads encountered. To return the recording pointers to the 1-G position, press the button on the lower left corner of the instrument.

MAGNETIC COMPASS

A magnetic compass is located on the forward part of the pilot's canopy. An on-off toggle light switch is located below and to the left of the compass.

COMMUNICATIONS AND AVIONIC EQUIPMENT

The communications and avionic equipment includes:

AN/AIC-18	Interphone system.
COMNAV-50	UHF communication and navigation system (UHF-1 and UHF-2).
AN/ARA-48	Automatic direction finder equipment, used with COMNAV-50.
AN/ARC-186(V)	VHF communication system.
618-T	HF radio.
ARC-190	HF radio.
51RV-1	Instrument landing system.
Wilcox 914X-2	IFF transponder.
G-Band Beacon	For radar tracking during special tests.
I-Band Beacon	For radar tracking during special tests.
AN/ARN-118(V)	TACAN system.

Refer to Section IV for descriptions of the Inertial and Astro-inertial Navigation systems, the sensor equipment, and the defensive systems. Some instruments used to display avionic systems navigation information are described under Flight and Navigation Instruments, this section.

Pilot's Microphone Switches

The pilot's microphone is connected to the radio transmitter selected on the interphone control panel when the pilot either depresses the microphone switch on the right throttle knob or selects the TRANS position of the microphone switch on the control stick grip. A transmission side tone is heard while the transmitter is keyed. The button and switch are spring loaded to the off position. When not transmitting, the pilot can receive communication signals selected on the interphone control panel. When the switch on the control stick is moved to INPH, the pilot's microphone is only connected to the interphone circuit. The Mission Recorder System records voice signals when either TRANS or INPH is selected or the interphone panel HOT MIC switch is on.

RSO's Microphone Switches

The RSO's radio transmit switch is located in the left footrest and the interphone switch is located in the right footrest. A quiet/listen (muting) switch is provided on the right side of the cockpit floor to inhibit all audio input to the RSO's headset except pilot's CALL signals. The quiet/listen switch operates only when the RSO's interphone selector knob is in the MUTE or INTER position. The Mission Recorder System records the RSO's voice signals whenever the aft cockpit radio transmit or interphone switch is depressed or the interphone panel HOT MIC switch is on.

INTERPHONE SYSTEM, AN/AIC-18

The interphone system provides crew intercommunication plus crewmember-to-ground-crew interphone and crewmember radio communication. Each crewmember's microphone and headset are connected to the communication equipment selected by the selector knob on individual interphone control panels. The system provides:

1. Either press-to-talk or HOT MIC interphone communication between cockpits.
2. HF, UHF-1, UHF-2, and VHF radio transmission and reception for the pilot and RSO.
3. A CALL button in each cockpit for emergency communication between cockpits.
4. Interphone communication with the ground crew by means of an external receptacle in the nosewheel well.

5. Landing gear system unsafe audio warning (pulsed tone) and low KEAS aural warning (steady tone) to pilot and RSO.

6. Monitoring of navigation radio audio signals.

7. Interphone communication with tanker during aerial refueling.

8. Recording of interphone system signals on the Mission Recorder System (MRS) whenever the pilot's or RSO's transmit or interphone switches are keyed, or when HOT MIC is selected.

Interphone Control Panel

An interphone system control panel is located on the pilot's right console and on the RSO's left console. Power is supplied by the essential dc bus through individual INTPH circuit breakers in each cockpit. Each panel contains a volume control, seven monitoring knobs, a HOT MIC knob, a six-position rotary selector, and a CALL button. See Figure 1-67.

Monitoring Switches

The seven combination volume-control and push-pull switches at the top of the interphone panel are labeled: INTER, UHF-1, HF, IFR COM, VHF, UHF-2, and TACAN/ILS. Pulling out a switch connects the output of the labeled equipment to the headset of the crewmember. Rotating a monitor switch knob varies the volume of the signal. The VOL (volume control) knob establishes the maximum strength of all seven monitor switches. The in-flight refueling communications (IFR COMM) monitor switch is used to monitor the tanker interphone system after refueling contact.

The TACAN and ILS systems share a single monitor button. TACAN signals are available except when the DISPLAY MODE SEL switch is in ILS or ILS APPROACH (when ILS signals become available).

Selector Knob

The rotary selector knob, labeled INTER, UHF-1, HF, UHF-2, VHF, and MUTE, connects the microphone and headset of the crewmember in that cockpit to the communication systems as follows:

INTER - Provides normal voice communication with the opposite cockpit and with the ground crew (if connected).

UHF-1 - For transmission and reception using the UHF radio in the forward cockpit.

HF - For transmission and reception using the HF radio.

UHF-2 - For transmission and reception using the UHF radio in the aft cockpit.

VHF - For transmission and reception using the VHF radio.

MUTE - Provides the RSO with a quiet/listen (muting) capability. When this position is selected and the quiet/listen switch on the aft cockpit floor is depressed, all interphone signals are muted except those from the pilot's CALL. This position is not normally used unless isolation from communication system distractions is required.

Hot Microphone Switch

The push-pull HOT MIC switch permits continuous interphone conversation between crewmembers without depressing a microphone button. Two-way hot microphone operation results if each crewmember pulls out the HOT MIC knob and sets the rotary selector switch to any position except INTER. (HOT MIC transmissions are not possible when the rotary selector in that cockpit is set to INTER.) In one-way operation, the HOT MIC signals are not received unless the other crewmember has the interphone monitor switch pulled up and/or the rotary selector switch set to INTER.

INTERPHONE CONTROL PANEL

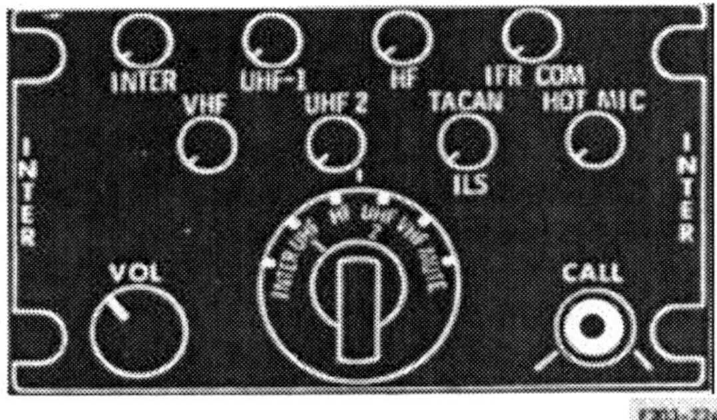

Figure 1-67

Volume Control

The volume (VOL) control knob adjusts the maximum audio level for all monitored signals.

Call Button

Depressing the spring-loaded CALL button permits interphone conversation to override any signals being received by the other crewmember. The calling volume is preset at a higher level and cannot be adjusted. A microphone switch does not have to be depressed to use CALL.

INTERPHONE SYSTEM NORMAL OPERATION

Normal communication between cockpits is accomplished by depressing the INPH switch (Pilot's Control Stick Grip and RSO's right side communication button on aft cockpit footrest). The crewmember in the opposite cockpit can hear if the interphone (INTER) monitor switch is pulled out and set to normal volume.

COMNAV-50 UHF RADIO

The COMNAV-50 UHF radio provides voice transmission and reception on any of 7000 channels in the P-Band frequency range. A direction-finding capability is also supplied. The radio is conventional except for a capability to operate in either an "Internal" mode (compatible with conventional UHF radio equipment), and an "External" mode (not compatible with other types of UHF radios). In the external mode, coded communication is only possible with other COMNAV-50 (or equivalent) radios. This mode has high resistance to jamming, allows message privacy, and has a range measuring capability.

Two independent UHF radios are provided, designated UHF-1 (front cockpit) and UHF-2 (aft cockpit). They can be used independently, within limits, for internal mode communication. A modulator/demodulator (Modem) control (COMM) panel in the aft cockpit controls coding of external mode signals and discrete selection of the ranging partner. The Modem controls can be switched by the RSO to become a part of either UHF system and give either UHF-1 or UHF-2 an external mode operating capability. Internal mode voice communication capability (without ADF function) is maintained by the opposite system; however, external mode transmissions can interfere with reception by the system operating in the internal mode if proper frequency separation is not maintained. (See Figure 1-68).

Transmitter Power Output

Transmitter power output in the internal mode is adjustable in five steps from 8.0 microwatts to 30 watts in the frequency range from 225 to 399.975 MHz. Power positions 5 and 6 have the same transmitter power output in the internal mode.

WARNING

ILS reception can be affected by UHF transmission at high power settings.

NOTE

When making an ILS approach, set power 4 or lower.

The transmitter power output in the external mode is adjustable in six steps from 8.0 microwatts to 100 watts in the frequency range from 230 to 394.975 MHz. Transmitter power should be kept as low as practical to reduce the possibility of detection. The following tables list recommended transmitter power levels for air-to-air communication in the external mode.

For Voice Communication and Ranging

Power Level	Estimated Distance Capability
6	300 plus nm
5	100 to 300 nm
4	10 to 100 nm
3	1 to 10 nm
2	less than 1 nm
1	less than 0.1 nm

For Direction Finding

Power Level Of Other Transmitter	Distance To Other Transmitter
6	100 to 200 nm
5	30 to 100 nm
4	3 to 30 nm
3	0.3 to 3 nm

External Mode Signal Characteristics

In the external mode, the transmitted signal is encoded to appear as noise to all receivers except those equipped with a compatible decoding device. The "mission code" feature prevents intelligible reception by stations which possess the necessary equipment but do not have the code. Ranging can occur only between two stations with the address code. Automatic direction finding (ADF) can also be accomplished with an addressed station in the external mode concurrently with ranging. Voice/ranging communication and ADF operation have distinct range differences for the same power level. Range measurements can be accomplished only in the external mode.

UHF Radio Equipment Location and Power Supplies

The UHF radio units, modulator/demodulator (Modem coding unit), and the ARA-48 automatic direction finding (ADF) equipment are located in the radio bay and cooled by the environmental control system. Power is furnished by the essential ac bus and the monitored dc bus.

CAUTION

If either canopy is open, the aft canopy latch handle must be in the aft position or the cockpit air handle must be in the forward (off) position for adequate equipment cooling. Otherwise, most of the cooling air would exit through the cockpit openings instead of the bays.

UHF Control Transfer Switch

The push-on/push-off control transfer switch, labeled UHF TRANS, is located on the left side of the RSO's instrument panel. The switch determines which UHF radio has ADF and EXT mode capabilities. When the UHF TRANS switch is on (illuminated), UHF-2 is connected to the ADF antenna, the UHF Modem, and the forward UHF blade antenna; and UHF-1 is connected to the aft UHF blade antenna (INT mode voice communication capability only). Depressing the UHF TRANS switch when it is illuminated extinguishes the light and reverses UHF-1/UHF-2 capabilities. ADF and EXT mode communication/ranging can only be accomplished by the UHF radio connected to the ADF antenna, UHF Modem, and the forward antenna by the UHF TRANS switch. INT mode voice communication is always possible on either UHF radio. Refer to Figure 1-69.

NOTE

- **Either crewmember can monitor or transmit on either the UHF-1 or UHF-2 system.**
- **The RSO always controls the Modem panel.**

COMNAV-50 UHF COMMAND RADIO CONTROL PANELS AND INDICATORS

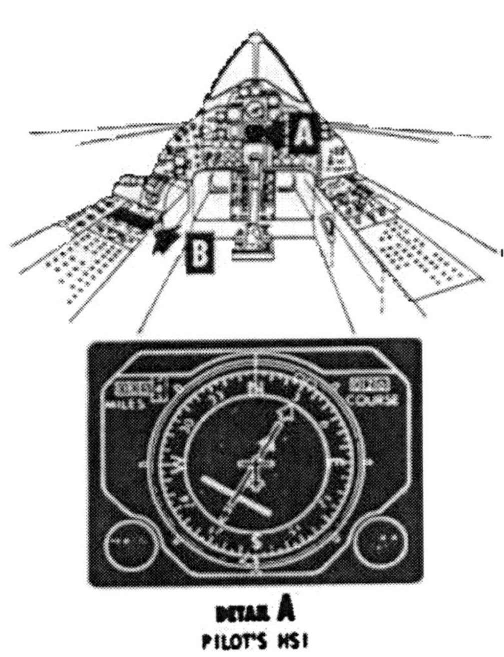

DETAIL A
PILOT'S HSI

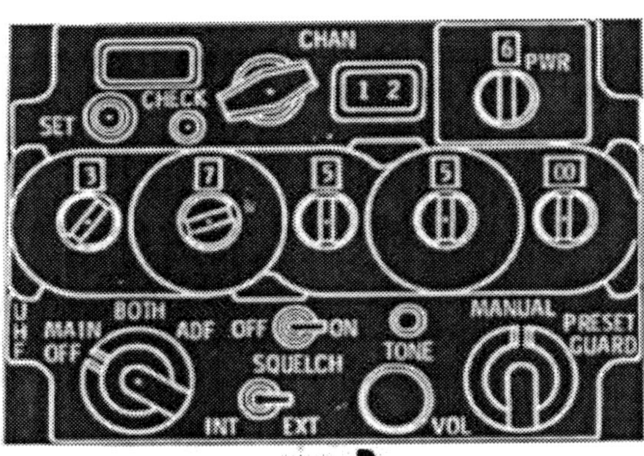

DETAIL B
Pilot's UHF-1 and RSO's UHF-2
Comnav-50 Radio Control Panel

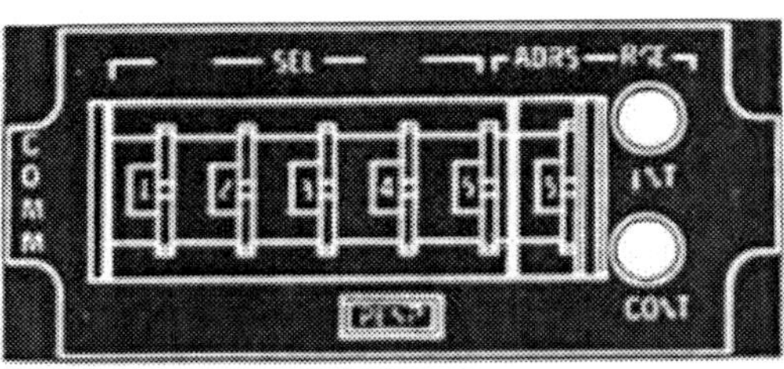

DETAIL C
RSO's Modulator/Demodulator
(Modem) Control Panel

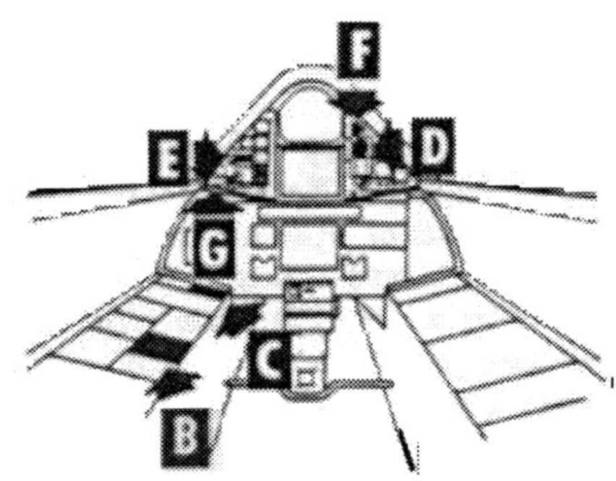

DETAIL D
RSO'S BDHI

DETAIL E
RSO'S FREQUENCY INDICATOR

DETAIL F
RSO'S DISTANCE INDICATOR

DETAIL G
RSO'S CONTROL TRANSFER SWITCH

F203-290(a)

Figure 1-68

UHF RADIO CONTROL PANELS

A UHF radio control panel, labeled UHF, is located in each cockpit. Each crewmember can independently control operation of a UHF transmitter and its guard channel receiver. The ADF function on a radio is only operative when the RSO has selected control of the ARA-48 direction finding equipment (and the forward UHF antenna) for that radio, using the UHF TRANS switch.

Function Selector Switch

A four-position rotary function selector switch turns the radio on and OFF and selects MAIN, BOTH or ADF. ADF is operable with UHF-1 if the UHF TRANS switch light is off, and with UHF-2 if the UHF TRANS switch light is on.

The UHF radio is not energized when the function selector switch is OFF. In MAIN, only the transmitter and main receiver operate. In BOTH, the transmitter, and main and guard receivers operate. The Modem unit (used for external operation) is in standby when the selector is not OFF. In ADF, the ARA-48 is energized, and the main receiver and transmitter operate. Directional signals from the ARA-48 can be displayed on the forward cockpit HSI bearing pointer and the aft cockpit BDHI No. 1 needle.

NOTE

The guard receiver is inoperative in the ADF position.

Manual-Preset-Guard Selector

The MANUAL-PRESET-GUARD switch controls frequency selection. In MANUAL, the manual frequency selector switches are functional. In PRESET, the preset channel selector switch is functional. In GUARD, guard channel frequency (243.0 MHz) is set on the main receiver and transmitter.

Preset Channel Selector Switch

The preset channel selector switch, labeled CHAN, selects one of twenty preset frequencies when the MANUAL-PRESET-GUARD selector is in PRESET. The channel number appears to the right of the selector.

Manual Frequency Selector Switches

Five rotary switches located across the middle of the UHF radio panel selects any one of 7000 frequencies when the MANUAL-PRESET-GUARD selector knob is in MANUAL. The frequency is displayed above the switches.

INT-EXT Mode Switch

The two-position transmitting mode selector switch is labeled INT-EXT. In INT, the UHF radio transmits and receives narrow-band AM signals. This position is used for conventional UHF transmitting and receiving. In EXT, the radio and modulator/demodulator (Modem) are used together to receive and transmit the wide-band pseudonoise encoded signal. Range information in EXT is displayed on the distance indicator in the aft cockpit. Direction-finding using the ARA-48 ADF can be done with the switch in INT or EXT. The UHF radio which is <u>not</u> controlling external mode operation has <u>only</u> internal mode communication capability and no ADF function.

Power Selector Switch and Indicator

A rotary PWR switch controls transmitter output power in the internal and external modes. A digit above the knob indicates the relative power output, from "1" through "6". Power output can be set from a maximum of 30 watts (position 6 and 5) to a low of 8.0 microwatts (position 1) in the INT, or narrow band mode. In the EXT, or wide band mode, power is set from a maximum of 100 watts (position 6) to approximately 8.0 microwatts (position 1).

Volume Control

Clockwise rotation of the VOL knob increases receiver volume. The normal setting is nearly full clockwise so that the interphone controls can be set to make UHF volume compatible with other signals.

Tone Button

The TONE push-button generates a 1020-cycle tone for audio checking or homing. The tone can be generated in either the internal or external mode; however, the tone is inoperative in the external mode when the CONT switch light is illuminated.

Squelch Switch

The two-position SQUELCH switch provides ON and OFF positions for the control of receiver background noise. OFF position allows reception of weak signals and noise.

Frequency Set Switch

Pressing the SET button changes the frequency of that preset channel to the frequency set by the manual frequency selector switches.

Check Switch

Pressing the CHECK button displays the manual frequency corresponding to the selected preset channel in the display window above the switch.

REMOTE FREQUENCY INDICATOR

A UHF remote frequency indicator on the RSO's instrument panel shows the setting selected in the forward cockpit UHF. When the MANUAL-PRESET-GUARD switch on the UHF-1 radio control is in MANUAL, the selected frequency is displayed numerically on the indicator. The selected preset channel number is displayed when in PRESET. The indicator displays "G" when in GUARD.

MODULATOR/DEMODULATOR (MODEM) CONTROL PANEL

The modulator/demodulator (Modem) coding equipment provides the UHF radio with the following capabilities:

1. Communications with discreetly selected partners.
2. Message privacy.
3. Ranging (semi-automatic or automatic).
4. Combined direction finding and ranging.

The Modem control (COMM) panel is located on the forward portion of the RSO's left console. The Modem controls only function in the EXT mode of the UHF radio (UHF-1 or UHF-2) that is using the forward UHF antenna (as selected by the RSO's UHF TRANS switch).

Code Selector Switches

Five rotary selector switches, labeled SEL, have positions from 0 through 7. These switches establish the signal code in the external mode. A code of all zeros cannot be used. Stations must have identical code settings to communicate in the external mode.

Range Address Switch

The right rotary selector switch, labeled ADRS, provides selective ranging. It has eight settings: 0 through 5 plus "A" and "T". The "0" position is an off position which prevents another terminal from ranging, although discrete voice communication capability is retained. Positions 1 through 5 provide range addresses. The "A" position allows a range measurement on any terminal which can respond (having the same synchronization code setting), regardless of its range address code. This is considered an emergency code. "T" is a test position for checking indicator lights on the Modem control panel.

Interrogate Switch and Indicator Light

Depressing the interrogate (INT) push-button switch initiates range and bearing interrogations in the external mode. The INT switch illuminates while the range (or direction and range) is obtained. The switch light extinguishes after three seconds. If the UHF radio function switch is in MAIN, BOTH, or ADF the one-time range measurement in nautical miles and tenths of miles is displayed on the RSO's Distance Indicator. If the UHF radio function switch is in ADF, a one-time bearing is provided to the pilot's HSI and RSO's BDHI. The INT switch is also used with the continuous ranging (CONT) switch to establish automatic ranging and direction finding.

Continuous Ranging Switch

A continuous ranging (CONT) switch and indicator light at the right of the Modem panel initiates continuous ranging. Illumination of the CONT switch light shows that the Modem is in continuous ranging or in continuous ranging and direction finding. The mode depends on setting the UHF radio function selector switch to MAIN, BOTH, or ADF, as discussed above. The CONT switch illuminates at both stations while in continuous ranging or continuous ranging and direction finding; however, the station addressed must have activated its continuous ranging switch to maintain the sequence. The RSO's Distance Indicator is updated each five seconds in the continuous mode. The automatic cycle terminates if the CONT switch is depressed, if a microphone switch at either terminal is depressed for at least three seconds after tone terminates, or the INT/EXT mode switch is placed to INT. The distance indicator at the transmitting station will then reset to zero and the indicator at the receiving station will retain the last distance value. After the cycle is broken by depressing a microphone switch, the operator of either station can reestablish the mode by depressing the INT pushbutton.

Response Light

A respond (RESP) indicator light at the bottom of the Modem control panel illuminates while the other UHF system operating in the external mode is performing a range measurement.

DISTANCE INDICATOR

A distance indicator, on the RSO's instrument panel, displays the distance in nautical miles and tenths of miles between two COMNAV-50 (or equivalent) radio systems ranging in the external mode. The maximum indication is 999.9 miles. A negative contact indicates 000.0.

UHF ANTENNAS

Two fixed UHF blade antennas and a flush-mounted direction finding (ADF) antenna are provided. The forward (No. 1) blade antenna is under the left chine, abeam the cockpit. The aft (No. 2) antenna is under the right chine by the wing leading edge. The ADF antenna is on the centerline of the fuselage, below the aft cockpit.

The RSO's UHF TRANS switch controls access to the forward UHF blade and ADF antennas. See UHF Control Transfer Switch under COMNAV-50 UHF Radio, this section.

UHF RADIO OPERATION

The pilot controls UHF-1. The RSO controls UHF-2, the UHF TRANS switch and the Modem (COMM) panel. When the UHF TRANS switch is on (illuminated), only UHF-2 can use the direction finding and/or external communication modes. To use ADF and/or EXT mode functions on UHF-1, the RSO must select UHF TRANS off (not illuminated). The RSO controls the Modem for EXT operation with either radio. See Figure 1-69.

Operations in Internal Mode

Normal Operation:

UHF radio panel:

1. **Mode switch - INT.**
2. **Volume control - Nearly full clockwise.**

   Use the interphone panel controls for volume adjustments. If necessary, decrease the UHF volume control level to maintain compatibility with the range of adjustments available in the interphone panel.

COMNAV-50 UHF AND ARC-186(V) V RADIO CONTROL AND SIGNAL CHANNELS
UHF TRANS on, RSO on UHF-2, Pilot on UHF-1, VHF off.

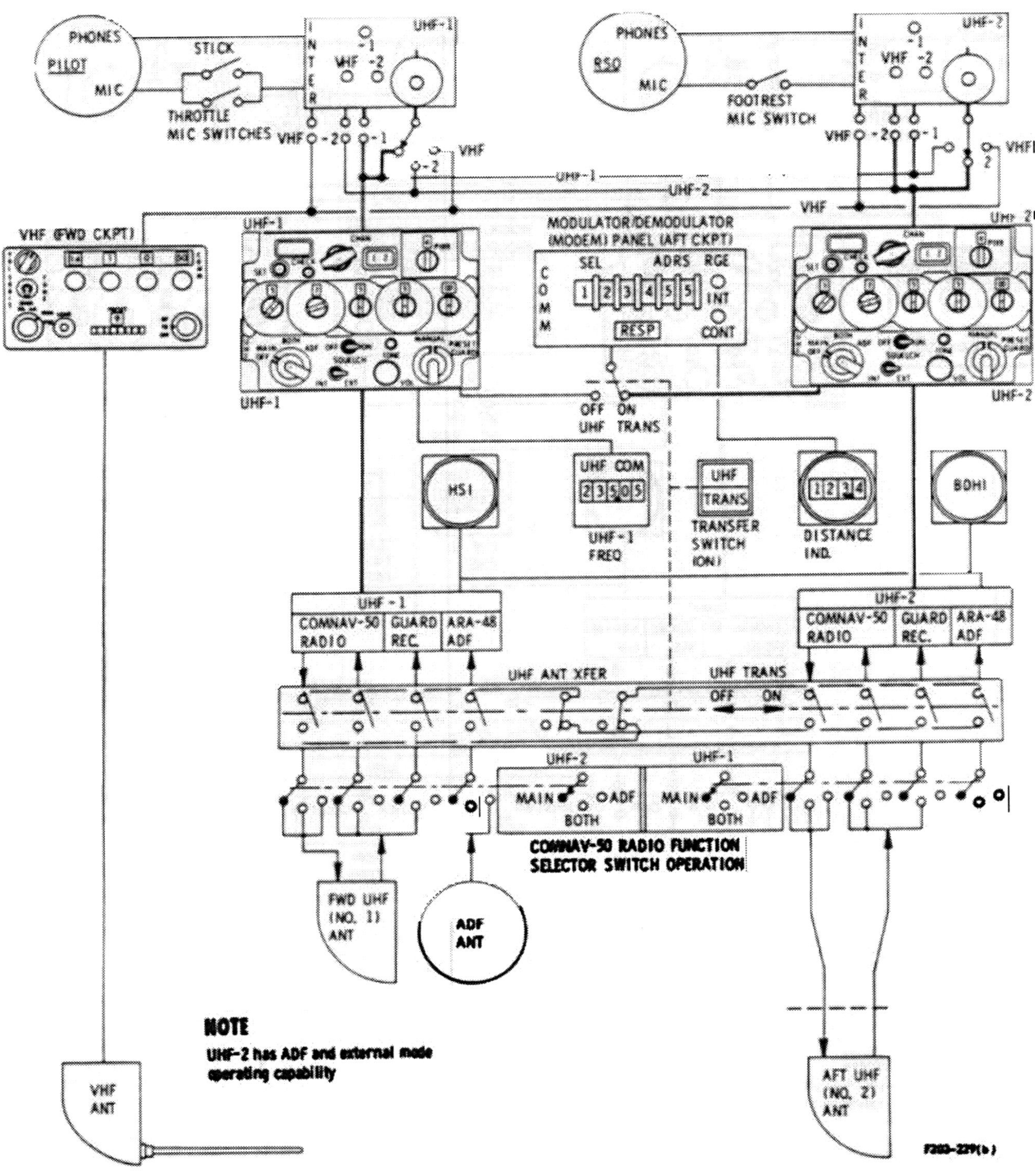

Figure 1-69 (Sheet 1 of 3)

COMNAV-50 UHF AND ARC-186(V) VHF RADIO CONTROL AND SIGNAL CHANNELS
UHF TRANS off, RSO on UHF-2, Pilot on UHF-1, VHF off.

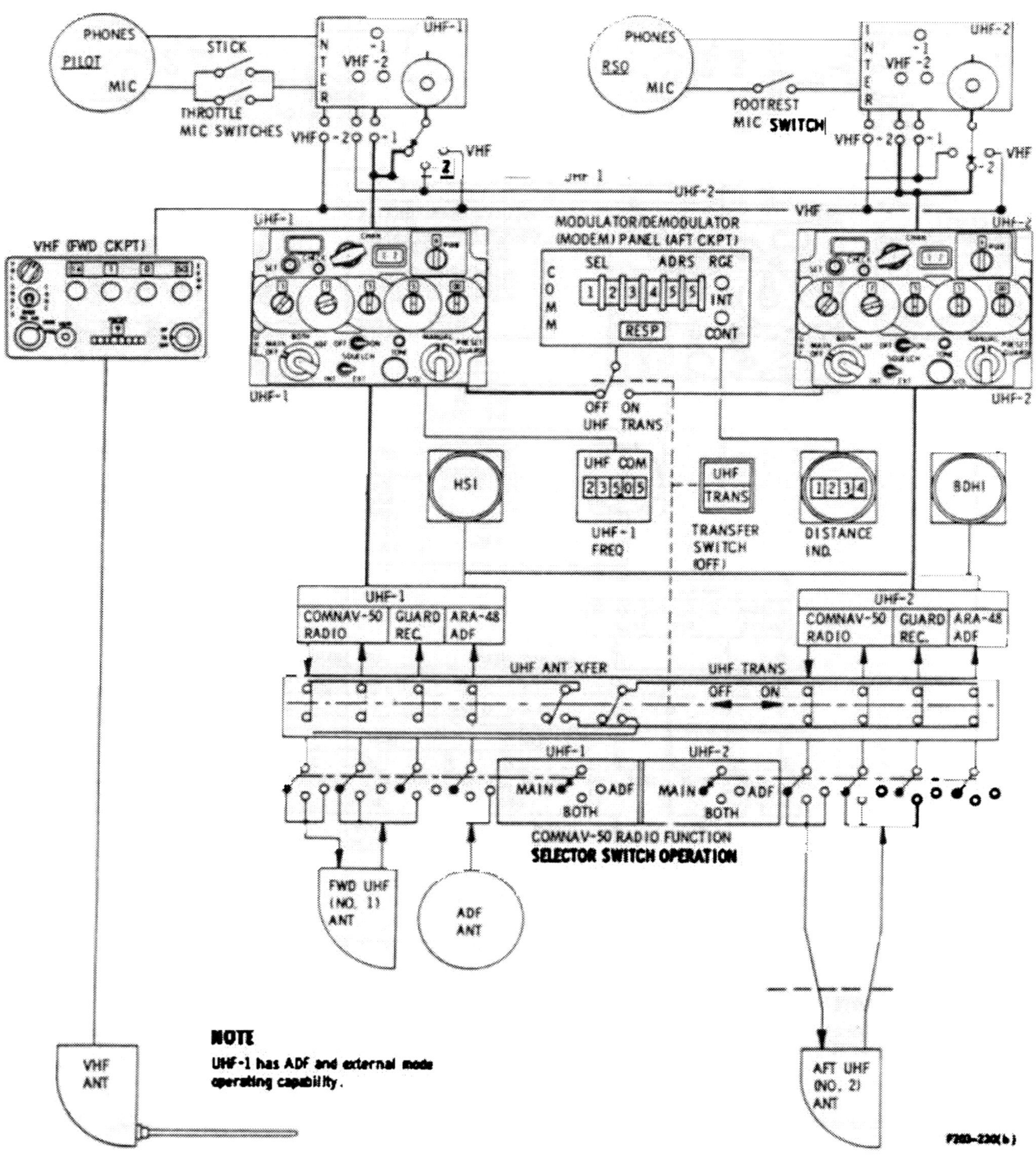

Figure 1-69 (Sheet 2 of 3)

COMNAV-50 UHF AND ARC-186(V) VHF RADIO CONTROL AND SIGNAL CHANNELS
Pilot on VHF, RSO on UHF-1, UHF-2 off, UHF TRANS off.

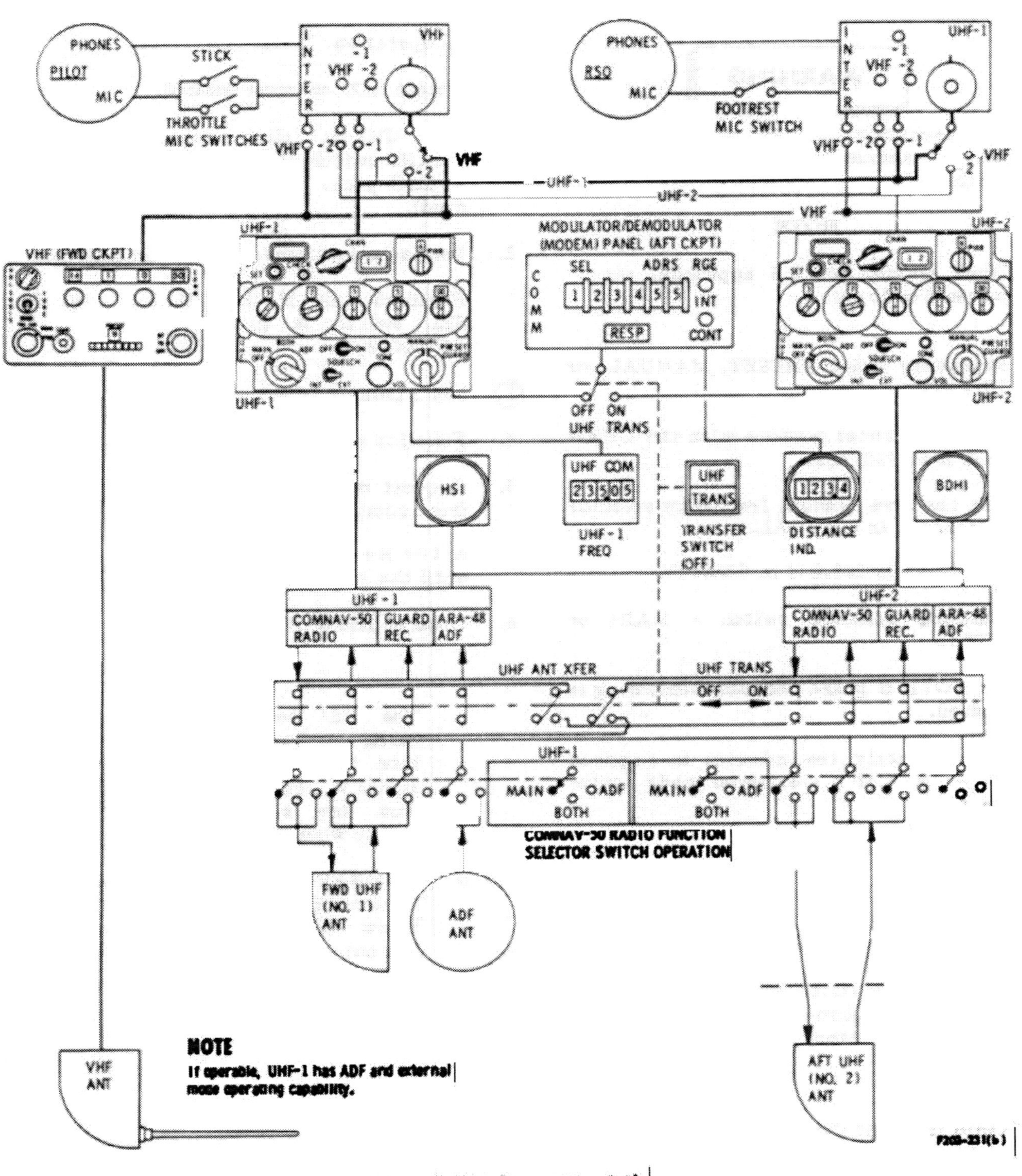

Figure 1-69 (Sheet 3 of 3)

3. **Power switch - Set.**

Position 6 is normally used.

WARNING

ILS reception can be affected by UHF tranmission at high power settings.

NOTE

When making an ILS approach, set power 4 or lower.

4. **Frequency - Set PRESET, MANUAL, or GUARD.**

Set the channel number with the CHAN knob if in PRESET.

Use the five manual frequency selector switches if in MANUAL.

Frequency is 243.0 if in GUARD.

5. **Function selector switch - MAIN or BOTH.**

Set BOTH if guard channel monitoring is desired.

Approximately ten minutes is required for warmup of the external mode equipment.

For voice communications:

6. **Interphone controls - Set.**

Select UHF-1 or UHF-2 with the interphone panel rotary selector knob and pull appropriate monitor button. Adjust volume by interphone volume control and the monitor button.

To transmit:

7. **Radio transmit switch - Hold depressed.**

For internal mode direction-finding:

Accomplish steps (1), (2), (3), and (4) as for normal operation. Then:

1. **Obtain ADF antenna control.**

   For UHF-1, pilot coordinate UHF TRANS switch off with RSO. For UHF-2, RSO press UHF TRANS on (illuminated).

2. **Bearing select switch - TACAN ADF.**

   Required only if the DISPLAY MODE SEL switch is in INS, ILS or ILS APPROACH.

3. **No. 1 pointer select switch - ADF.**

4. **Function selector switch - ADF.**

5. **Request continuous transmission or tone from communication station.**

   A 100 Hz chopping noise will be present until the ADF antenna seeks a null.

6. **Observe HSI or BDHI bearing.**

NOTE

- **The HSI bearing pointer and BDHI No. 1 pointer rapidly oscillate 5 to 10 degrees when a signal is received. The indicators are stationary or drift slowly when no signal is present.**
- **COMNAV-50 range and bearing information may be lost when below the tanker if the tanker has not switched to lower antenna.**

Operations in the External Mode

Normal Operation:

1. **Obtain UHF TRANS control.**

The RSO can operate the UHF-2 system independently when the UHF TRANS switch is on (illuminated). If UHF-1 is to be operated in the external mode, the UHF TRANS switch must be off and the RSO must operate the Modem panel. When operating one UHF in the external mode, the other UHF can only be operated in the internal mode. External mode transmissions by one radio can interfere with reception by the other radio.

UHF radio panel:

2. Mode switch - EXT.

NOTE

Communication in the external mode is possible only with another station having the same capability.

3. Volume control - Nearly full clockwise.

Use the interphone panel controls for volume adjustments. If necessary, decrease the UHF volume control level to maintain compatability with the range of adjustments available in the interphone panel.

4. Power switch - Set.

Position 6 is normal; however, position 5 is normally the maximum within the U.S.

5. Frequency - Set PRESET or MANUAL.

Set the channel number with the CHAN knob if in PRESET.

Use the manual frequency selector switches if in MANUAL.

With GUARD selected the main receiver and transmitter are switched to the internal mode. GUARD has priority over the external mode. ADF is functional.

6. Function selector switch - MAIN or BOTH.

Set BOTH if guard channel monitoring is desired. Guard channel monitoring is not functional in ADF mode.

Approximately ten minutes is required for warm-up of the external mode equipment.

Modem (COMM) panel:

(T7) Code selector switches - Set as briefed.

Set 0 to 7 on each of the five code selector switches.

(T8) Range address switch - Set 0, or as briefed.

The zero setting prevents ranging (unless the other station's ADRS switch is in A) but does not restrict voice communication capability.

For voice communication:

9. Interphone controls - Set.

Select UHF-1 or UHF-2 with the interphone panel rotary selector knob and pull appropriate monitor button. Adjust volume by interphone volume control and the monitor button.

To transmit:

10. Radio transmit switch - Hold depressed.

A one-second tone will be heard. Begin transmission after tone.

For semiautomatic (one-time) ranging or range and bearing:

Accomplish steps (1) through (8) above, then:

For bearing display:

1. Bearing select switch - TAC/ADF

Required only if the DISPLAY MODE SEL switch is in INS, ILS or ILS APPROACH.

(2.) **No. 1 pointer select switch - ADF**

3. **Function selector switch - ADF**

Guard channel monitoring is not available in ADF.

To initiate ranging:

(T4.) **Interrogate (INT) push-button - Depress momentarily.**

The light in the INT switch illuminates for approximately three seconds. A range indication appears on the RSO's distance indicator when the light goes out and bearing will be indicated on the HSI and BDHI if in ADF mode.

To update range or range-and-bearing indications:

(T5.) **INT push-button - Depress momentarily.**

To communicate with range partner:

6. **Microphone switch - Press.**

Wait for tone to mute.

For automatic (continuous) ranging or range and bearing:

Accomplish steps (1) through (8) as for Operations in the External Mode. Then:

For bearing display:

1. **Bearing select switch - TAC/ADF.**

Required only if the DISPLAY MODE SEL switch is in INS, ILS, or ILS APPROACH.

(2.) **No. 1 pointer select switch - ADF.**

3. **Function selector switch - ADF.**

Guard channel monitoring is not available in ADF.

To initiate automatic ranging cycle:

(T4.) **CONT push-button - Depress.**

(T5.) **INT push-button - Depress.**

NOTE

If ranging stops or can not be initiated, but voice communication is satisfactory, it may be due to marginal signals or to a temporary condition. Attempt to resume ranging by pressing the INT switch. If ranging does not resume, increase power at one or both stations.

(T6.) **CONT light - Check illuminated.**

(T7.) **INT and RESP lights - Check alternating illumination.**

Both stations will update readings: each 5 seconds if ranging only; each 8 seconds if ranging with one way ADF; or each 12 seconds if ranging with two way ADF.

NOTE

- **The equipment will automatically reinterrogate once if a ranging interrogation cycle is not completed after continuous ranging has been established. The digital range indication will be held for approximately 10 seconds and then reset to zero, if ranging is not reestablished.**
- **An erroneous range may occasionally appear, but the proper value will be updated during the next interrogate cycle.**

To terminate cycling:

8a. **Microphone switch - Press.**

Wait for tone to mute. A tone will be heard: from 0 to 5 seconds (ranging only), 0 to 8 seconds (one way ADF), or 0 to 12 seconds (two way ADF), depending on progress of the cycle.

NOTE

If ranging only, and a transmission is begun within 1.5 seconds after muting, that transmission will be to the ranging partner only. Subsequent transmissions will be heard by all stations having identical code selections.

OR

b. INT/EXT mode switch - INT.

OR

(T C) CONT push-button - Depress

Check that CONT light goes out.

As an alternate method of receiving ADF bearing during air refueling rendezvous, the UHF radio may be used to provide instantaneous direction finding (without ranging) in the external mode. This procedure is advantageous at close range.

The RSO should request the tanker to discontinue continuous ranging and transmit a tone in the external mode for five seconds every 15 seconds. The SR-71 UHF radio should be in the external mode with ADF selected. When the tone is transmitted, the ADF bearing will lock on steady and not tend to oscillate. Ranging with the tanker can be reestablished intermittently by requesting the tanker to reengage continuous ranging at one minute intervals.

AN/ARA-48 AUTOMATIC DIRECTION FINDER (ADF)

The AN/ARA-48 ADF antenna is used with the UHF system. When the UHF internal (INT) mode is used, the direction finder will point to emissions from any standard UHF radio transmitting on the same frequency. When the UHF external (EXT) mode is used, the ADF function is only compatible with other COMNAV-50 (or equivalent) equipment.

The ADF antenna is located under the RSO on the aircraft centerline. ADF is selected by the UHF radio function selector switch. Directional information can be displayed by the bearing pointer of the pilot's HSI and by the No. 1 pointer of the RSO's BDHI. Power is obtained from the essential ac and monitored dc buses. (See UHF Radio System Operation, this section).

AN/ARC-186(V) VHF RADIO

The ARC-186(V) VHF radio provides AM transmission and reception from 108.000 to 151.975 MHz. Frequency spacing is 25 KHz and 20 channels can be preset in addition to the preset guard frequency (121.5 MHz). Either narrowband or wideband operation is available, but must be preset by maintenance personnel. The FM capability of this radio is not operative.

The receiver/transmitter is located in the right forward chine bay. The antenna is located on the lower left fuselage, opposite the UHF-2 antenna. The radio control panel (Figure 1-70) is on the pilot's right console. Electrical power is provided by essential dc bus number 2.

Mode Select Switch

The rotary mode select switch has three positions:

OFF - Removes power.
TR - Applies power.
DF - Inoperative.

Frequency-Control/Emergency-Select Switch

The rotary frequency-control/emergency-select switch has four positions:

EMER FM - Inoperative.

EMER AM - Selects guard frequency (121.5 MHz).

MAN - Enables manual frequency selection.

PRE - Enables preset channel selection.

VHF CONTROL PANEL

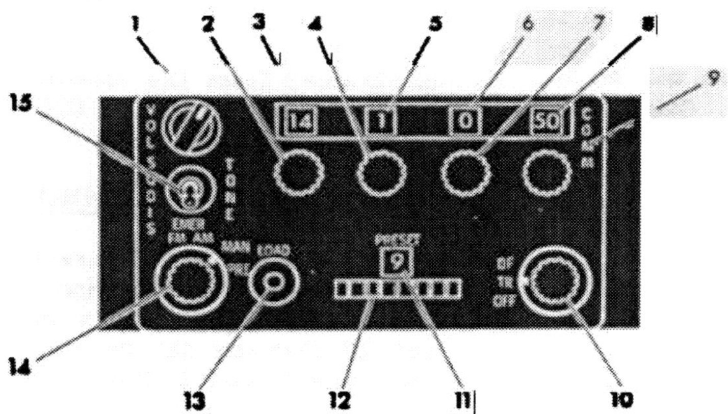

1	VOLUME CONTROL	9	0.025 MHz SELECTOR
2	10.0 MHz SELECTOR	10	MODE SELECT SWITCH
3	10.0 MHz INDICATOR	11	PRESET CHANNEL INDICATOR
4	1.0 MHz SELECTOR	12	PRESET CHANNEL SELECTOR
5	1.0 MHz INDICATOR	13	LOAD SWITCH
6	0.1 MHz INDICATOR	14	FREQUENCY CONTROL/EMERGENCY SELECT SWITCH
7	0.1 MHz SELECTOR	15	SQUELCH DISABLE TONE SELECT SWITCH
8	0.025 MHz INDICATOR		

F203-225

Figure 1-70

Squelch-Disable/Tone-Select Switch

The squelch-disable and tone-select toggle switch has three positions:

Center Position	-	Enables squelch.
SQ DIS	-	Disables squelch.
TONE	-	A spring-loaded-to-center position that transmits a tone at 1000 Hz for audio checking or homing.

Volume Control

Clockwise rotation of the volume control knob, labeled VOL, increases volume. The normal setting is nearly full clockwise so that the interphone controls can be set to make VHF volume compatible with other signals.

Manual Frequency Selector Switches

Four rotary switches are used to manually select a frequency when the frequency-control/emergency-select switch is in MAN. The windows above each switch display the selected frequency.

Preset Channel Selector

A rotary preset channel selector switch is used to select one of 20 preset frequencies when the frequency-control/emergency-select switch is in PRE. The selected channel number is displayed above the switch.

Load Switch

Pressing the recessed LOAD push-button switch changes the preset channel to the frequency set by the manual frequency selector switches.

VHF OPERATION

The VHF radio control panel is in the forward cockpit.

1. Mode select switch - TR.
2. Frequency-control/emergency-select switch - PRE or MAN.

   Set the channel number with the preset channel selector if in PRE.

   Use the manual frequency selector switches if in MAN.
3. Volume control - Nearly full clockwise.
4. Interphone controls - Set.

   Select VHF with the interphone panel rotary selector switch and pull VHF monitor button. Adjust volume by interphone volume control and the monitor button.
5. Squelch-disable/Tone-select switch - Center position.

To transmit:

6. Transmit switch - Hold depressed.

HF RADIO SYSTEM, 618-T

The 618-T HF radio is a long-range voice communications transceiver. The modes of transmission are single sideband (SSB) and amplitude modulation (AM). The frequency range is 2 to 30 MHz, tunable in 1-KHz steps. Because of the nature of the antenna (comprising the pitot boom and the insulated forward portion of the aircraft nose) it is not advisable to transmit at frequencies below 4 MHz. The equipment includes a transceiver and semiautomatic antenna coupler/coupler control, mounted in the radio bay in the right chine. The HF radio control panel is on the RSO's left console (Figure 1-71). Electrical power is supplied by the essential ac and essential dc buses.

Long-range HF communication is highly sensitive to hourly and seasonal variations in propagation conditions. For best results, frequency assignment planning should be based on HF propagation predictions.

Function Switch

The four-position function selector switch, labeled OFF, USB, LSB, and AM, energizes the equipment and selects the desired operating mode. In USB (upper sideband), only the upper sideband signal is transmitted or received. This is the sum of the voice signal and the radio frequency (rf) signal. In the LSB (lower sideband) position, only the lower sideband signal is transmitted or received. This signal is the difference of the voice signal and the rf signal. In the AM position, the signal is amplitude modulated and both sidebands and the original rf signal are transmitted and received.

Frequency Selector Knobs

Four rotary switches manually select a frequency. The windows in the middle of the panel display the frequency selected.

RF Sensitivity Knob

The RF sensitivity knob, labeled RF SENS, adjusts the receiver sensitivity level to control the signal-to-noise ratio.

Normal Operation:

1. Interphone controls - Set.

   Select HF with the interphone panel rotary selector switch and pull HF monitor button. Adjust volume by interphone volume control and the monitor button.
2. Function switch - USB, LSB, or AM.
3. Frequency selector knobs - Set.

   The muting of sound in the headset indicates the receiver is tuning to the new frequency.

HF CONTROL PANEL - 618T

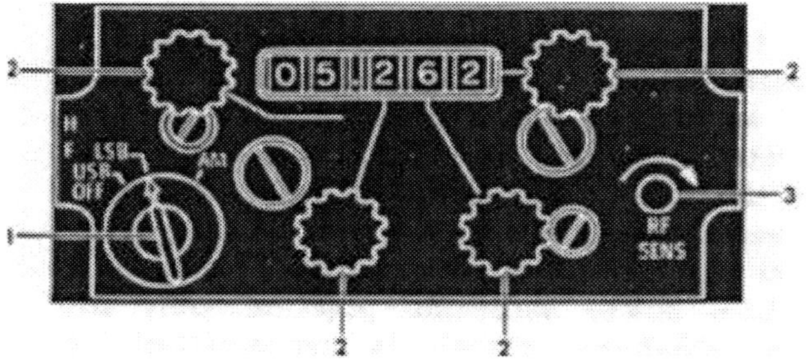

1 FUNCTION SWITCH
2 FREQUENCY SELECTOR SWITCHES
3 RF SENSITIVITY SWITCH

Figure 1-71

When background sound is again heard:

4. RF sensitivity knob - Adjust.

Turn the RF SENS knob clockwise until a distinct crackling background noise is heard. At this setting, the receiver is at maximum sensitivity and further rotation of the knob will not improve signal reception. Adjusting the RF SENS knob until there is no distinct background noise lowers receiver sensitivity and incoming signals may not be received. The background noise level varies with locality and propagation conditions, and small adjustments may be necessary to maintain the optimum sensitivity setting.

5. Transmit switch - Depress. Wait for the equipment to tune.

A 1000 Hz tone will be heard until tuning is complete. Tuning may require 38 seconds.

When the tone ceases, the transmitter is tuned. Adjust the interphone HF monitor switch volume.

NOTE

The HF radio should be retuned after takeoff to match the antenna inflight impedance condition.

WARNING

Rf energy from the HF radio during tuning or transmission has caused erroneous light and instrument indications.

Emergency Operation

If an overload exists in the power supply output, a protective circuit turns off the HF equipment. Attempt to restore normal operation as follows:

1. Function switch - OFF, then back to desired operating mode.

If the antenna coupler makes several consecutive tuning cycles, a thermal relay de-energizes the equipment. To restore operation:

1. Function switch - OFF. After 2 minutes the thermal relay will cool.

2. Function switch - Set to desired operating mode.

HF RADIO SYSTEM, AN/ARC-190 (V)

The ARC-190 receives and transmits on 280,000 frequencies in a band from 2 to 29.9999 MHz spaced at 100 Hz. Frequencies below 4 MHz should not be used due to the nature of the antenna (comprising the pitot boom and the insulated forward portion of the aircraft nose). Modes of operation are upper and lower sideband, amplitude modulation and continuous wave. System components are: a receiver-transmitter (R/T) and antenna coupler located in the R bay;

HF CONTROL PANEL - ARC 190

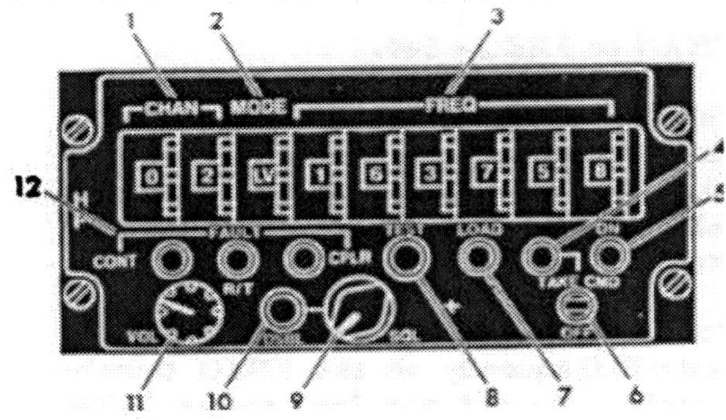

1 CHANNEL SELECT THUMBWHEELS
2 MODE SELECT THUMBWHEEL
3 FREQUENCY SELECT THUMBWHEELS
4 TAKE COMMAND LIGHT
5 ON LIGHT
6 TAKE COMMAND/OFF SWITCH
7 LOAD PUSHBUTTON
8 TEST PUSHBUTTON
9 SQUELCH SWITCH
10 SQUELCH ENABLE/DISENABLE PUSHBUTTON
11 VOLUME SWITCH
12 FAULT INDICATOR LIGHTS

Figure 1-72

antenna and antenna loading coil in the nose; and the HF control panel on the RSO's left console (Figure 1-72). Essential ac power is supplied through a circuit breaker in the C bay.

Automatic tuning occurs in both receive and transmit. Receive tuning requires less than 10 milliseconds. Transmit tuning requires 35 milliseconds on any of the 30 preset channels and 1 second on a manually selected frequency. The first transmission on a new frequency, even a momentary transmission, initiates the transmit tune cycle. A 1000-Hz audio tone is heard until tuning is complete.

Long-range voice HF communication is highly sensitive to hourly and seasonal variations in propagation conditions. For best results, frequency assignment planning should be based on HF propagation predictions.

Channel (CHAN) Switches

Two thumbwheel switches used to select 30 preset channels 00 through 29.

Mode Switch

An 8-position thumbwheel switch used to select modes of operation: LV, lower sideband voice; UV, upper sideband voice (must be set when selecting manual frequencies); LD, lower sideband data (not used); UD, upper sideband data (not used); CW, continuous wave; AM, amplitude modulation; P, preset (must be set when using the 30 preset channels); A, undefined (CONT FAULT will illuminate with A selected).

Frequency (FREQ) Switches

Six thumbwheel switches used to manually select frequencies.

ON Light

Illuminates when radio is on.

TAKE CMD/OFF Switch

Three position switch, spring-loaded to center position used to turn the radio on and off. Momentary TAKE CMD (forward) position turns the radio on. Momentary OFF (aft) position turns the radio off.

LOAD Switch

Pushbutton used to load preset channels in memory.

TEST Switch

Pushbutton used to initiate self-test. When pressed, a receive self-test cycle is initiated and all FAULT lights (3) illuminate. When released, all FAULT lights extinguish unless a component fault is present where indicated. A transmit self-test is initiated by depressing the transmit microphone switch (pilot or RSO with HF selected on interphone panel). A fault is indicated by illumination of the fault light(s).

NOTE

The transmit self-test can only be initiated after completion of a receive self-test.

FAULT Lights

CPLR - Indicates a coupler malfunction.

R/T - Indicates a receiver-transmitter malfunction or illuminates when an unloaded preset channel is selected with P selected in the MODE switch.

CONT - Indicates a control panel malfunction, FREQ switches set below 02.0000 MHz, or the MODE switch is set to A.

Squelch (SQL) Switch

Rotary switch that provides squelch threshold in 3 preset levels. Squelch is disabled in the fully counterclockwise position.

Disable (DSBL) Switch

Pushbutton that alternately enables and disenables the squelch (SQL) switch.

Volume (VOL) Switch

8-position rotary switch sets transmit-receive audio and transmit audio sidetone at 7 preset levels.

Normal Operation

1. Interphone rotary selector knob - HF.

2. Interphone HF monitor switch - Pull and rotate to approximately 12 o'clock position.

3. TAKE CMD/OFF switch -TAKE CMD.

   Momentarily position the switch to TAKE CMD to turn the radio on. The ON and TAKE CMD lights will illuminate.

4. MODE switch - Set.

   Set P for preset channel.

   Set UV for manually selected frequency.

5. CHAN or FREQ - Set.

   To use a preset channel, select the desired channel on the CHAN thumbwheels. If an unloaded preset channel is selected, the R/T FAULT light illuminates.

   To manually select a frequency, set the desired frequency on the FREQ thumbwheels. Do not use frequencies below 04.0000. If a frequency below 02.0000 is set, the CONT fault light illuminates.

6. VOL switch - Nearly full clockwise.

   Use the interphone panel controls for volume adjustments. If necessary, decrease the HF volume to maintain compatibility with the range of adjustments available in the interphone panel. The VOL switch also controls the transmit audio sidetone.

7. SQL and DSBL switches - Set.

   Rotate the squelch switch fully clockwise. If background noise is audible in the full clockwise position, depress the DSBL pushbutton to enable squelch. Rotate the SQL counterclockwise until background noise is audible. This position allows reception of any audible signal but with some continuous background noise. For normal operation rotate the switch one position more clockwise to set the threshold of reception just above the background noise.

8. TEST switch - Press.

   Check all three fault lights illuminate. When the switch is released, check no fault lights remain illuminated.

To transmit:

9. Microphone switch - Depress.

A 1000-Hz tone is heard during transmit tuning on the first transmission on a new frequency. Tuning is complete within one second.

WARNING

Rf energy from the HF radio during tuning or transmission has caused erroneous light and instrument indications.

To program a preset channel:

1. **Mode switch - UV.**
2. **FREQ. - Set.**
3. **CHAN - Set.**
4. **LOAD pushbutton - Depress.**

   The select channel will be programmed to the selected mode and frequency.

   Repeat as required for other channels.

To turn system off:

1. **TAKE CMD/OFF switch - OFF.**

   Momentarily position the switch to OFF. The ON and TAKE CMD lights will extinguish.

Emergency Operation

If the R/T FAULT light illuminates while receiving or transmitting:

1. **Cycle a FREQ thumbwheel (if in UV mode) or a CHAN thumbwheel (if in P mode) at least three times.**

Illumination of the R/T FAULT light when an unloaded preset channel is selected is normal.

If the R/T FAULT light remains on:

2. **Attempt a short transmission.**

CAUTION

If the ARC-190 HF transmits normally with the R/T FAULT light illuminated, use short transmissions sparingly to prevent overheat.

If transmission and/or reception is inoperative:

3. **TAKE CMD/OFF switch - OFF.**

   Attempt to restore normal operation by cooling the system at least two minutes. Reattempt normal operation as desired.

INSTRUMENT LANDING SYSTEM (ILS), 51RV-1

An ILS receiver supplies signals to the bank and pitch steering bars and glide slope indicator on the ADI, and to the course deviation indicator (CDI) on the HSI. Refer to Attitude-Director Indicator, Horizontal Situation Indicator, this section, and to Figure 1-66.

An invalid localizer or glide slope signal is indicated by a red warning flag appearing behind the bank steering bar or glide slope indicator, respectively.

WARNING

ILS reception can be affected by UHF transmission at high power settings.

The receiver operates on 20 frequencies. Localizer frequencies range from 108.1 to 111.9 MHz, and glide slope frequencies range from 329.3 to 335.0 MHz. The proper glide slope frequency is automatically tuned when the localizer frequency is selected.

ILS CONTROL PANEL

An ILS control panel (Figure 1-73), labeled NAV, is located on the pilot's right console. The panel contains concentric power (outer

ring) and frequency (inner knob) switches. The power switch has two positions: OFF and PWR. The panel also has concentric volume control (outer ring) and frequency (inner knob) switches. Turning the VOL (outer ring) clockwise increases localizer identification volume. The TACAN/ILS interphone switch must be pulled and the DISPLAY MODE SEL switch must be in ILS or ILS APPROACH to hear the ILS identifier and/or marker beacon. A window, labeled MC, displays the localizer frequency (selected by rotating the inner knobs of the concentric switches). The ILS TEST buttons, labeled UP/L and DN/R, test the ILS system (excluding the antenna). The VOR button is not operable.

MARKER BEACON

A conventional 75 MHz ILS marker beacon receiver illuminates the amber MARKER BEACON light on the pilot's instrument panel and generates an audio signal when the aircraft is over an ILS marker. Power for the receiver is furnished by the essential dc bus.

NOTE

> The marker beacon antenna is located inside the right nosewheel door and is operational only when the nosewheel is down.

ILS Operation

1. Interphone ILS monitor switch - Pull and set volume.
2. DISPLAY MODE SEL switch - ILS or ILS APPROACH.
3. Power switch - ON.

   Allow 90 seconds for warmup.
4. Volume - Nearly full clockwise.

To self-test the ILS:

1. Select any localizer frequency.
2. Align the HSI course arrow with the lubberline.
3. DISPLAY MODE SEL switch - ILS APPROACH.
4. Press UP/L.

   The localizer and glide slope warning flags disappear, (the INS must be in NAV for the localizer warning flag to disappear), the glide slope indicator moves 1-dot up, the localizer moves 1-dot left, and the pitch and bank steering bars move half-scale up and left, respectively.
5. Press DN/R.

   The same actions occur as in step 4, but the directions are down and right.

ILS CONTROL PANEL

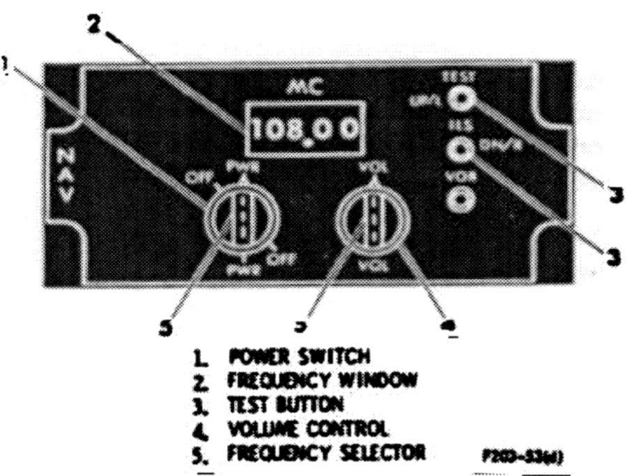

Figure 1-73

AN/APX-108(V) IFF (W/SB R-2668) WILCOX 914X-2 IFF (W/O SB R-2668)

The IFF transponder responds to radar interrogation. The system includes altitude reporting, selective identification, and emergency reporting features. The Mode 4 function provides an encrypted IFF capability.

IFF CONTROL PANEL

The IFF control panel is located on the RSO's instrument panel. The controls for Mode 4 are on the left side of the IFF control panel, and the controls for Modes 1, 2, 3A and C are on the right side. See Figure 1-74.

Master Selector Switch

The rotary master selector switch has five positions:

OFF - Removes power. The switch must be pulled out before it can be rotated to OFF.

STBY - Standby. Applies power, but transmission is inhibited.

LOW - Only responds to strong (local) interrogations. Used at the request of a controller.

NORM - Normal operation. All modes selected have full sensitivity.

EMER - Emergency. Responds to interrogations in Modes 1, 2 and 3A. Mode 3A responds with code 7700. Mode C and 4 operate normally if selected. The switch must be pulled out before it can be rotated to EMER.

Mode 1, 2, 3A and C Controls

Mode 1, 2, 3A and C Control Switches

Four three-position toggle switches select Modes 1, 2, 3A and C. The ON (center) position places the corresponding mode in operation and the OUT (down) position disables the corresponding mode. The momentary TEST (up) position is used to test each mode. Continuous illumination of the TEST indicator light, while a mode control switch is held in TEST and the master selector switch is in NORM, indicates successful self-test of that mode.

Mode C responds with altitude information pulses in addition to the framing pulses (for tracking) sent in Mode 3A.

RAD-OUT-MON Switch

The three-position RAD-OUT-MON switch is provided for testing and monitoring the IFF system. The spring-loaded momentary RAD position is used with ground equipment for maintenance test. In the MON (monitoring) position, the TEST light will illuminate intermittently when responding to radar interrogations in the mode(s) selected. In the OUT position, the TEST light will only respond to self-test inputs using the mode control switches and will not indicate response to radar interrogation.

Mode 1 and 3A Code Selector Switches

Six thumb-wheel switches select the codes for Mode 1 and Mode 3A. The two switches on the left select Mode 1 codes and the other four switches select Mode 3A codes. The left Mode 1 switch is numbered 0 through 7, the other is numbered 0 through 3 on each half of the drum. The Mode 3A switches are numbered 0 through 7.

Identification-of-Position Switch

The three position identification-of-position (I/P) switch controls transmission of I/P pulse groups. The I/P timer is energized for thirty seconds when the switch is momentarily held in spring-loaded IDENT,and I/P replies will be made if a Mode 1, 2, or 3A interrogation is recognized within thirty seconds. The I/P pulse group is not transmitted when the I/P switch is in OUT. The MIC position is inactive.

Test Indicator Light

A rotate-to-dim, press-to-test, green TEST light indicates satisfactory operation of the transponder. With the master selector switch

IFF CONTROL PANEL

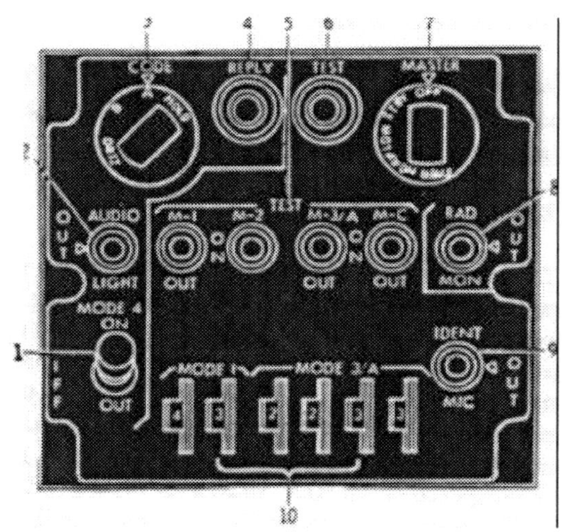

1 Mode 4 On-Out Switch
2 Mode 4 Indication Switch
3 Mode 4 Code Select Switch
4 Mode 4 Reply Light
5 Mode Control Switches
6 Test Indicator Light
7 Master Selector Switch
8 Rad-Out-Mon Switch
9 Identification-of-Position Switch
10 Code Selector Switches

F20J-173(-)

Figure 1-74

Mode C Altitude Reporting and Identification Capability Schedule

Mode Selector Switch Settings		Response	
Mode 3A	Mode C	Mode 3A	Mode C
ON	ON	Normal	Normal (Altitude Code)
ON	OUT	Normal	*Framing Pulses Only
OUT	ON	None	Normal (Altitude Code)
OUT	OUT	None	None

*Response indicates aircraft with Mode C altitude reporting capability to interrogating station.

Notes:

Mode 3A selector switch enables Mode 3A and Mode C decoder.

Mode C selector switch enables Mode C decoder and information pulses.

Mode C altitude reporting information is enabled by the DAFICS M computer. If the M computer fails, Mode C will continue to report the altitude at the time the computer failure occurred.

Figure 1-75

in NORM, the light illuminates when a mode switch (1, 2, 3A, or C) is placed in TEST if the self-test is satisfactory. When the system is operating (master selector switch in LOW or NORM), and the RAD-OUT-MON switch is in MON, the light blinks when the IFF responds to interrogation in the mode(s) selected.

Mode 4 Controls

Mode 4 is also controlled by the master selector switch.

Mode 4 Code Select Switch

The Mode 4 code select switch is labeled ZERO, B, A, and HOLD. Placing the switch in ZERO cancels (zeroizes) both the code settings. The switch must be pulled out before it can be rotated to ZERO. When the switch is in A or B, the transponder will respond to Mode 4 interrogation sources using the same code. The spring-loaded HOLD position is used (before power is removed from the transponder) to retain codes when the aircraft is on the ground.

NOTE

Weight must be on the nosegear before the HOLD position is functional.

To retain code settings, the switch must be held in HOLD for at least 5 seconds and transponder power must be left on for an additional 15 seconds. Otherwise, the code settings may not be mechanically latched and will zeroize when the master selector switch is turned off or power is disconnected.

Mode 4 Reply Light

A rotate-to-dim, press-to-test, green REPLY light illuminates when the Mode 4 indication switch is in either AUDIO or LIGHT, if transponder Mode 4 replies are satisfactory. Press-to-test is only operative in the AUDIO or LIGHT position.

Mode 4 Indication Switch

A three-position AUDIO-OUT-LIGHT switch controls Mode 4 indications.

W/O SB R-2668, with the indication switch in AUDIO (up), an audio tone is heard in the aft cockpit only if a proper Mode 4 interrogation is received, and the REPLY light illuminates when Mode 4 generates a reply.

With SB R-2668, with the indication switch in AUDIO (up) an audio tone is heard in the aft cockpit if Mode 4 does not respond to a proper interrogation and the REPLY light will not illuminate.

NOTE

The RSO's IFR COM knob controls the volume of the Mode 4 audio.

In LIGHT (down), the REPLY light illuminates without audio when Mode 4 replies are transmitted. In OUT (center), both light and audio indications are inoperative.

Mode 4 On-Out Switch

A two-position toggle switch controls Mode 4 operation. A Mode 4 response cannot be transmitted unless: ON (up) is selected; the Master Selector Switch is in LOW, NORM, or EMER; and the Mode 4 code has not been zeroized. The switch must be pulled out to be moved to OUT (down).

Mode 4 IFF Caution Light

A rotate-to-dim, amber IFF CAUTION light, is located on the RSO instrument panel. The light illuminates each time the transponder fails to reply to a proper interrogation. If the Mode 4 codes are zeroized, or if self-test detects a system fault, the light will be on steady. Pressing the LAMP TEST switch checks the light .

IFF NORMAL OPERATION

Modes 1, 2, 3A and C

1. Master Selector Switch - STBY.

   Three minutes are required for warm-up.

2. Master Selector Switch - NORM.

should only be used at the request

3. Mode 1 and 3A Code Selector Switches - Set.

NOTE

The Mode 2 Code is preset on the ground.

4. Mode Switch(es) - As required.

To test operation of individual Modes:

1. RAD-OUT-MON switch -OUT.

   If the switch is in MON, the TEST light will illuminate for both self-test and monitor functions.

2. Individual mode switch - TEST.

   Illumination of the TEST light indicates the corresponding Mode is operational. Repeat for each mode.

NOTE

Test modes individually.

3. Mode switches - As required.

To monitor operation of individual Modes:

1. RAD-OUT-MON switch - MON.
2. Individual Mode switch - ON.

   The TEST light will blink to indicate responses to interrogation. Repeat for other modes.

To transmit identification of position:

1. Identification switch - IDENT.

Mode 4

1. Mode 4 ON-OUT switch - ON.
2. Code select switch - A or B, as required.

   Code A is the code for a prescribed 24-hour period; Code B is for the next 24-hour period.

3. Audio/Light indicator switch - As desired.

To zeroize the codes:

1. Mode 4 code select switch - ZERO.

   The codes can also be zeroized by turning the master selector switch OFF, if the HOLD function has not been used.

NOTE

Once zeroized, Mode 4 is inoperative until the codes are reinserted on the ground.

To retain the codes after landing:

1. Mode 4 code select switch - HOLD.

   Place the switch in the HOLD position for 5 seconds, then wait another 15 seconds before turning equipment OFF.

To turn IFF off:

1. Master selector switch - OFF.

IFF EMERGENCY OPERATION

Modes 1, 2, and 3A

1. Master selector switch - EMER.

NOTE

In EMER, Mode 4 and Mode C replies are normal.

Mode 4

Illumination of the IFF CAUTION light indicates the transponder will not respond to Mode 4 interrogations.

With IFF CAUTION light illuminated:

1. Master selector switch - Check NORM.

2. Mode 4 ON-OUT switch - Check ON.

3. Mode 4 code select switch - Check.

   Check A or B.

G-BAND BEACON

The G-band (formerly C-band) beacon is a radio frequency transponder used to aid radar tracking of aircraft during special tests. The beacon responds to interrogation in the 5400 to 5900 MHz range. The transmitter radiates with at least 400 watts peak power at 500 pps. Receive and transmit frequencies are displaced at least 50 MHz for protection of the beacon receiver. Transmitter operation can not be adjusted while in-flight.

The beacon is controlled by two two-condition push-button switches located left of the viewsight controls on the RSO's instrument panel. Power is controlled by the BEACON switch. Actuating the switch illuminates an ON legend in the lower half of the switch. A warm-up period of at least three minutes is desired. A 15-minute warm-up period results in maximum signal stability. Another switch actuation turns the beacon off and extinguishes the ON legend. Two antennas are provided. Actuating the ANT (antenna) switch illuminates either a B (bottom antenna) or T (top antenna) legend in the lower quarters of the switch when the beacon is on.

Power for the beacon is from the Essential DC Bus through the BEACON circuit breaker on the RSO's right console.

I-BAND BEACON

The I-band beacon is provided when Mission Kit 4AT1030 is installed. The antenna can be installed in the left EIP door, right EIP door, left TECH door, or right TECH door.

The I-band beacon is a radio frequency transponder used to aid radar tracking of aircraft during special tests. The beacon responds to interrogation in the 8500 to 9600 MHz range. Receive and transmit frequencies are displaced at least 50 MHz for protection of the beacon receiver. Transmitter operation can not be adjusted while in flight.

The beacon is controlled by the OOC LH OPER power switch if the antenna is installed on the left EIP door or the left TECH door or by the OOC RH OPER power switch if the antenna is installed on the right EIP door or the right TECH door. Actuating the switch illuminates the ON legend in the lower half of the switch. System warm-up is from 20 seconds to two minutes. Another switch actuation turns the beacon off and extinguishes the ON legend. (The FAIL light is not functional and will not illuminate.)

Power for operation of the beacon is provided from the Monitored DC Bus through the left or right OOC CONT AND PWR circuit breaker in the C-bay.

TACAN SYSTEM AN/ARN 118(V)

The Tactical Air Navigation (TACAN) system operates with ground stations and cooperating aircraft. Continuous slant range and bearing information is obtained from ground stations. Range and bearing are also obtained through mutual transponding with cooperating aircraft; however, bearing information can not be transmitted by the SR-71 to the other aircraft. Operational range can be over 300 nautical miles at high altitudes.

126 X-mode and 126 Y-mode channels are available.

Transmitter interrogation frequencies range from 1025 to 1150 MHz with 1-MHz separation. Receiver frequencies range from 962 to 1024 MHz and 1151 to 1213 MHz when operating in the air-to-ground (T/R) mode.

NOTE

> TACAN reception can be affected by UHF transmissions at high power settings in external mode.

In the air-to-air mode, receiver frequencies used are the same as for transmitter interrogation; however, a pair of channels with 63 MHz separation is required. Since the air-to-air modes can operate at or near the IFF system frequencies (1090 and 1030 MHz for transmit and receive), IFF interference can cause unreliable TACAN operation on channels 1-11, 58-74, and 121-126.

The IFF receiver is suppressed during TACAN transmission and the TACAN receiver is suppressed during IFF transmissions, to protect the receivers. If no signal is received when a strong TACAN or IFF signal is expected, a malfunction of the suppression circuits can be identified by momentarily turning off one of the systems.

When TACAN channels are changed, acquisition time for the new channel is less than one second. No more than three seconds are required for bearing lock-on.

If the received TACAN distance signal is invalid, the distance displays on the HSI and BDHI are covered. If the DISPLAY MODE SEL switch is in TACAN, the steering bar warning flag appears if TACAN bearing is not valid. The identifier of the interfering station is purposely garbled if co-channel interference occurs in T/R. If signal loss occurs, velocity memories keep the bearing and range indications tracking for up to 3 and 15 seconds, respectively, or until the signal is reacquired. An automatic self-test is performed after each signal loss and the TEST light illuminates if the test fails.

Power for TACAN operation is provided by the essential ac bus and by the monitored dc bus through circuit breakers in the C-bay.

The TACAN antenna is a high temperature annular slot type (identical to the IFF antenna) located on the bottom centerline of the fuselage forward of the nosewheel well.

TACAN CONTROL PANEL

A TACAN control panel and a TACAN control transfer switch are provided in each cockpit (Figure 1-76).

Channel Selectors

Two channel selector knobs and an X-Y mode selector, control TACAN frequency. The channel number (01 through 126), selected by rotating the tens and units channel selector knobs, is displayed on the panel. Rotating the ring which surrounds the base of the units channel selector knob selects X or Y mode.

Mode Selector Switch

The rotary mode selector switch has five positions:

OFF - Removes power.

REC - Bearing to selected ground station. Range is not available.

NOTE

- After tuning a new station, the bearing pointer may slew to a bearing 90° greater than the actual bearing and remain there for about two seconds. This is normal.
- After signal loss and re-acquisition, the bearing pointer will slew to 270° and remain there for about seven seconds during the automatic self-test. After self-test, the bearing pointer may slew to a bearing 90° greater than the actual bearing and remain there for about two seconds.

T/R - Bearing and slant range to a selected ground station.

A/A REC - Bearing to a suitably equipped cooperating aircraft (also in A/A with 63 channel separation). Range is not available.

A/A T/R - Bearing and slant range to a cooperating aircraft.

AN/ARN-118(V) TACAN CONTROLS

TACAN CONTROL PANEL
PILOT RIGHT CONSOLE
RSO LEFT INSTRUMENT PANEL

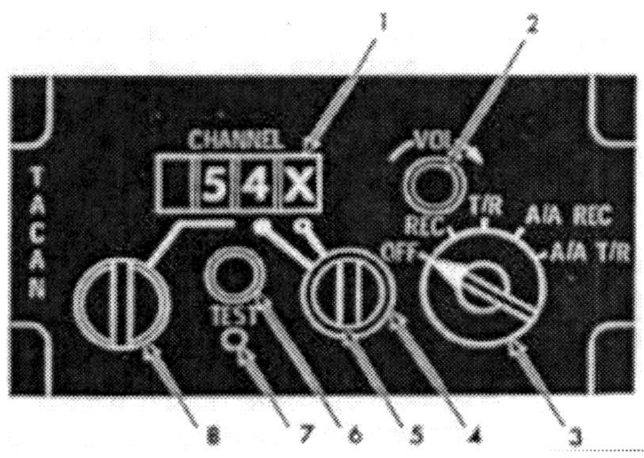

1 CHANNEL DIGITAL DISPLAY
2 STATION IDENTIFICATION VOLUME CONTROL
3 MODE SELECTOR CONTROL
4 X-Y CHANNEL SELECTOR CONTROL
5 UNITS CHANNEL SELECTOR CONTROL
6 R/T STATUS INDICATOR LIGHT
7 TEST SWITCH
8 TENS CHANNEL SELECTOR CONTROL
9 TACAN CONTROL TRANSFER SWITCH - PILOT'S AND RSO'S INSTR PANELS

F200-201(b)

Figure 1-76

NOTE

Some aircraft, including the SR-71, cannot transmit bearing. Range is available to both aircraft.

Volume Control Knob

Rotating the volume (VOL) control clockwise increases the TACAN identification signal audio level. The TACAN/ILS interphone switch must be pulled and the DISPLAY MODE SEL switch must be in ANS, INS, or TACAN ADF to hear the TACAN identifier.

Self-Test Push-button

A self-test of the TACAN equipment and its interface with the HSI and BDHI is initiated by actuating the TEST push-button. The adjacent indicator light flashes momentarily when the TEST switch is pressed.

NOTE

If the TEST light illuminates other than at the start of self-test, a TACAN fault exists.

A self-test can be terminated immediately by rotating a channel selector or the mode selector switch to another position.

NOTE

- o Bearing and/or distance indications may still be present when the TEST light is on. Such indications may be accurate, but are unreliable.
- o Be prepared for failure of TACAN equipment if the TEST light illuminates.

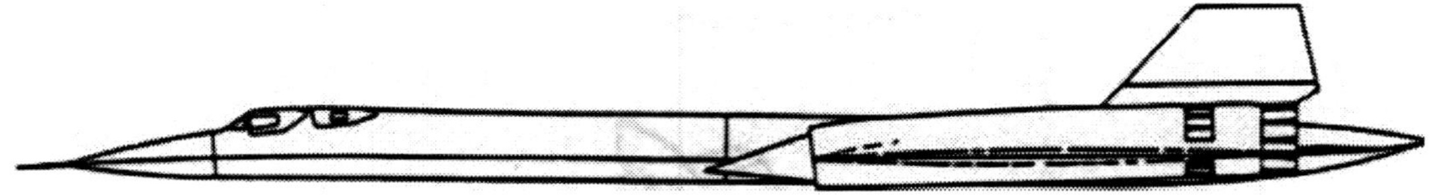

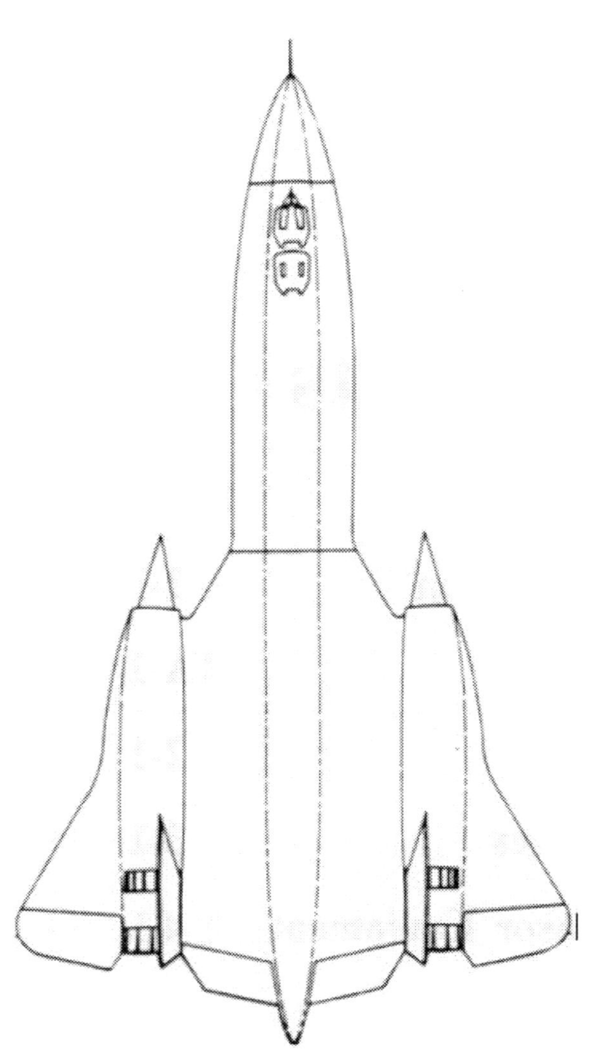

THIS PAGE INTENTIONALLY LEFT BLANK OR STILL CLASSIFIED.

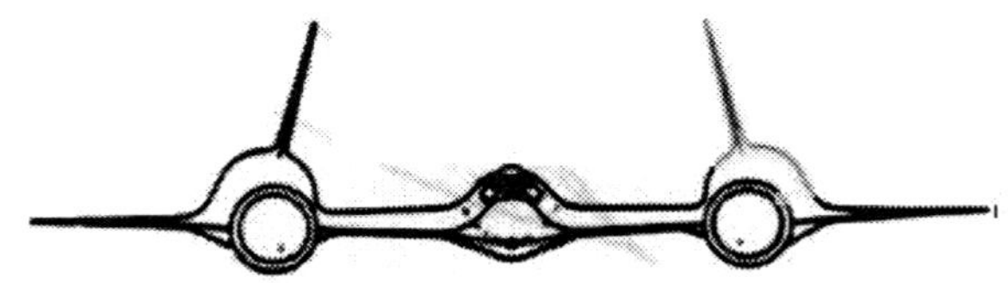

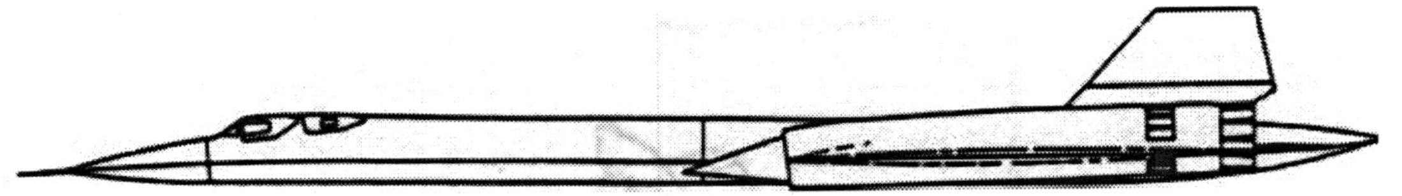

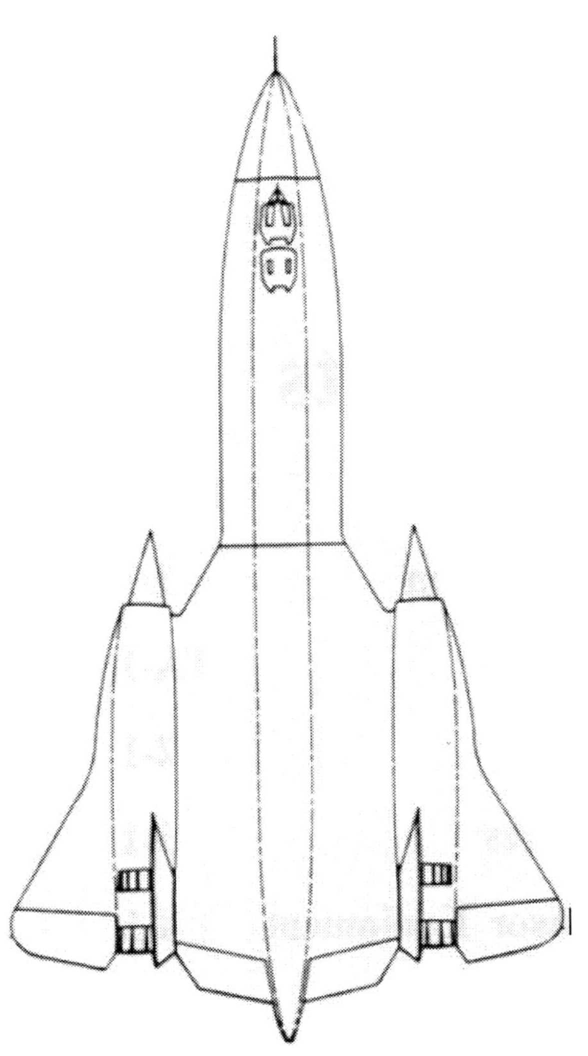

THIS PAGE INTENTIONALLY LEFT BLANK OR STILL CLASSIFIED.

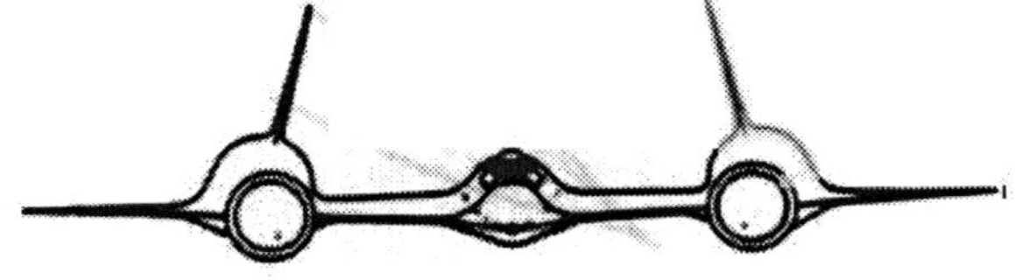

Small recesses on the outside of each side of the canopies provide lifting points so the canopies can be opened from outside. A latch on top of the nitrogen counterbalance cylinder engages when the canopy is fully open and holds the canopy in that position until the latch is released by pressing the latch release lever. The two canopies are independent in operation, except for the external jettison feature.

CAUTION

- Canopies shall be opened or closed only when the aircraft is stopped. To prevent wind forces from shearing the canopy hinge pins, hold canopy securely when opening
- Maximum taxi speed with the canopy open is 40 knots. Gusts should be included as part of the 40 knot limit.

Canopy Latching Mechanism

Each canopy is latched closed by four hooks (two in each canopy sill). The canopy is latched and unlatched by a handle at the forward right side of each cockpit. Moving the handle rotates a transverse torque tube behind the seat to simultaneously position all four hooks. With the canopy closed, forward movement of the handle latches and locks the canopy, while aft movement releases the hooks and unlocks the canopy. The ejection or canopy jettisoning sequence unlatches the canopy by gas pressure from the canopy unlatch thruster behind the seat.

The aft canopy latching torque tube is mechanically connected to a cockpit air shutoff valve in the air-conditioning system. To conserve cooling air for electronic equipment, this valve shuts off air flow to both cockpits when the canopy latch handle is in the aft position. This valve is also operated by the RSO's Cockpit Air handle. The forward (off) position of the RSO's Cockpit Air handle will shut off air to both cockpits even if the aft canopy latch handle is forward. Refer to Environmental Control System Controls, Cockpit Air Valve, this section.

CAUTION

If either canopy is open, the aft canopy latch handle must be in the aft position or the cockpit air handle must be in the forward (off) position for adequate equipment cooling. Otherwise, most of the cooling air would exit through the cockpit openings instead of the bays.

Canopy Counterbalance System

Normal opening and closing of the canopy is assisted by gaseous nitrogen pressure from the canopy counterbalance cylinder at the left rear of the seat. Without counterbalance assistance, the canopy is difficult to raise (requiring approximately 112 pounds of force), and can drop with sufficient force to injure personnel. A counterbalance system pressure gage is located above the nitrogen cylinder and the pointer should indicate in the green if system pressure is normal.

WARNING

To avoid injury, verify that counterbalance pressure indication is normal before pushing the canopy latch release.

CAUTION

Do not rotate the canopy upward beyond its normal full-open position or canopy shear pins may be severed.

Canopy Seal System

An inflatable seal is installed along the edge of each canopy frame. The seal is inflated by engine bleed pressure to provide an airtight seal between the canopy and the canopy sills and windshield. A pressure regulator and control valve is provided in each cockpit. The seal selector lever is located in the right forward corner of each cockpit and is used to inflate or deflate the seals. Seal pressure is regulated to no more than 23 psi.

CANOPIES, CANOPY CONTROLS and

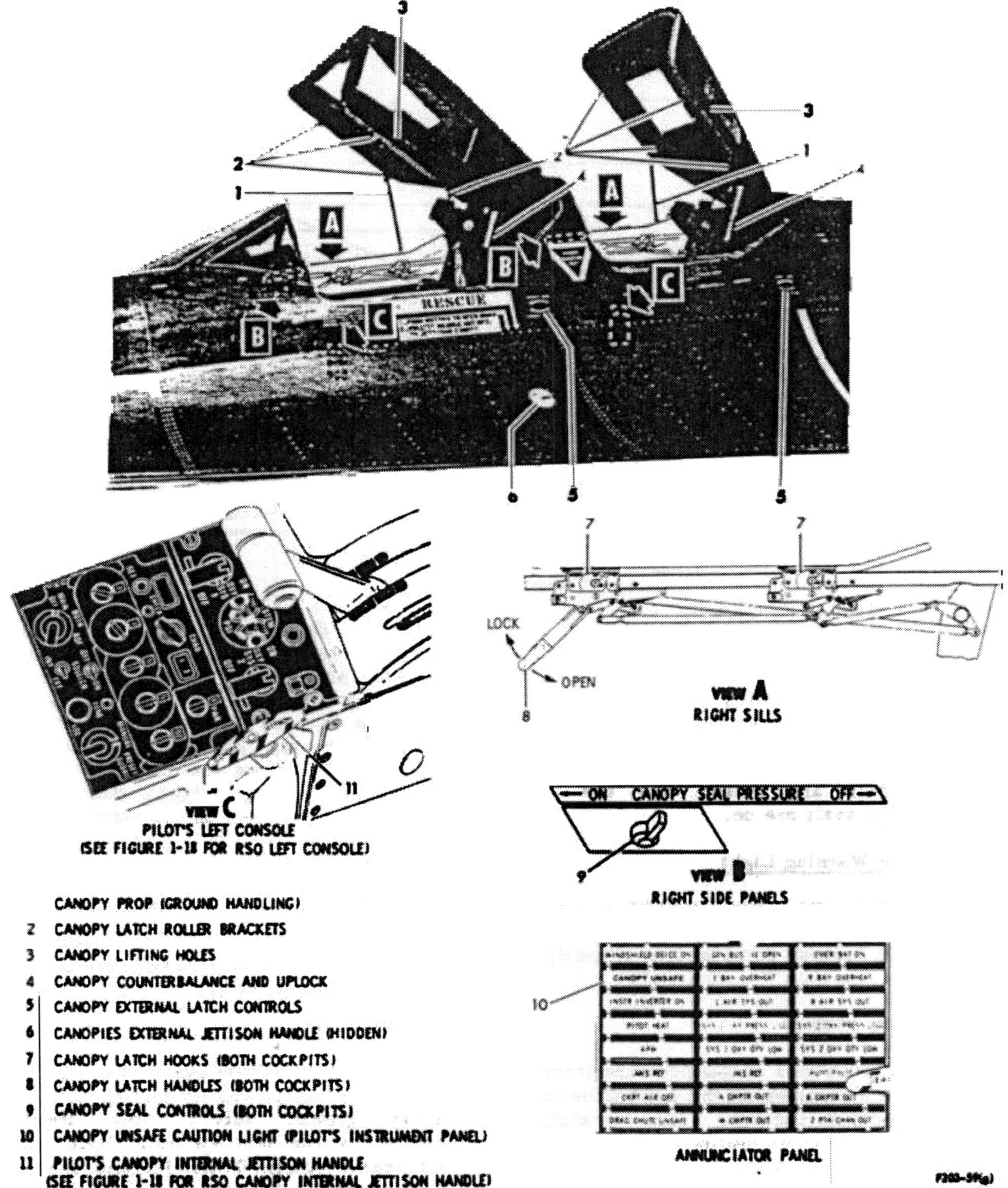

RESCUE
LOCK
OPEN
VIEW A
RIGHT SILLS
VIEW C
PILOT'S LEFT CONSOLE
(SEE FIGURE 1-18 FOR RSO LEFT CONSOLE)
ON CANOPY SEAL PRESSURE OFF
VIEW B
RIGHT SIDE PANELS
ANNUNCIATOR PANEL
CANOPY PROP (GROUND HANDLING)
2 CANOPY LATCH ROLLER BRACKETS
3 CANOPY LIFTING HOLES
4 CANOPY COUNTERBALANCE AND UPLOCK
5 CANOPY EXTERNAL LATCH CONTROLS
6 CANOPIES EXTERNAL JETTISON HANDLE (HIDDEN)
7 CANOPY LATCH HOOKS (BOTH COCKPITS)
8 CANOPY LATCH HANDLES (BOTH COCKPITS)
9 CANOPY SEAL CONTROLS (BOTH COCKPITS)
10 CANOPY UNSAFE CAUTION LIGHT (PILOT'S INSTRUMENT PANEL)
11 PILOT'S CANOPY INTERNAL JETTISON HANDLE
(SEE FIGURE 1-18 FOR RSO CANOPY INTERNAL JETTISON HANDLE)

Figure 1-77

REAR VIEW PERISCOPE

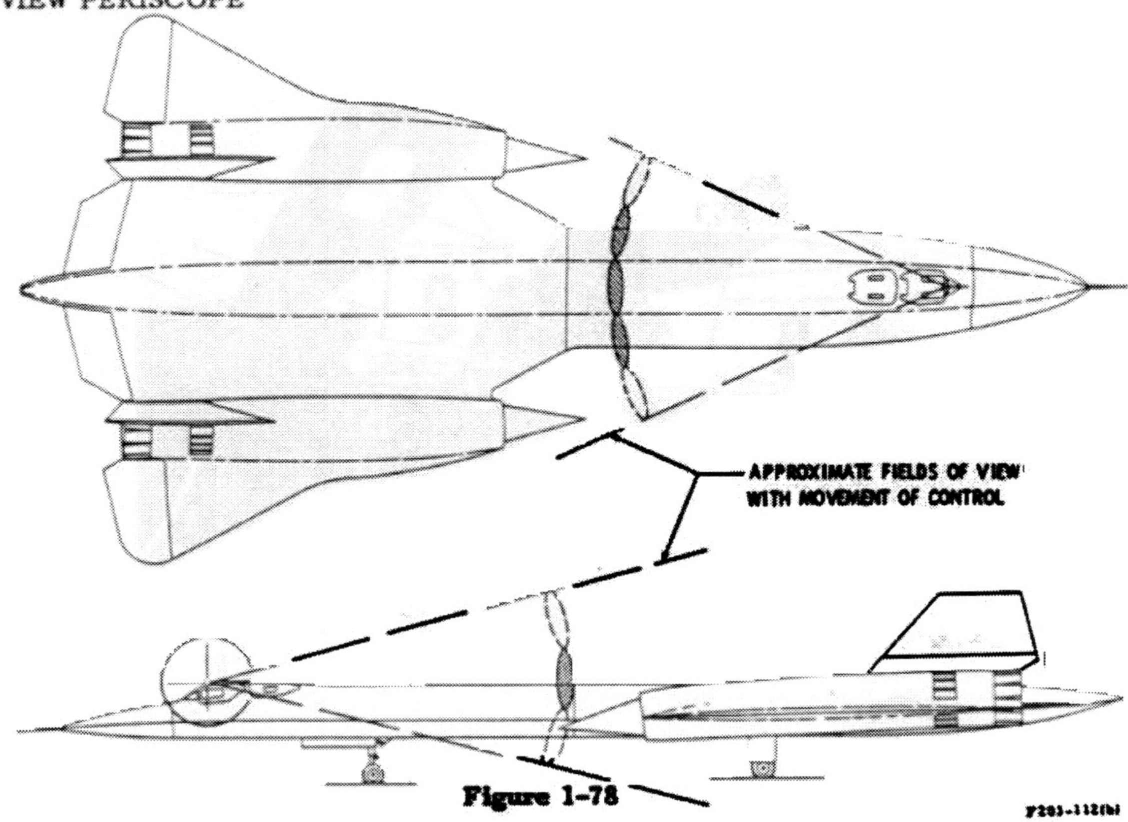

Figure 1-78

CAUTION

- Do not inflate seals unless the canopy is latched closed, or damage to seals may occur.
- The OBC must be off unless both canopies are closed and the canopy seals are on.

Canopy Unsafe Warning Light

Illumination of the CANOPY UNSAFE annunciator caution light indicates that one or both canopies is not latched down and/or properly sealed.

Canopy Jettsion System

The canopy is jettisoned by gas pressure from a canopy unlatch thruster on the aft bulkhead of the respective cockpit. The unlatch thruster may be fired by pulling the ejection seat D-ring, or the CANOPY JETTISON T-handle located outboard of the left forward corner of each ejection seat.

When the unlatch thruster fires, it rotates the canopy hooks to the unlatch position and releases the canopy. The thruster charge is then ported to the canopy-removal thruster and seal-hose cutter. The seal-hose cutter severs the canopy seal hose, and the canopy-removal thruster forces the canopy up and aft, shearing the hinge attach points and thrusting the canopy clear of the aircraft.

The canopy can also be removed in-flight by manually unlatching it and pushing it up into the airstream. If the cockpit is pressurized when the canopy latches are released, the canopy will be blown upward into the airstream by cockpit pressure.

WARNING

Do not enter or leave the cockpit unless ground safety pins are installed in the seat ejection D-ring, the seat ejection T-handle, and the canopy jettison T-handle.

Canopy External Latch Controls

Individual external latch release fittings are flush-mounted on left side of the fuselage, just below the canopy hinge points. Fittings accept a 1/2-inch-square bar extension to open the canopies from outside the aircraft. The canopy must be raised manually after releasing the latch hooks externally.

Canopy External Jettison System

Both canopies may be jettisoned on the ground by the external jettison system. An external jettison handle is located under an access panel on top of the left chine, at the aft end of the forward cockpit. When the external jettison handle is pulled, both canopies jettison in sequence, the forward canopy first, the aft canopy 1 second later. A long jettison cable allows the person pulling the handle to be well clear of the fuselage during jettison.

Cockpit Sunshades (Bat Wings)

Two 12.5" x 8" sunshades (bat wings) in the front cockpit are used to block the intense sunlight prevalent at high altitude, and to reduce cockpit light reflections at night. A sunshade is located on each canopy side-frame, below the side windows, and is attached to a rod mounted 5" aft of the forward end of the canopy. Each sunshade is individually adjustable: the sunshade can be extended along the rod, the rod elevated, and the sunshade rotated. They may be joined together to form an extension of the glare shield (refer to Night Flying, Interior Light Reflections, Section VII). A lever to adjust rotational friction is located on the top, inboard portion of the left bat wing; similar levers at both rod attachment points adjust elevation friction and allows the bat wings to be joined together. The right bat wing may be folded lengthwise and stowed above the PVD projector. Both bat wings have extension panels, located on the lower side, which increase the area by 50% when pulled out. The altimeter correction card is attached to the outboard side of the right bat wing.

REAR-VIEW PERISCOPE

A manually extended rear-view periscope is mounted in the top of the pilot's canopy. It is moved by using a white nylon handle mounted on the aft side of the viewing tube. Pushing the handle left unlocks the tube, allowing the periscope to be extended. Pushing the tube upward to a spring-detented position makes the rear view available. Cockpit pressure assists extension and resists retraction. The cone of view is approximately 10 degrees across; however, head movement extends the viewing cone to approximately 30 degrees total angle. (See Figure 1-78.) When extended, the periscope can be rotated horizontally to move the center of the viewing arc up to 10 degrees from the aft centerline. The lens provides a 2 to 1 reduction ratio.

MAP PROJECTORS

A strip-film projector is installed in each cockpit to provide a strip map of the route to be flown with mission data and emergency information. (See Figure 1-79.) Film strips are provided for each mission. The pilot and RSO projectors are of different design and are independently controlled.

PILOT'S MAP PROJECTOR

The pilot's map projector is located in the bottom center of the instrument panel. Projector controls are located on the panel borders of the 4-1/4 by 4-1/4 inch display and on the map projector control panel on the left console, forward of the throttle quadrant. The projector can display up to 25 feet of 35 mm film.

Illumination Control

The illumination control slide switch at the bottom edge of the navigation map display is a combination power on-off switch and illumination rheostat. When the control is in the left-hand detent, the projector is deenergized; moving the control down and sliding it to the right energizes the projector lamp, blower, and film-drive motors. Moving the

MAP PROJECTORS

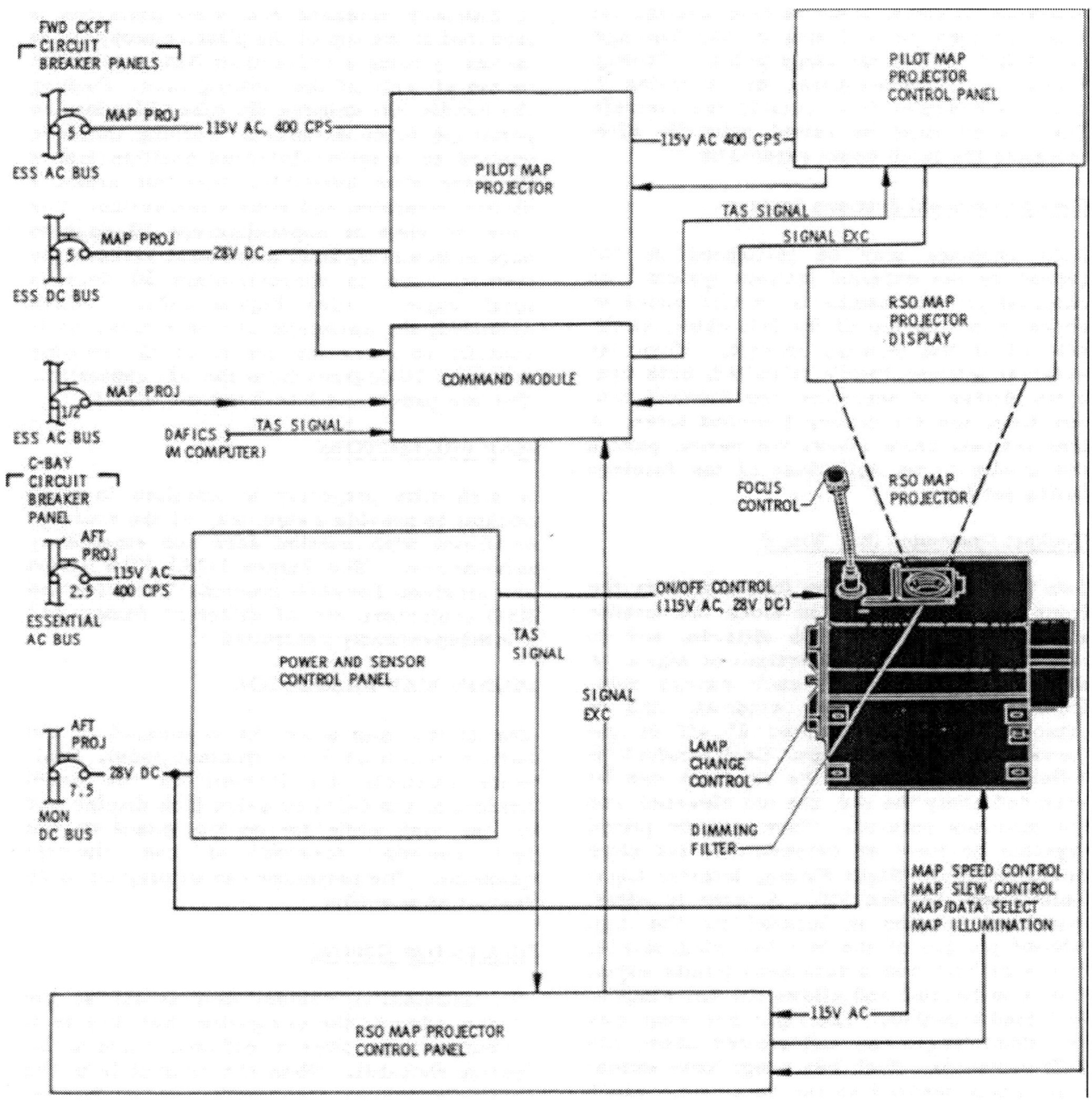

Figure 1-79

control from DIM towards BRT increases the display brightness.

Projector Lamp DIM-BRT Switch

A three-position projector lamp dimming toggle switch is provided on the bottom edge of the navigation map display, to the right of the illumination control. In DIM, low voltage is applied to the navigation map projector lamp and the range of image brightness is reduced. In BRT the maximum range of image brightness is available. The unmarked middle position provides medium brightness.

Lamp Change Control

The lamp change control on the right border of the nav map display is labeled LAMP CHG. The lever switch has detent positions (labeled 1, 2, and 3) to provide a choice of three bulbs. If a bulb burns out, another may be selected by moving the lever to the right, out of detent, and repositioning it to one of the other two marked positions.

Slew Control Switch

The rotary slew control switch on the map projector control panel, is spring-loaded to the center (off) position. Counterclockwise rotation of the switch causes reverse slew and clockwise rotation causes forward slew of the film strip. The rate of film movement is variable in both directions. Increased rotation of the switch increases the slew speed.

NOTE

> To return to a specific position, reverse slew, then slew forward and stop on the point of interest. This eliminates any delay in automatic film drive advance.

Drive Switch

The drive switch on the map projector control panel, selects either manual rate control (MANUAL MACH), automatic map rate synchronized to true airspeed (AUTO), or OFF (film stop).

Map Rate Control and Indicator

The map rate control on the map projector control panel, is a single-turn potentiometer calibrated in Mach. This control permits manual synchronization between map film speed and aircraft groundspeed when the drive switch is in the MANUAL MACH position. The control has a dial face with seven major outer divisions, representing Mach, and a single pointer. The inner O marking with + and - arrows provides a means of biasing the map rate when the drive switch is in AUTO. For normal operation the pointer should be in the O position to provide zero bias. Counterclockwise rotation from the O position decreases map film speed; clockwise rotation increases map film speed. The control has no effect when the drive switch is in OFF. The projector is intended for use with map films scaled at 365 nautical miles per inch.

CAUTION

> Attempting to turn the control clockwise past 35 or counterclockwise past zero will damage internal stops and cause a loss in dial reference.

PILOT'S MAP PROJECTOR OPERATION

1. Illumination control - Slide right to turn on and obtain desired brillance.
2. Map DRIVE switch - AUTO or MANUAL MACH, as desired.
3. RATE control - O position for AUTO or as required for MANUAL MACH.

   Adjust + or -, as necessary.
4. SLEW control - REV or FWD, as required.

RSO's MAP PROJECTOR

The RSO's map projector system includes the projector, projector controls, and a 9 by 9-inch display screen. Some map projector controls are located on the viewsight control panel (Figure 4-33 without S/B R-2538, Figure 4-33A with S/B R-2538). When the RCD is installed, the map projector is mounted on the floor and projects onto a hinged viewing screen below the RCD. When the RCD is not installed, the projector is mounted horizonally at a higher position on the bottom of the instrument panel, and projection is then onto the display screen mounted in place of the RCD.

NOTE

> The map film continues to move when data film is selected for viewing, even though the film is not projected. The rate of movement always corresponds to 250 nautical miles per inch.

Projector Power Switch

The projector power switch is located on the PWR & SENSOR control panel. A white illuminated legend PROJ appears in the top half of the switch. A green ON legend, that illuminates alternately when the pushbutton is depressed, and a red non-functional FAIL legend occupy the lower quarters of the switch.

Lamp Change Control

A rotary lamp change control is located on the side of the projector case. The control has three detented positions for different bulb selections. When the RCD is installed, the control is located on the right side of the projector case and may be operated by the RSO's foot when a bulb burns out. When the RCD is not installed, the control is located on the left side of the projector case and can be reached by hand.

Projector Focus Control

Map projector focus is adjustable by turning the focus control knob flexible stem on top of the map projector assembly.

Map Drive Switch

A map drive switch on the viewsight control panel provides selection of either manual map rate control (MAN-TAS), or automatic map rate (AUTO) synchronized to true airspeed.

Map Rate Control and Indicator

The map rate control on the viewsight control panel is a two-turn potentiometer, calibrated to true airspeed, which permits synchronization between map film speed and aircraft groundspeed when the map DRIVE switch is in MAN - TAS. The control has a dial face with 10 peripheral divisions labeled 1 through 0 (10), a long and a short pointer, and a movable knurled bezel to move the pointers. The short pointer indicates thousands of knots and the long pointer indicates hundreds of knots. The inner O marking with + and - arrows provides a reference to bias the map rate when the DRIVE switch is in AUTO. For normal automatic operation both the short and long pointers should be on the O inner scale position (1 on outer scale) to provide zero bias. Counterclockwise rotation decrases map film speed; clockwise rotation increases map film speed.

Slew Control Switch

A rotary control switch, labeled SLEW, is located on the viewsight control panel. It controls the direction and speed of the film strip that has been selected by the film-select switch. Counterclockwise rotation of the switch toward REV (REW with video viewsight) causes reverse slew and clockwise rotation toward FWD causes forward slew. Increasing switch rotation causes an increasing rate of film movement in the indicated direction up to 0.6 ips for map film and 0.9 ips for data film. At the end of switch travel an electrical contact is made to cause rapid slewing of 3.5 ips for map film and 6 ips for data film. (Image speed across projector screen is nine times the film speeds.)

NOTE

> To return to a specific position, reverse slew, then slew forward and stop on the point of interest. This eliminates any delay in automatic film drive advance.

Film Select Switch

A two-position film-select toggle switch, labeled SELECT, is located on the viewsight control panel. The DATA and MAP positions select the type of film to be projected.

Illumination Control Switch

A three-position illumination control switch, labeled ILLUM, is located underneath the right end of the optical viewsight control panel and on the front of the video viewsight panel. The switch is springloaded to the center off position. A motor closes or opens an iris in the projector to change the degree of illumination; when held in the forward (DIM) position or the aft (BRT) position on the optical viewsight panel, or in the down (DIM) position or up (BRT) position on the video viewsight panel.

Dimming Filter

A manually positioned dimming filter can be placed over the projector lens to reduce brightness during night operation.

RSO MAP PROJECTOR OPERATION

1. PROJ power switch on PWR & SENSOR panel - Press to illuminate ON.
2. ILLUM switch - BRT or DIM, as desired.
3. SELECT switch - DATA or MAP, as desired.
4. DRIVE switch - AUTO or MAN-TAS as desired.
5. MAP rate knob - as desired.

If no display:

1. ILLUM switch - BRT as required.
2. Lens dimming filter - Check open.
3. Projector slew control - Slew to assure dark leader is not blocking light.
4. Lamp change control - Rotate to new position.

If image is out of focus:

1. Focus adjustment - Set.

LIGHTING EQUIPMENT

EXTERIOR LIGHTING

Landing/Taxi lights and Switch

A 1000 watt landing light and a 450 watt taxi light are mounted on the nose gear strut. The lights are controlled by a luminous three-position landing and taxi lights switch on the pilot's left instrument side panel. The switch positions are LAND LT (up), TAXI LT (down), and OFF (center). The switch is ineffective when the nose gear is retracted, as the uplock switch prevents power being applied to the lights. Power for the lights is obtained from the essential ac bus through the LDG LT and TAXI LT circuit breakers on the pilot's right console.

CAUTION

Use the taxi light if lighting is required during ground operation or after landing. The landing light burns out without airstream cooling.

Anti-Collision/Fuselage Lights and Switch

Two combination retractable anti-collision and fuselage lights are located at the top and bottom of the fuselage, near the middle. The lights are controlled by a three-position toggle switch on the pilot's lighting panel. The switch positions are: ANTI COLLISION (forward), FUS (aft), and OFF (center). In ANTI COLLISION, the lights extend, illuminate red, and rotate at 45 rpm (which produces 90 flashes per minute). In FUS (fuselage), the lights are retracted and illuminate white. The lights are retracted and off when the switch is in OFF. Three-phase essential ac powers the lights through the ANTI COLL LTS circuit breaker on the pilot's right console.

Tail Lights and Switch

A white tail light is located on the top and bottom of the tail cone. The lights are controlled by a three-position toggle switch, labeled TAIL LT, on the pilot's lighting panel. In STEADY (aft), the lights illuminate continuously. In FLASH (forward), the lights flash 85 times per minute. The center position is off. Power is furnished by the essential ac bus through the TAIL light circuit breaker on the pilot's right console.

Fuselage and Tail Lights Intensity Switch

A FUS & TAIL lights intensity switch is located on the pilot's lighting panel. The positions are BRT (forward) and DIM (aft). The switch controls the light intensity of the tail lights and also the anti-collision lights when they are operated as fuselage lights.

FORWARD COCKPIT LIGHTING

Integral and edge lighting for the instruments and consoles is controlled by the lighting panel on the left console. The four instrument lights circuit breakers on the lighting panel (LH, RH, FUEL CONT, and FUEL CONT TEST) are only operative if the ac hot bus INSTR light circuit breaker on the right console is closed. Front cockpit interior lighting also includes: one floodlight located above each side console, and one floodlight located above the circuit breaker panel on each side; a thunderstorm light located on the canopy on each side; a movable spotlight secured on the console on each side; flex point lights mounted on the canopy sill on each side; two emergency instrument lights located on the instrument panel glareshield; and one combination floodlight/emergency instrument light, above the right console, which illuminates the right instrument side panel.

ADI Light Rheostat

The ATT DIR IND rheostat on the lighting panel, controls light intensity of the attitude director indicator only. Power is from the ac hot bus through the FUEL CONT TEST instrument lights circuit breaker.

Instrument Lights Rheostat

The INSTR LTS rheostat on the lighting panel, controls instrument panel light intensity. Power is from the ac hot bus through the LH and RH instrument lights circuit breakers.

Console Lights Rheostat

The CONSOLE LTS rheostat on the lighting panel, controls light intensity of the right and left consoles. When the rheostat is not OFF, all caution and warning lights (including annunciator panel lights) are dimmed, except the nacelle fire warning lights and the landing gear handle warning light. Power is from the essential ac bus through the PANEL lights circuit breaker on the right console.

Floodlights Rheostat

The FLOOD LTS rheostat on the lighting panel, controls light intensity of the floodlights. Power is from the essential ac bus through the FLOOD lights circuit breaker on the right console.

Thunderstorm Lights Switch

The THUNDERSTORM lights switch on the lighting panel has two positions: ON (forward) and OFF (aft). Power is from the essential dc bus through the SPOT light circuit breaker on the left console.

Spotlights

The spotlights (utility lights) incorporate a rheostat on the aft end to control light intensity. A push-button switch bypasses the rheostat to provide maximum intensity. Power is from the essential dc bus through the SPOT light circuit breaker on the left console.

Flex Point Lights

The bodies of the two flex point lights are fixed to the left and right canopy sills. A flexible shaft, attached to each light body, contains a hood used to vary the size of the light beam. The hood is covered with velcro pile for stowing the flexible shaft under the canopy sill when not in use. A rheostat at the aft end controls light intensity. A push-button switch bypasses the rheostat to provide maximum intensity. Power is from the essential dc bus through the SPOT light circuit breaker on the left console.

Emergency Instrument Lights

The emergency instrument lights are automatically energized if ac hot bus power is lost or the ac hot bus INSTR light circuit breaker is open. Power is from the essential dc bus through the EMER INSTR light circuit breaker on the left console.

AFT COCKPIT LIGHTING

Integral and edge lighting for the instruments and consoles are controlled from the lighting panel on the left console. The left console PNL and LGD lighting circuit breakers on the lighting panel are only operative if the essential ac PNL L circuit breaker on the right console is closed. The right console TEST & BRT, PNL, and LGD lighting circuit breakers on the lighting panel are only operative if the essential ac PNL R circuit breaker on the right console is closed. Aft cockpit interior lighting also includes: one floodlight located above the left console; two floodlights above the right console; two floodlights above the circuit breaker panel on the right side; a floodlight on the canopy on each side; a spotlight secured on the left console; a spotlight secured on the right console; and flex point lights mounted on the canopy sill on each side.

Instrument Lights Rheostat

The INSTR rheostat on the lighting panel controls instrument panel light intensity. Power is from the essential ac bus through the INST lights circuit breaker on the right console.

L and R Console Lights Rheostats

The L CONSOLE and R CONSOLE rheostats on the lighting panel control legend light intensity for the left and right consoles, respectively. Power is from the essential ac bus through the respective PNL and LGD circuit breakers.

Floodlights Rheostat

The FLOOD rheostat on the lighting panel controls light intensity of the floodlights. Power is from the essential ac bus through the FLD lights circuit breaker on the right console.

Spotlights

The spotlights (utility lights) are identical to the spotlights in the forward cockpit. Power is from the essential dc bus through the SPOT LTS circuit breaker on the right console.

Flex Point Lights

The flex point lights are identical to the flex point lights in the forward cockpit. Power is from the essential dc bus through the SPOT LTS circuit breaker on the right console.

ENVIRONMENTAL CONTROL SYSTEM

These aircraft operate in an extremely adverse speed and altitude environment. Ram air temperatures may exceed 400°C at design airspeed, and ambient static air pressure can be less than 1/3 psi near the limit altitude. The external skin surfaces are painted black to radiate heat. Special insulating materials are used extensively to minimize temperature build-up within the aircraft. Pressurized and conditioned air must be supplied to the crew and equipment to provide a suitable environment. Both crewmembers wear full pressure suits for protection against cockpit depressurization or bailout at high altitude.

Air-conditioning includes two identical and parallel air-cycle compressor turbine refrigeration systems. Each is supplied by ninth-stage bleed air from one engine. The bleed air is regulated to no more than 26 psi, and is cooled by air-to-air and air-to-fuel heat exchangers before it is ducted to the air-cycle units. A temperature of about 200°F is maintained while below 44,000 feet. At higher altitudes, the air receives the maximum possible cooling. The bleed air supply is shut off if pressure can not be maintained above 5 psi. Cold air manifolds from the refrigeration turbines supply the chine and mission bays and the electrical, radio, and navigation bays. Conditioned air, which is temperature-controlled by the pilot, is used to pressurize and ventilate the cockpits and pressure suits. Cockpit exhaust air cools the nose and radar equipment and supplements cooling of the chine and equipment bays. Regulated bleed air supplies windshield deicing and, when mixed with conditioned air, windshield defogging. Regulated bleed air is also used to pressurize the canopy seals, some of the nose radar equipment, and the windshield rain removal system reservoir. The inlet control pressure ratio transducers and the inlet forward bypass actuator LVDT's, which are located in high temperature environments in the nacelles, are also cooled by regulated bleed air. Engine inlet ram air and engine fuel are used as heat sinks for the air-conditioning system. The ram air is exhausted within the nacelles after passing through the air-to-air heat exchangers. Heat sink fuel is either returned to the engine fuel supply system or diverted to tank 4 after use in the air-to-fuel heat exchangers. Refer to Figure 1-80, Environmental Control System, and Figure 1-39, Fuel Heat Sink System, for schematic diagrams.

Either of the two refrigeration systems can maintain a suitable cockpit environment and cool the ANS and essential radio and flight equipment in the E and R Bays. However, if one of the refrigeration systems fails at high supersonic speeds, cooling air for the nose, chine, and mission bays must be turned off and use of most of the equipment in these spaces should be discontinued.

The air pressure available at the chine and bay equipment boxes is substantially the same as that provided in the cockpits; however, pressure is reduced by orifices in individual equipment supply or bay vent manifolds to regulate flow to individual equipment. Normally, the cooling air temperature is automatically controlled to -30°F when at supersonic flight altitudes. Selecting FULL COLD on the manifold temperature switch can result in much lower supply temperatures in some cases. With the manifold temperature switch in AUTO, equipment cooling air temperature is approximately +37°F when below 41,000 to 44,000 feet, to prevent ice formation in the cold air manifold.

Water separators are provided in the supply manifolds to the ANS and bay areas. Although they have sufficient capacity for normal flight operations, they may not be adequate for sustained periods of ground operation when high humidity conditions exist. Suitable ground support equipment can be connected when prolonged ground operation is anticipated. The cockpit air supply is not dehumidified.

COCKPIT PRESSURIZATION SCHEDULES AND TEMPERATURE SELECTION

The crewmembers operate within a sealed and specially insulated compartment which contains the two cockpits. The compartment can be pressurized at either a 10,000 FT (foot) or 26,000 FT schedule. Each of the schedules can be used without restriction; however, the 26,000 FT schedule is usually preferred since it allows more airflow through the cockpits and enhances cockpit and bay cooling. An automatic control restricts the maximum rate of pressure change to approximately 5000 fpm when the selection is altered. Refer to Environmental Control Systems Controls, Cabin Pressure Switch, this section.

The cockpits remain essentially unpressurized while below 26,000 to 28,000 feet pressure altitude with the 26,000 FT schedule. Cockpit pressure is then maintained at 26,000 feet at all higher flight altitudes. With the 10,000 FT schedule selected, the cockpits remain substantially unpressurized while below 10,000 feet. They maintain approximately 10,000 feet until aircraft altitude exceeds

28,000 feet, then a pressure 5 psi greater than ambient at higher altitudes. Refer to Figure 1-81. Note that at 25,000 feet aircraft altitude, for example, cockpit pressure altitude would be only slightly less than the flight altitude with the 26,000 FT pressure schedule, and would equal 10,000 feet with the 10,000 FT schedule. The 10,000 FT schedule can be used at subsonic flight altitudes if a crewmember wishes to open the helmet faceplate temporarily.

Cooling capability of the air-conditioning system will be reduced somewhat during descents from supersonic cruise if the 10,000 FT schedule is used. This is due to increased back pressure at the cooling turbines and the resultant decrease in cooling capability. To minimize the effect of reduced cooling, the cockpit AUTOTEMP control should be turned toward COLD before descending from high supersonic cruise.

NOTE

> Use of the 26,000 FT schedule is recommended if air-conditioning system difficulties result in high cockpit temperatures.

A safety relief valve opens automatically to maintain 5.4 psi differential pressure if the normal pressure regulator valve malfunctions. The safety valve can also be fully opened by the PRESS DUMP switch, on the pilot's left instrument side panel. Operation of this solenoid dumps cockpit pressure to ambient static pressure very rapidly.

The pilot controls cockpit air temperature by adjustment of his automatic controls or by manually selecting the proportion of regulated hot air that is mixed with the cockpit cold air supply. Suit ventilation air temperature is also controlled by the pilot, but suit vent flow is controlled individually. Individual controls are provided for helmet faceplate heat settings and for oxygen system selection. The pilot controls the refrigeration system shutoff valves and the cockpit pressure dump valve. The RSO controls the cockpit air shutoff valve.

ENVIRONMENTAL CONTROL SYSTEM CONTROLS

Refer to the following discussions in this section: Windshield, for a description of the Windshield Hot Air De-Icing/Rain Removal System; Canopy, for the canopy latching system and the canopy seal; and Life Support Systems, for the suit and helmet protective systems (including suit and faceplate heat control).

Refrigeration Switches

The two-position REFRIG switches, on the pilot's left instrument side panel, control the engine bleed air supply valves for each refrigeration system. The valves are located downstream from the air-to-air and air-to-fuel heat exchangers in the "heat sink packages" in each wing, inboard of the nacelles. See Figure 1-80.

The ON (up) position of either the L (left) or R (right) refrigeration switch deenergizes the respective control circuit and allows the corresponding system shutoff valve and pressure regulator to open.

Each valve is spring-loaded closed and when its control circuit is deenergized, the valve opens fully if at least 5 psi bleed air pressure is available from its engine. The OFF (down) position of either switch energizes a solenoid in the corresponding shutoff valve and closes the valve, stopping all of the air supplied for that pressurization and air conditioning system. The solenoid is also energized automatically by action of either of two thermal switches if temperatures exceed 365 $\pm 15^\circ$F at the compressor inlet or intercooler outlet positions of the systems air cycle refrigeration machine. The shutoff valve reopens automatically when lower temperatures are sensed. An L or R AIR SYST OUT caution light on the annunciator panel illuminates when the corresponding shutoff valve is closed.

The shutoff valve solenoids are powered by the essential dc bus through the AIR SOV CONT circuit breaker on the pilot's left console.

ENVIRONMENTAL CONTROL SYSTEM

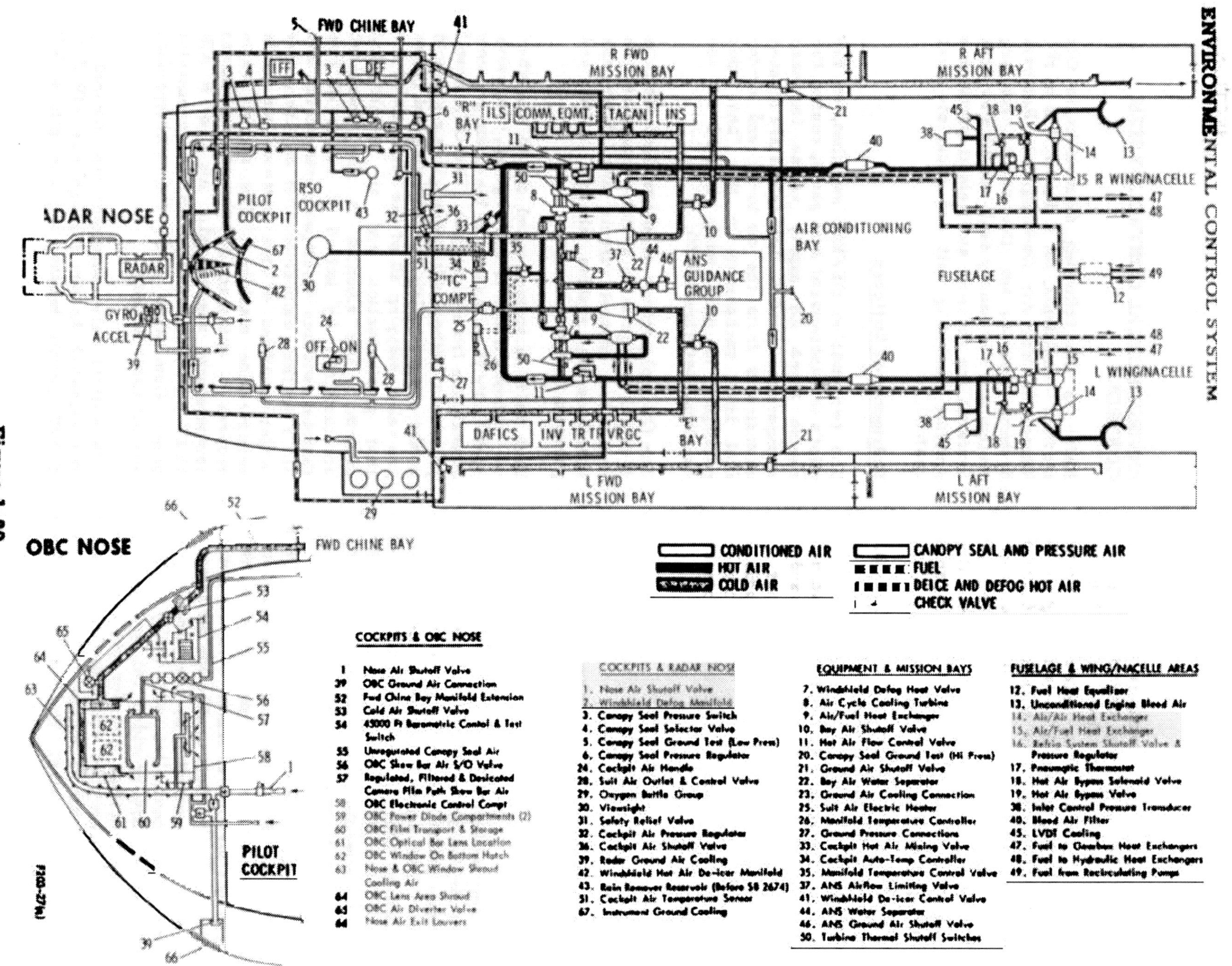

Figure 1-80

COCKPIT PRESSURIZATION SCHEDULE

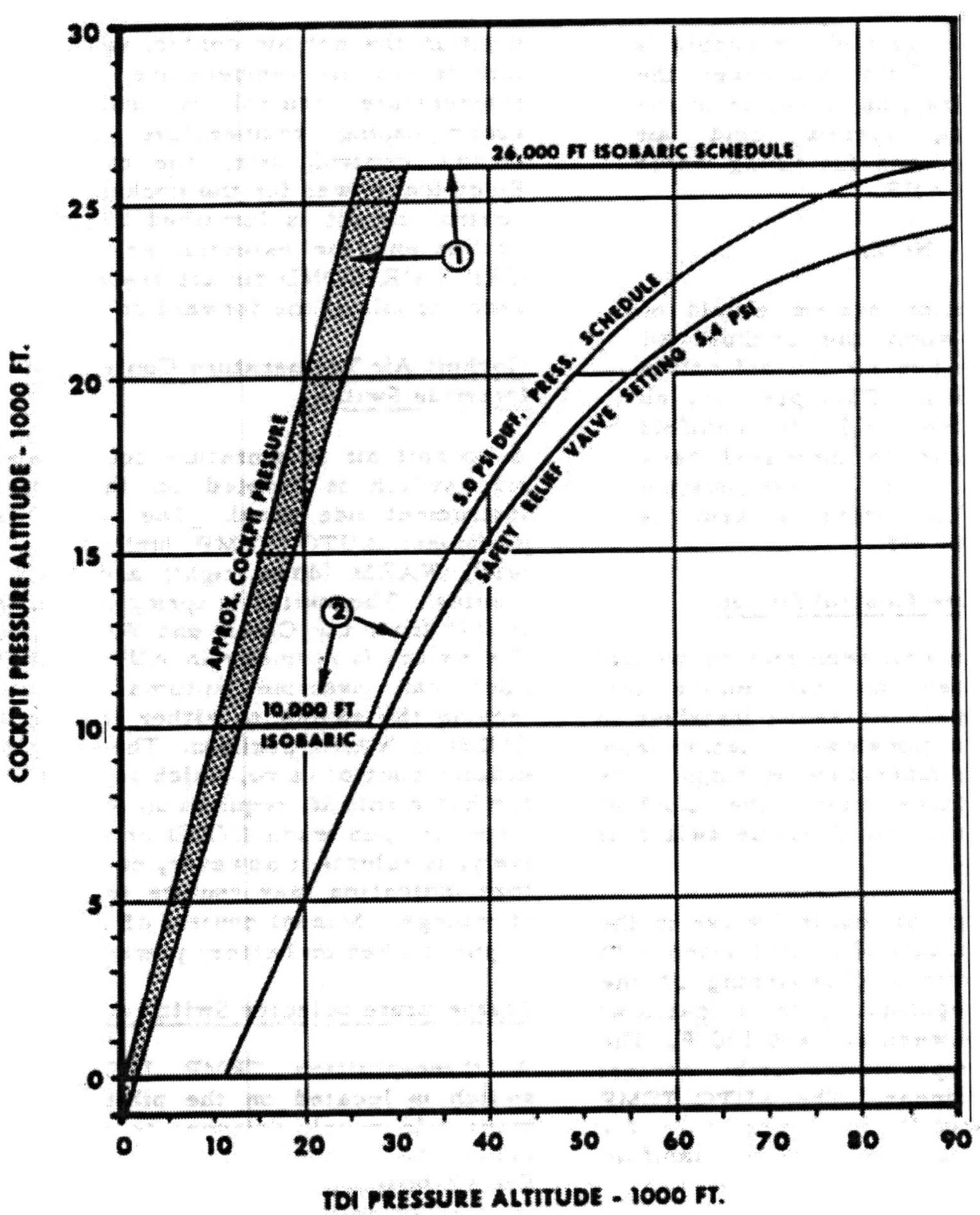

Figure 1-81

CAUTION

Both refrigeration switches must be OFF during ground operation until just before the ground air supply is disconnected. This minimizes the possibility of trapping moisture in the air-conditioning system cold air supply manifold and sustaining water damage to the ANS.

NOTE

One refrigeration system should be turned OFF when the cockpit air valve and the bay air shutoff valves are both closed. This prevents an increase in the cold air manifold temperature due to increased back pressure at the refrigeration turbines. The right system is normally turned off.

Cockpit Temperature Control Switch

An AUTO TEMP cockpit temperature control rheostat is located on the pilot's left instrument side panel. An arrow inscribed in the panel indicates clockwise rotation from COLD to WARM temperature settings. The control is ineffective unless the Cockpit Temperature Control and Override switch is in AUTO TEMP.

The setting of a hot air control valve in the cockpit air supply manifold is modulated with AUTO TEMP selected. Positioning of the control rheostat regulates cockpit (outflow) air temperature between 40° and 100°F. The variation of temperature with control position is nearly linear. The AUTO TEMP function automatically compensates for changes in hot and cold air manifold temperatures.

A duct high temperature limit sensor operates with AUTO TEMP selected to limit cockpit air to a maximum of 126° ± 12°F. A high limit switch is also provided which limits cockpit air to a maximum 155° ± 5°F in auto or manual control. Both limit controls reduce temperature by reducing flow from the regulated hot air duct.

Because a ratcheting technique is used to position the hot air control valve and modulate mixed air temperature, the automatic temperature control is much slower in accomplishing temperature changes than manual control, using the override switch. Electrical power for the cockpit temperature control circuit is furnished by the essential dc bus and the essential ac bus through a CKPT AIR COND circuit breaker located on each console in the forward cockpit.

Cockpit Air Temperature Control and Override Switch

A cockpit air temperature control and override switch is located on the pilot's left instrument side panel. The switch has four positions: AUTO TEMP (up), COLD (down left), WARM (down right) and MAN HOLD (center). The switch is spring loaded to MAN HOLD from the COLD and WARM positions. The switch is normally in AUTO TEMP. The pilot can override automatic control by moving the switch to either the momentary COLD or WARM position. The cockpit temperature control valve, which supplies air from the hot manifold, requires up to 10 seconds to close or open when COLD or HOT, respectively, is selected; however, cockpit temperature indication may require several minutes to change. Manual control of temperature is required when on battery power.

Temperature Selector Switch and Indicator

A three-position TEMP IND SELECTOR switch is located on the pilot's left instrument side panel, adjacent to a temperature indicator. Selection of the R-BAY, CKPT, or E-BAY position displays the air temperature in the corresponding compartment on the temperature indicator. Sensors for the E-BAY and R-BAY temperatures are located at the aft outboard walls of these bays. The sensor for the cockpit temperature indication is located near the cockpit pressure regulator and outflow valve, aft of the RSO's seat. The

Indicator displays temperature in 5°F increments from 30°F to 160°F. Temperature indications below 30°F are normal for E-Bay or R-Bay while above 44,000 feet. Power for the indicator and thermoresistor elements is from the essential dc bus through the CKPT AND BAY TEMP circuit breaker on the pilot's left console.

NOTE

The pressure suits may keep the crewmembers relatively comfortable in a cockpit environment that is too warm. The temperature gage allows anticipation of conditions that might eventually become too hot.

Manifold Temperature Control Switch

A two-position MANIFOLD TEMP switch, on the pilot's left instrument side panel, selects either a full cold or an automatic refrigeration system operation. The AUTO (up) position is normally selected. In AUTO, the cold air manifold temperature is regulated to +37°F when below 44,000 feet and to a minimum of -30°F at higher altitudes. During descents from above 44,000 feet, cold air manifold temperature regulation to +37°F begins at 41,000 feet.

In FULL COLD (down), air entering the cold air manifold receives the maximum cooling from the air cycle refrigeration machines and their intercoolers. As a result, cold air manifold temperature varies with bleed air temperature, fuel temperature in the intercoolers and heat exchangers, altitude, and system back pressure. Temperatures can be considerably lower than -30°F in some cases. Therefore, FULL COLD is not recommended unless automatic air-conditioning regulation is not satisfactory.

Power for the automatic manifold temperature control is from the essential ac bus. Power for the manifold temperature control valves and hot air bypass control systems is from the essential dc bus. Circuit breakers for both circuits are labeled MANF TEMP and are located in the forward cockpit.

At altitudes where +37°F is provided in the cold air manifold, air in the hot air supply lines is mixed with unconditioned (hot) engine bleed air and heated to about 200°F after it has passed through the nacelle air-to-air and air-to-fuel heat exchangers. This regulation and mixing is accomplished by a pneumatic thermostat and hot air bypass valve in each of the hot air supply lines. Bypass valve positions are controlled by barometric switching similar to that used for control of temperature level selection in the cold air manifold. They are not affected directly by manifold temperature control switch settings; however, both bypass valves close and hot air supply line temperatures are minimized if dc power from the MANF TEMP circuit breaker is interrupted after FULL COLD manifold temperature is selected.

Bay Air Switch

A two-position BAY AIR switch is provided on the pilot's left instrument side panel. In the ON (up) position, pressurized air from the cold air manifold is directed to the chine bays and chine bay equipment boxes. It also allows conditioned air from the cockpits to exhaust through a sonic venturi to the nose and nose radar compartments. The switch must be pulled out before it can be moved OFF (down). In OFF, shutoff valves in the chine bay distribution manifolds terminate direct flow from the cold air manifold. The spring-loaded-closed nose air shutoff valve actuator is deenergized and air flow to the nose from the cockpits is stopped. Air continues to flow to the cockpits, pressure suits, ANS equipment, and E-Bay and R-Bay equipment. It is exhausted through the chine bays for secondary cooling.

The BAY AIR switch should be on during normal flight and ground operations.

CAUTION

The BAY AIR switch should always be positioned OFF if one refrigeration system is turned off, manually or automatically, while at high speeds (high CIT).

Bay air may be off or on, to suit requirements for cockpit cooling, if one refrigeration system is off while CIT is low (subsonic). Bay air should not be turned off when cockpit air is shut off unless one REFRIG switch is also turned off, as cold air manifold temperatures could increase due to back pressure at the refrigeration turbines, and all of the cooling flow would then be directed toward and exceed the supply requirements for the ANS and E-Bay and R-Bay equipment boxes.

Power for the bay air shutoff valves and for opening the nose air shutoff valve is from the essential dc bus through the AIR SOV CONT circuit breaker on the pilot's left console.

CAUTION

With the Bay Air switch off, the chine bay equipment will overheat during subsonic or supersonic flight if electrical power to the equipment is not turned off. If supersonic, the E and/or R BAY OVERHEAT caution lights may illuminate even with the equipment off.

Cabin Pressure Switch

A two-position lift-lock CABIN PRESS switch is on the pilot's right console and on the RSO's instrument panel. The 26,000 FT position is normally preferred on flights involving high altitudes and air refueling as cockpit pressure remains relatively constant and cockpit and bay cooling is more effective. The switches in both cockpits must be in the 26,000 FT position to select the 26,000 FT pressure schedule. If either crewmember selects the 10,000 FT position, the 10,000 FT pressure schedule is selected. Figure 1-101 depicts the pressure schedules.

NOTE

- At low power settings, cockpit altitude may exceed 10,000 feet while below 28,000 feet altitude with the 10,000 FT schedule selected due to reduced bleed air supply pressure.
- Suit and cockpit cooling may be reduced due to increased back pressure with the 10,000 FT schedule selected.

Cockpit Air Valve

A cockpit air shutoff valve is located in the duct which supplies conditioned air to the cockpits. It is operated manually, either by the cockpit air handle under the RSO's left canopy sill, or by the RSO's canopy latch handle. The valve is open and air is furnished to the cockpits when the cockpit air handle is aft <u>and</u> the canopy latch handle is forward (locked). The valve is closed and conditioned air flow to the cockpits is shut off when the cockpit air handle is positioned forward to its OFF detent <u>or</u> when the canopy latch handle is aft (unlatched). Cockpit air is normally shut off for landing to prevent fog entering the cockpits from the air-conditioning system. The pilot's CKPT AIR OFF annunciator panel caution light illuminates when the cockpit air valve is closed.

WARNING

In high humidity conditions, do not resume use of cockpit air while at low altitude or after landing until certain that onset of fog will not endanger the aircraft.

When the cockpit air valve (36, Figure 1-80) is closed, the hot air mixing valve (33, Figure 1-80) is also closed automatically and the cockpit air supply temperature becomes full cold. If the cockpit temperature control switch is in AUTO TEMP, the hot air mixing valve reopens when the flow of cockpit air is resumed, and modulation at the selected automatic temperature level continues after a short delay. If the air temperature control is in MAN HOLD, the air supply temperature remains full cold when flow of cockpit air is resumed. The pilot must reset the temperature if a warmer temperature is desired.

Movement of the RSO's cockpit air handle has no effect when the canopy latch handle is in the aft (unlatched) position. Movement of the RSO's canopy latch handle has no effect on the cockpit air valve when the RSO's cockpit air handle is in the forward (off) position. In both cases, the cockpit air valve remains closed.

CAUTION

If either canopy is open, the aft canopy latch handle must be in the aft position or the cockpit air handle must be in the forward (off) position for adequate equipment cooling. Otherwise, most of the cooling air would exit through the cockpit openings instead of the bays.

Unless operating with a ground air supply connected, closing the cockpit air valve may increase the cold air manifold temperature slightly due to an increase in back pressure at the refrigeration turbines.

CAUTION

Do not operate with the cockpit air and bay air shutoff valves closed while above 10,000 feet unless one refrigeration system is also shut off. The increase in back pressure can result in decreased refrigeration turbine efficiency, higher than normal cold air manifold temperatures, and ANS equipment overheating.

The cockpit air temperature controls are not effective with cockpit air shut off, but hot air from the windshield defog system may be used to raise cockpit air temperature.

Cockpit Pressure Dump Switch

A two-position PRESS DUMP switch is on the pilot's left instrument side panel. It allows the pilot to dump cockpit pressure to ambient. In OFF (down), the cockpit pressure safety relief valve is closed and the cockpits are pressurized. The switch must be pulled out before it can be moved to ON (up). With ON selected, a solenoid in the safety relief valve fully opens the valve and the cockpit compartment depressurizes rapidly to air conditioning bay pressure (approximately ambient). The pressure suits inflate if cabin altitude exceeds 35,000 feet. As pressure in the cold air manifold is not affected substantially by the loss of cockpit pressure, airflow to equipment boxes in the bays and chines is not materially reduced by depressurizing the cockpits. Hot air remains available for defogging and windshield deicing. Power for the dump valve solenoid is from the essential dc bus through the PRESS DUMP circuit breaker on the pilot's left console.

WARNING

Cockpit pressure dump and/or repressurization occurs very rapidly.

Cabin Altimeter

A single-revolution cabin altimeter is located on the pilot's left instrument side panel. It is vented to cockpit static pressure and indicates cockpit altitude in increments of 1,000 feet from sea level to 50,000 feet.

Air System Caution Lights

The following caution lights on the annunciator panels are associated with the environmental control system.

Forward cockpit:
- L AIR SYS OUT
- R AIR SYS OUT
- E BAY OVERHEAT
- R BAY OVERHEAT
- CKPT AIR OFF
- WINDSHIELD DEICE ON
- CANOPY UNSAFE
- ANS REF
- BAY AIR OFF

Aft cockpit:
- ANS FAIL

The L or R AIR SYS OUT light illuminates when its corresponding refrigeration system shutoff valve is closed, either manually (using the refrigeration switches) or automatically because of high air temperature at the refrigerator compressor inlet or intercooler outlet.

The E or R BAY OVERHEAT light illuminates when air temperature exceeds 150°F at the aft outboard wall of the corresponding compartment.

The CKPT AIR OFF light illuminates when the cockpit air valve is closed either by aft (unlocked) positioning of the RSO's canopy latch handle or by setting the RSO's cockpit air handle to OFF (forward).

The WINDSHIELD DEICE ON light illuminates when the windshield deicing switch is in ON DE-ICE. The need for hot air deicing should be monitored, as the quantity of air available to the air contitioning and pressurization systems is reduced by deicing. Temperature of the air supply is approximately 200°F when below 44,000 feet.

The CANOPY UNSAFE light illuminates if either canopy is not closed properly on its sill, if either canopy latch handle is not full forward (latched), or if the canopy seals in either cockpit are not fully inflated.

The forward cockpit ANS REF and aft cockpit ANS FAIL lights illuminate if cooling air flow to the ANS is inadequate (less than 2.5 pounds per minute), if ANS component temperatures are too high or too low, or if the ANS is not operating normally. The RSO's ANS MAL and TEMP LIMIT/TEMP TOLR lights indicate abnormal air temperature and /or quantity. (Refer to Astroinertial Navigation System, Section IV.)

The BAY AIR OFF light illuminates if the BAY AIR switch is OFF or fails, or if the cap for the ground air receptacle is improperly installed, or if the microswitch in the cap for the ground air receptacle fails.

LIFE SUPPORT SYSTEMS

AIRCRAFT OXYGEN SYSTEM

One standby and two normal oxygen systems are provided. The normal system has two 10-liter liquid-oxygen-converter containers, and the standby system has one. Each of the three containers, located in the left chine opposite the aft cockpit, have a buildup circuit which maintains pressure at 65-100 psi until the oxygen supply is virtually depleted. Pressure above 120 psi opens a relief valve and vents overboard. The liquid oxygen from each converter is warmed and converted into a gas by passing through the supply line tubing (See Figure 1-82.)

Oxygen Control Panels

Control panels for both the normal and standby systems are located on the pilot's left console. A control panel for the normal system only is located on the RSO's left console. Two ON-OFF levers, labeled SYS 1 (system 1) and SYS 2 (system 2), located on each panel, manually control oxygen shutoff valves. When a lever is ON (full forward) a mechanical latch prevents moving the lever to OFF (aft). The normal control panels have a dual-reading pressure gage (0 to 140 psi) and the standby control panel has a single pointer pressure gage (0 to 140 psi). The normal oxygen gages indicate pressure only when the individual levers are ON, while the standby system pressure gage indicates pressure regardless of the position of the levers on the standby panel. Moving any lever to

ON opens a valve and permits oxygen to flow to the respective pressure suit helmet or oxygen mask of the crewmember. Check valves are provided at the seat disconnect for each of the supply lines, to prevent free flow of oxygen if a system valve is turned ON while the suit connections are disengaged.

NOTE

If standby system pressure is higher than the normal system pressure, standby system oxygen pressure is displayed on the normal oxygen control panel gage(s) when the system(s) (1 and/or 2) levers are ON for both the normal and standby control panels.

Liquid Oxygen Quantity Gage

A LIQUID OXYGEN quantity gage is on the pilot's left instrument side panel and the RSO's instrument panel. The gages have dual pointers that indicate the volume in liters of the normal system or standby system, depending on the position of the liquid oxygen system quantity switch. The gages are calibrated in 1/2-liter increments from 0 to 10 liters.

Liquid Oxygen System Quantity Switch

A two-position switch, labeled LOX QTY, is on the pilot's left console. In SYS 1 IND 1 (aft), the liquid oxygen quantity indicators in both cockpits display the volume of liquid oxygen in systems 1 and 2. In STANDBY IND 1 (forward), needle 1 of both cockpit liquid oxygen quantity indicators display the volume of liquid oxygen in the standby system.

NOTE

The volume of liquid oxygen in system 2 is always displayed by pointer 2 regardless of the position of the switch.

Oxygen System Caution Lights

Four oxygen system caution lights are installed on the annunciator panel in each cockpit. The SYS 1 OXY QTY LOW and SYS 2 OXY QTY LOW lights illuminate when the quantity of the respective system is less than 1 liter. If the No. 1 pointer is selected to indicate the standby system quantity, the SYS 1 OXY QTY LOW light indicates that the standby system quantity is less than 1 liter. The SYS 1 OXY PRESS LOW and SYS 2 OXY PRESS LOW lights illuminate when the pressure in the supply lines of the respective system is less than 50 (± 3) psi.

EMERGENCY OXYGEN SYSTEM

Each crewmember has two independent emergency oxygen bottles in the ejection seat survival kit. Each 45 cubic-inch capacity cylinder is pressurized to 2000 psi. Both emergency oxygen bottles are supplied automatically to the helmet upon ejection. Both oxygen systems can be activated manually by pulling the "green apple". Once the emergency oxygen system is actuated, it cannot be shut off. The emergency oxygen system should be actuated if the aircraft is not delivering the desired amount of oxygen from the ship systems, or hypoxia or noxious fumes are suspected. When actuated, check-valves prevent oxygen flow into the aircraft system. To prevent emergency oxygen flowing if aircraft system pressure is available, emergency oxygen system regulated pressure is slightly lower than the aircraft system pressure. Because emergency oxygen system regulated pressure is lower than the aircraft system pressure, turn the normal oxygen system supply levers to OFF after actuating emergency oxygen if contamination of the aircraft system is suspected. Pressure in the emergency system containers is indicated by gages on the forward left side of the survival kit (Figure 1-84) container.

Oxygen Consumption

The rate of oxygen consumption varies inversely with cockpit altitude. For a normal mission profile with the cockpit pressure switch in 26,000 FT, the average rate of oxygen consumption for two persons is 1 liquid liter per hour; with the switch in 10,000 FT, this average rate increases to 1.3

OXYGEN SYSTEM

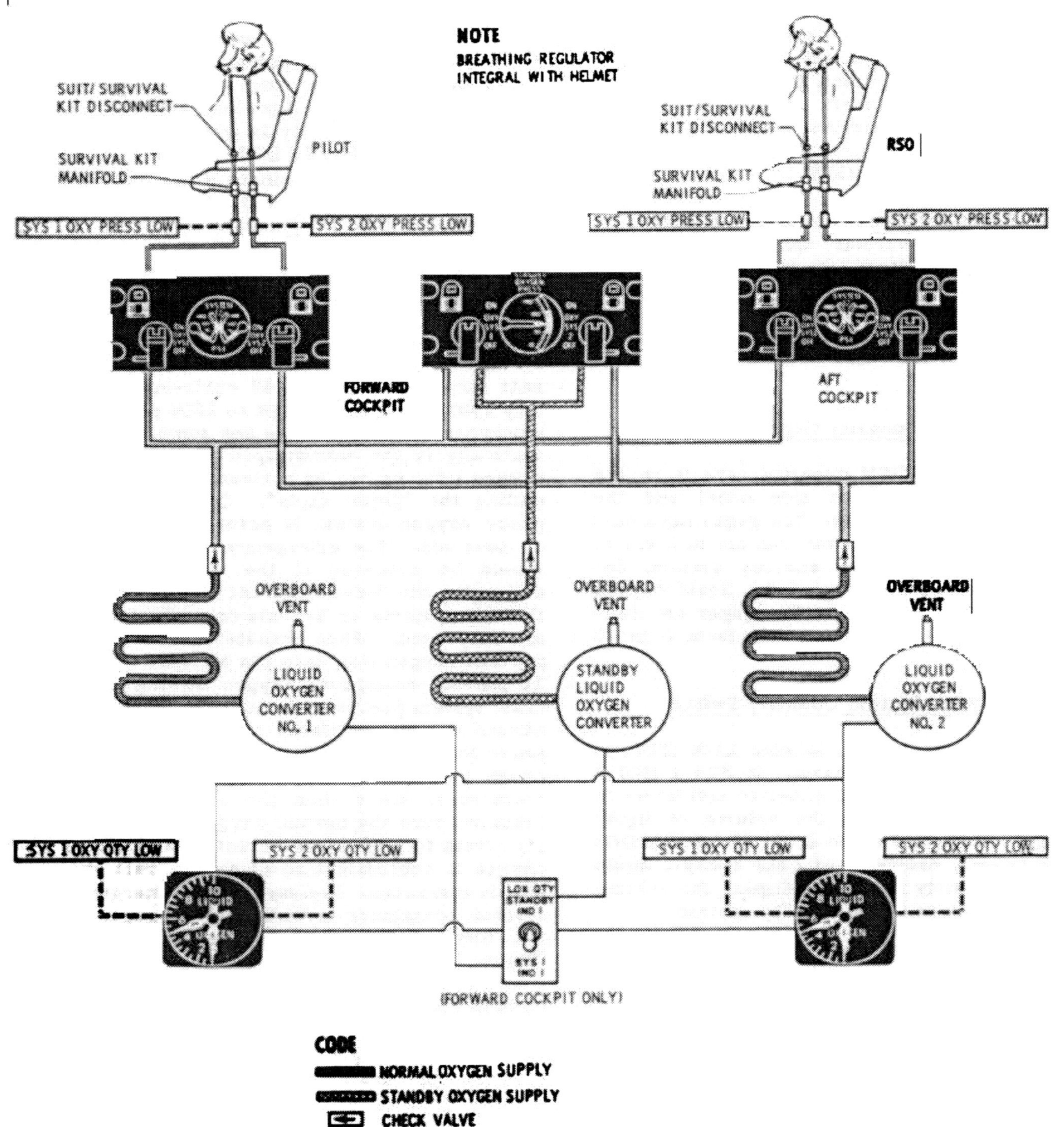

Figure 1-82

liquid liters per hour. The rate of consumption for individual crewmembers may vary between 50% and 150% of average consumption rates. Emergency oxygen duration is approximately 15 minutes.

FULL-PRESSURE SUIT

The model 1030 full-pressure suit provides the crewmember a safe environment, regardless of cockpit pressure. The suit has six layers: internal comfort liner, vent duct, bladder, exposure garment, link net restraint, and exterior cover. The ventilation layer allows vent air to circulate between the crewmember's cotton underwear and the bladder layer. The airtight bladder holds pressurized air. The link net is a mesh used to shape the suit to the crewmember's body. The outer garmet provides some protection from high temperatures and wind. A vertical entry zipper is locate in the back.

Thermal Protective Garment

The double-walled, dual-function gas container/exposure garment contains the pressurized gas. The double-wall provides an interspace which can be orally inflated by a tube stored behind the left thigh pocket, to create an air-space, thermal-barrier for exposure protection. This double-walled bladder extends over most of the torso, arms, and legs (except for the hands and feet).

Pressure Suit Vent Air

Air for suit ventilation comes directly from the air conditioning cold air manifold. In cruise, the temperature of this air may be as low as -30°F with suit heat OFF and the MANIFOLD TEMP switch in AUTO. The air temperature can be substantially lower with FULL COLD selected. The air can be warmed by positioning the suit heat rheostat. Vent air and exhaled breathing air exhausts through the suit controller valve which controls suit pressure and vent flow rate.

Suit Heat Rheostat

A suit heat rheostat, on the pilot's left instrument side panel, controls an electric heater for the air supply to the pilot's and RSO's pressure suits. Individual comfort adjustment is accomplished by varying suit airflow through the ventilated air valve on each suit.

Suit Controller Valve

Air pressure in the suit is regulated by dual suit controller valves on the front of the suit just above the waist. Each controller valve meters airflow to keep internal suit pressure at 3.5 psi if the cockpit depressurizes. A press-to-test button on each controller valve bypasses the suit controller by closing the outflow valve, allowing the suit to inflate. A knurled knob may be rotated to check suit inflation or to partially inflate the suit for comfort.

Helmet

The helmet head area is divided into two sections by a rubber face seal. The front area receives oxygen from either the aircraft or emergency oxygen system through two independent regulators built into the helmet. Inhalation causes oxygen to flow across the visor and accomplishes some visor defogging before it is inhaled. The rear area receives ventilating air. The face seal may not be positive; however, oxygen pressure in the face area is slightly higher than in the rear area and prevents air from leaking forward. If the oxygen supply is interrupted or exhausted, an antisuffocation valve in the helmet senses the drop in pressure and allows ambient air to enter the helmet to prevent suffocation.

A transverse "Baylor" bar encircles half of the helmet. It pivots at each side attachment point. When raised, it lifts the visor and closes the helmet oxygen regulator valves. When lowered, it opens the helmet oxygen regulator valves and closes the visor. A latch below the face opening must be engaged by the bar and a lever on the bar rotated 90 degrees to lock and hold the visor sealed.

A microphone adjustment knob, below the helmet face opening, adjusts the fore-and-aft position of the helmet microphone.

CAUTION

Do not force the microphone adjustment knob after the microphone has reached full travel; otherwise, the helmet shell may be damaged.

The visor is raised and lowered by the Baylor bar. With the bar raised, the visor is held up by a spring, uplock plunger, and cam on each side of the helmet. A transparent dark sunshade also pivots at each side connection. A friction clutch holds the sunshade in any position from fully open to closed. An external knob at the rear of the helmet adjusts the face seal. External helmet connections include two oxygen supply hoses and an electrical lead which provides power for visor heat, microphone, and earphone connections.

<u>Face Heat Switches</u>

A FACE HEAT rheostat switch is on the pilot's left instrument side panel and the RSO's instrument panel. The discrete switch positions are numbered 1 thru 7 with additional labels: OFF, LOW, MED and HIGH. Both oxygen flow and face heat defog the visor. Power for the switches is from the essential dc bus through individual FACE HTR circuit breakers in each cockpit.

CAUTION

- Do not use the HIGH face heat position when equipped with the PPG (glass) visor except for emergency heating. Continuous use of the HIGH heat position may delaminate the visor.
- The face heat switch should not be set above 5 with the visor raised, or the faceplate may be damaged.

<u>Boots</u>

A boot bladder is fastened to the suit by a zipper near the calf. The boot bladder will retain pressure. Flying boots are worn over the pressure suit boot bladder.

<u>Gloves</u>

Gloves fasten onto the suit at the wrist rings. The inner bladder of the glove is similar to the suit inner bladder and will retain pressure. The outer glove palm is leather.

TORSO HARNESS

The parachute harness is part of a torso harness worn over the pressure suit. The torso harness contains a built-in, dual-cell flotation vest with inflators to provide bouyancy in water after ejection. The two flotation sections are contained within removable velcro patches.

Each of the two cells is inflated by a water sensitive inflator which actuates within 0.5 second after immersion. Each automatic immersion inflator is a battery powered, electrically fired explosive squib which pierces the CO_2 bottle and inflates the vest. Each cell may also be manually actuated by a lanyard. Each cell may be inflated (or deflated) by an oral inflation tube.

The torso harness is also used for "shirt-sleeve" flights (flights below 50,000 ft. without a pressure suit). For shirt-sleeve flight, extra attachments are fastened on the harness to restrain the oxygen regulator. The left attachment (located where suit vent would be if a pressure suit were worn) has a pocket to hold the oxygen regulator. The right attachment (where the suit controller valve would be if a pressure suit were worn) restrains the oxygen line.

OXYGEN MASK AND REGULATOR

An oxygen mask assembly is used for "shirt-sleeve" flights below 50,000 ft. The mask assembly consists of a standard oxygen mask, a special oxygen regulator, an antisuffocation valve, and oxygen personal leads with connectors for both aircraft and emergency oxygen systems. If the regulator malfunctions or the oxygen supply is exhausted, the antisuffocation valve between the regulator and the mask senses a drop in pressure and permits ambient air to enter the mask to prevent suffocation.

EMERGENCY ESCAPE SYSTEM

The emergency escape system is comprised of an SR-1 stabilized ejection seat, canopy jettison system, and the operating controls and indicators in both cockpits. (See Figure 1-83.)

EJECTION SEAT

The SR-1 seat is usable from zero speed and altitude to the extremities of the flight envelope. The ejection seat is a rocket-propelled, upward-ejecting unit, which uses a drogue chute to stabilize seat descent until man-seat separation occurs automatically at approximately 15,000 feet. The personnel (main) chute is deployed automatically immediately after separation from the seat.

In the normal ejection sequence using the D-ring, the canopy is automatically jettisoned prior to seat ejection. If using the secondary ejection method (initiated by pulling the seat T-handle), the canopy must be jettisoned by separate action prior to pulling the T-handle.

WARNING

Do not pull the secondary ejection T-handle with the canopy still in place.

The survival kit, with emergency oxygen supply, and parachute are installed in the seat before the crewmember enters the cockpit. Hookup to the crewmember is by means of five quick-disconnect attachments.

Seat Vertical Adjustment

An electric motor, controlled by a 3-position switch on the right side of each seat bucket, moves the seat up or down. Seat movement is in the same direction as switch movement. The spring-loaded center position de-energizes the motor. Maximum vertical seat travel is 9 inches for the pilot's seat, and 6.75 inches for the RSO's seat. Power is from the essential dc bus through a SEAT ADJ circuit breaker in each cockpit.

Emergency Face Heat

A battery on the left side of the seat automatically provides face heat after ejection. The battery is hose-connected to the D-ring ballistic line, and a power cable is connected (by pull-to-release plug) to the helmet. During ejection, a squib in the battery fires to rupture diaphragms at both ends of the battery and forces electrolyte into the battery cells, providing a charged battery for visor heat. The battery cable plug is pulled loose from the helmet or seat during man-seat separation.

Inertia Reel

An inertia reel in the seat headrest maintains constant pressure on the shoulder harness straps to keep them taut and permit unrestricted movement of the crewmember. Placing the inertia reel control knob (on the left side of each seat) in the LOCK (forward) position manually locks the shoulder straps. When the inertia reel control knob is in the normal (UNLOCK) position, the reel will automatically lock with an instantaneous forward load of 2 to 3 g, and remains locked until released by moving the control knob to LOCK then back to UNLOCK.

Early in the ejection sequence, an initiator fires to roll up and lock the shoulder straps. Later in the ejection sequence, the shoulder straps are automatically severed when strap cutters fire. If the straps are not cut, they pull free of the inertia reel during man-seat separation. The strap cutters are fired manually by pulling the manual release (scramble) handle on the right side of the seat.

D-Ring Assembly

A D-ring, located on the forward part of the seat, is pulled to initiate ejection and also is a handhold for protection of the arms during ejection. A compression spring within the D-ring assembly acts as a shock absorber. During man-seat separation, the D-ring cable is automatically severed to release the D-ring. The D-ring cable cutter may be fired manually by pulling the scramble handle. A safety pin is inserted in a hole at the base of the D-ring (when on the ground) to prevent accidental firing of the ejection system.

T-Handle Assembly

A T-handle backup ejection control, is used if the D-ring initiator does not fire. The T-handle is located on the left side of the seat, under a cover shield which must be pushed away to gain access to the handle. A safety pin is inserted in a hole in the cover shield to prevent accidental firing of the ejection system when on the ground. Pulling the T-handle initiates only seat ejection; the canopy must be jettisoned by separate action prior to pulling the T-handle. The D-ring must be pulled before the T-handle can be actuated, since pulling the D-ring arms the T-handle system (secondary ejection sequence).

Scramble Handle

The manual release (scramble) handle on the right side of each seat is a quick release from the seat for bailout with the ejection seat inoperative, or for ground emergency egress. When a button at the forward end of the handle is depressed and the handle is pulled, the handle mechanically rotates the seat torque tube to release the lap belt and the parachute arming cable (the lap belt and cable remain attached to the parachute) and fires a no-delay initiator to release the crewmember from the seat through the following actions:

1. Foot retraction cables are cut.
2. Shoulder harness straps are severed.
3. Inertia reel is released (so shoulder straps pull free of reel if cutters fail).
4. D-ring cable is cut.
5. Main (personnel) chute lanyard is released from the seat.

NOTE

When pulling the scramble handle, expect a loud report from the initiator firing.

After a manual bailout using the scramble handle, the parachute must be manually deployed, using the shoulder D-ring.

Foot Retention System

The crewmember connects shoe spurs to foot retention fittings at the rear of the footrest. The fittings are cable-connected to reel assemblies which maintain the cables under slight spring tension to permit unrestricted foot movement. When ejection is initiated, the foot spur cables automatically retract to hold the crewmember's feet in the footrests. The spur cables are severed automatically during man-seat separation, or may be severed by pulling the scramble handle.

The spur is normally released from the cable by moving the foot aft (against the seat footrest) and raising the heel.

Parachute Beacon

A battery powered radio beacon is installed in the parachute. The beacon has a minimum operating life of 15 hours, and transmits an automatic signal on a frequency of 243.0 megacycles. During chute deployment, the beacon automatically turns on when a lanyard, attached to a chute riser, pulls a plastic plug from between the beacon control switch contacts. The beacon has a 22-inch telescoping antenna and a flexible removable antenna. Pulling a push-pull control knob, attached to the chute harness at the right front shoulder, disables the automatic beacon activation feature.

Seat Catapult

During ejection, the catapult gas charge is pressure-fired to initiate seat ejection. The catapult gas charge has a duration of 0.15 seconds, sufficient to raise the seat above the canopy sills, at which point the seat rocket motor automatically ignites. The seat rocket motor provides sufficient thrust and duration (0.5 second) to provide a seat elevation (relative to the aircraft) of approximately 300 feet.

When ejection is initiated by the D-ring, the seat catapult fires through a 0.3-second-delay initiator to provide time for canopy jettison prior to seat ejection. When the T-handle is used, the seat catapult fires immediately and no canopy ballistics are fired. The canopy must be jettisoned by separate action prior to pulling the T-handle.

Ejection Seat Drogue Parachute

A 6.5-foot-diameter drogue chute (stowed in the headrest) is deployed by a ballistic drogue gun, which fires 0.2 second after the seat catapult fires. The drogue (stabilizing) chute is connected to the seat at four points by a bridle and 10 feet of webbing. Ten seconds after drogue chute deployment (to permit seat deceleration), the lower two bridle lines are automatically severed to stabilize the seat, and the seat continues its descent upright. The drogue chute controls rate of descent and attitude until man-seat separation at approximately 15,000 feet.

Man-Seat Separation

Dual aneroid-actuated initiators located on the upper left and right sides of the seat, delay man-seat separation until 15,000 feet if ejection occurs at a higher altitude. Each unit contains a small dial, colored red and green. The dial is a leak indicator for the aneroid bellows and will indicate in the red band if the bellows leaks. When man-seat separation is initiated:

a. Foot retention cables are severed.

b. Drogue chute upper risers are severed 0.3 seconds after seat separation (to release drogue chute).

c. Lap belt is released.

d. Shoulder harness is severed and inertia reel is unlocked.

e. D-ring cable is severed.

f. Rotary actuator (butt-snapper) is fired.

When the rotary actuator is fired, the crewmember is forcibly separated from the seat as seat webbing is rapidly retracted by reel rotation. The seat webbing is secured to the front of the seat and passes under the seat pack, up to the rotary actuator in the headrest. Rapid tightening (retraction) of the seat webbing as the actuator fires, snaps the crewmember from the seat with a sling-shot action. The main chute is automatically deployed by a lanyard after man-seat separation.

Main Parachute

A 35-foot diameter main parachute in a backpack is deployed automatically after man-seat separation. The chute can also be deployed manually, after separation from the seat.

Automatic deployment is initiated by a drogue gun in the upper left corner of the pack. A lanyard connected to the seat fires the drogue gun as the man and seat separate. Accidental deployment is prevented by a lanyard housing which disconnects from the seat at separation.

The chute can be manually deployed by pulling the D-ring which is held in place by a Velcro patch on the suit, near the left shoulder harness. See Figure 1-83. The drogue gun is not fired when the chute is manually deployed.

NOTE

> The main chute must be deployed manually if the crewmember has used the scramble handle for manual bail-out, as the main chute lanyard is released from the seat when the scramble handle is pulled.

The chute has a lanyard with a 7-inch loop tacked with breakaway thread on each rear riser strap. After deployment, if the parachute is not damaged, each loop should be pulled downward with a sharp tug approximately 1-1/2 feet. This releases three pairs of suspension lines on each side (24 pairs

EJECTION SEAT

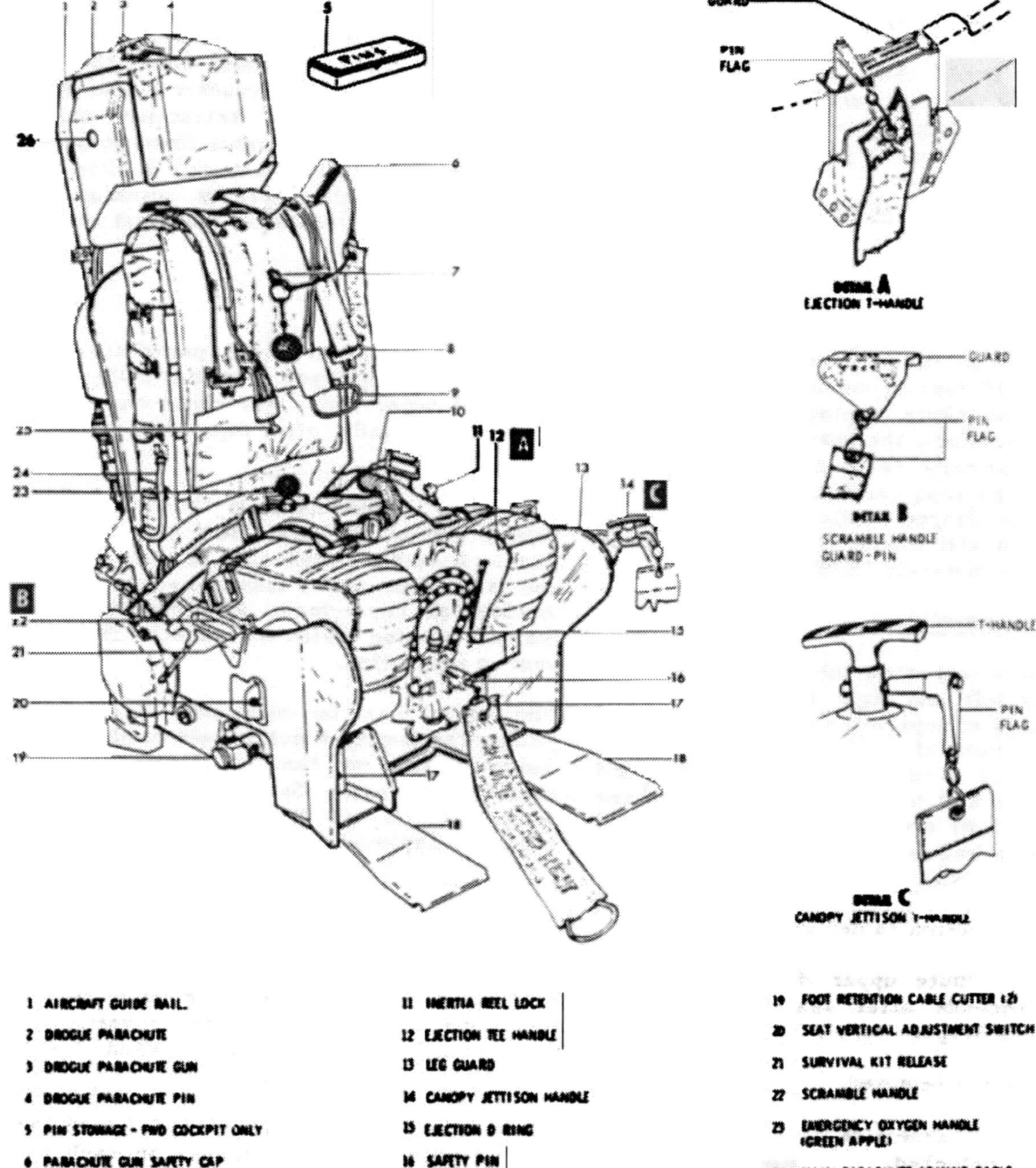

1 AIRCRAFT GUIDE RAIL
2 DROGUE PARACHUTE
3 DROGUE PARACHUTE GUN
4 DROGUE PARACHUTE PIN
5 PIN STOWAGE - FWD COCKPIT ONLY
6 PARACHUTE GUN SAFETY CAP
7 PARACHUTE GUN SAFETY PIN
8 SHOULDER HARNESS
9 MANUAL PARACHUTE DEPLOY RING
10 SUIT VENT AIR HOSE
11 INERTIA REEL LOCK
12 EJECTION TEE HANDLE
13 LEG GUARD
14 CANOPY JETTISON HANDLE
15 EJECTION D RING
16 SAFETY PIN
17 FOOT RETENTION CABLES
18 FOOT RAMPS
19 FOOT RETENTION CABLE CUTTER (2)
20 SEAT VERTICAL ADJUSTMENT SWITCH
21 SURVIVAL KIT RELEASE
22 SCRAMBLE HANDLE
23 EMERGENCY OXYGEN HANDLE (GREEN APPLE)
24 MAIN PARACHUTE ARMING CABLE
25 RADIO BEACON CONTROL
26 LEAK DETECTOR, ANEROID ACTUATED INITIATOR

Figure 1-83

remain), provides steering capability, and imparts a three to four ft/second forward speed to the chute.

WARNING

Do not pull either loop if the chute has sustained damage.

NOTE

Pull both loops. The chute will revolve continuously if only one set of lines is released.

For maximum effectiveness, start steering soon after chute deployment. A 180° turn requires 20 to 30 seconds.

The parachute may be equipped with a back cushion and/or a lowering device for descending from a tree. The lowering device consists of a let-down hook assembly, riser line section (colored blue), and 150 feet of nylon lead line packed in a special container in the parachute backpad. The last 25 feet of the lead line is colored differently (yellow, red, black, or white). See Figure 3-3.

Survival Kit

The survival kit is a hard-shell seat pack containing an emergency oxygen supply and survival equipment. Emergency oxygen quick disconnects are on the aft right corner of the kit, and two small oxygen quantity gages are at the forward left corner. (See Figure 1-84.) During ejection, emergency oxygen is turned on by a lanyard attached to the seat rail, or may be manually turned on by pulling the green apple. The kit may be released from the crewmember harness by pulling the kit release handle at the right side of the kit. The kit release handle should be pulled while descending in the main parachute, except when a tree landing is anticipated. When the release handle is pulled, the kit falls to the end of a 25-foot lanyard. The lanyard is disconnected from the harness if the release handle is pulled while seated.

WARNING

Pulling the kit release handle while seated disconnects both normal and emergency oxygen supplies.

PRIMARY EJECTION SEQUENCE

The D-ring is pulled to initiate the primary ejection sequence, which includes automatic canopy jettison. Pulling the D-ring fires an initiator on the front apron of the seat. Gas pressure from the D-ring initiator is ported directly to the foot retractor, the shoulder harness reel, and the canopy unlatch thruster. Pulling the D-ring also arms the secondary ejection sequence.

When the canopy unlatch thruster reaches full throw, gas pressure is ported to the canopy seal cutter and canopy removal thruster. Another initiator (pin-fired as the canopy is raised) ports pressure through the jettison valve to a 0.3-second delay initiator, which allows time for complete canopy jettison prior to firing the seat catapult initiator.

Gas pressure from the delay initiator provides pressure to fire the catapult initiator and arm the 0.2 and 1.4-second delay initiators in the drogue chute. The seat rocket motor ignites as the catapult raises the seat above the canopy sills.

The 0.2-second-delay initiator fires the drogue gun to deploy the drogue chute. The 1.4-second-delay initiator fires to arm the drogue chute lower riser cutters and the dual aneroid actuators. The lower riser cutters are armed through a 10-second delay, which provides time for seat deceleration before severing the lower riser lines. When the lower risers are severed, the seat is stabilized upright by the drogue chute. The dual aneroid actuators delay man-seat separation until 15,000 feet altitude if ejection occurs at a higher altitude. At completion of man-seat separation, the main chute is automatically deployed. See Figure 1-85.

SURVIVAL KIT

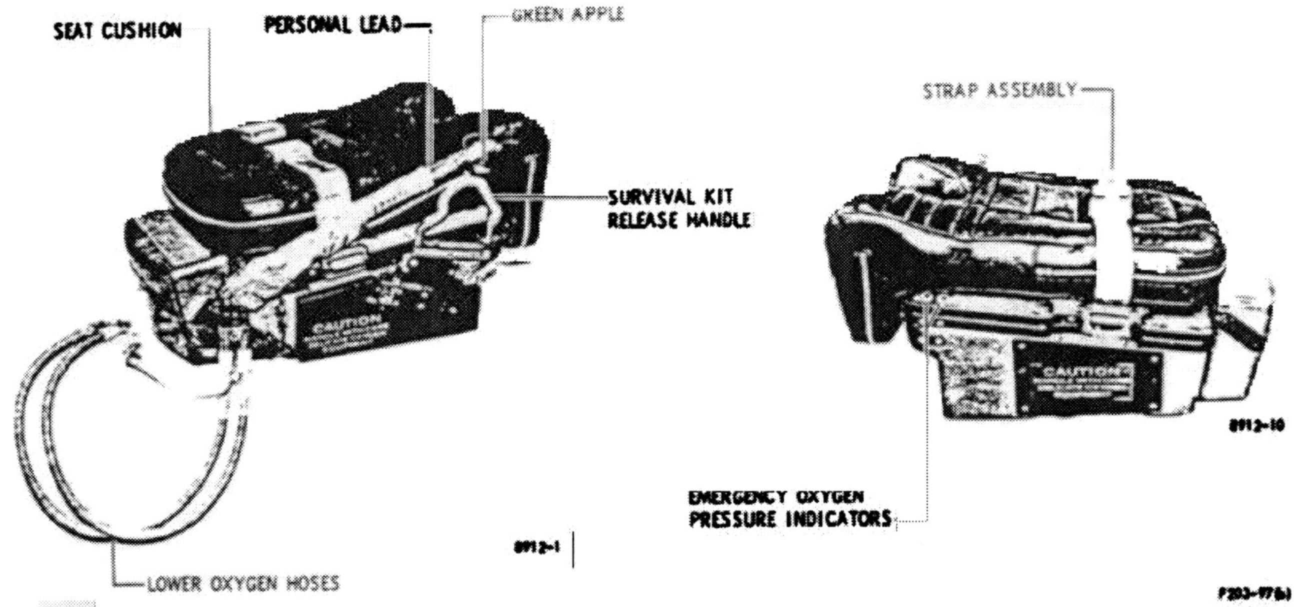

Figure 1-84

SECONDARY EJECTION SEQUENCE

If the primary ejection sequence malfunctions, the canopy must be jettisoned and then the ejection seat T-handle is pulled to initiate the secondary ejection sequence. The secondary sequence does not automatically jettison the canopy.

WARNING

Do not pull the secondary ejection T-handle with the canopy still in place.

NOTE

Secondary ejection sequence can not be initiated before the D-ring is pulled.

When the T-handle is pulled, the catapult initiator fires immediately and gas is ported to the two delay initiators in the drogue chute. Action downstream of the 1.4-second-delay initiator in the drogue chute is identical to the primary ejection sequence.

EGRESS COORDINATION SYSTEM

An egress coordination system supplements interphone communication. With this system, the pilot can issue and check compliance with an electrical bailout signal.

Egress Lights and Switches

A guarded RSO BAILOUT switch on the pilot's instrument panel has three positions: ALERT (down), OFF (center), and GO (up). Above this switch is a red RSO EJECTED warning light. The RSO instrument panel has an ALERT caution light, and PILOT EJECTED and BAILOUT warning lights. When the pilot actuates the RSO BAILOUT switch to ALERT or BAILOUT position, the corresponding light on the RSO panel illuminates. The ALERT light is a flashing amber light and the BAILOUT light is a steady red light. The forward cockpit RSO EJECTED light is operated by a switch on the aft cockpit ejection seat track. The aft cockpit PILOT EJECTED light is operated by a switch on the forward cockpit ejection seat

EJECTION TRAJECTORIES

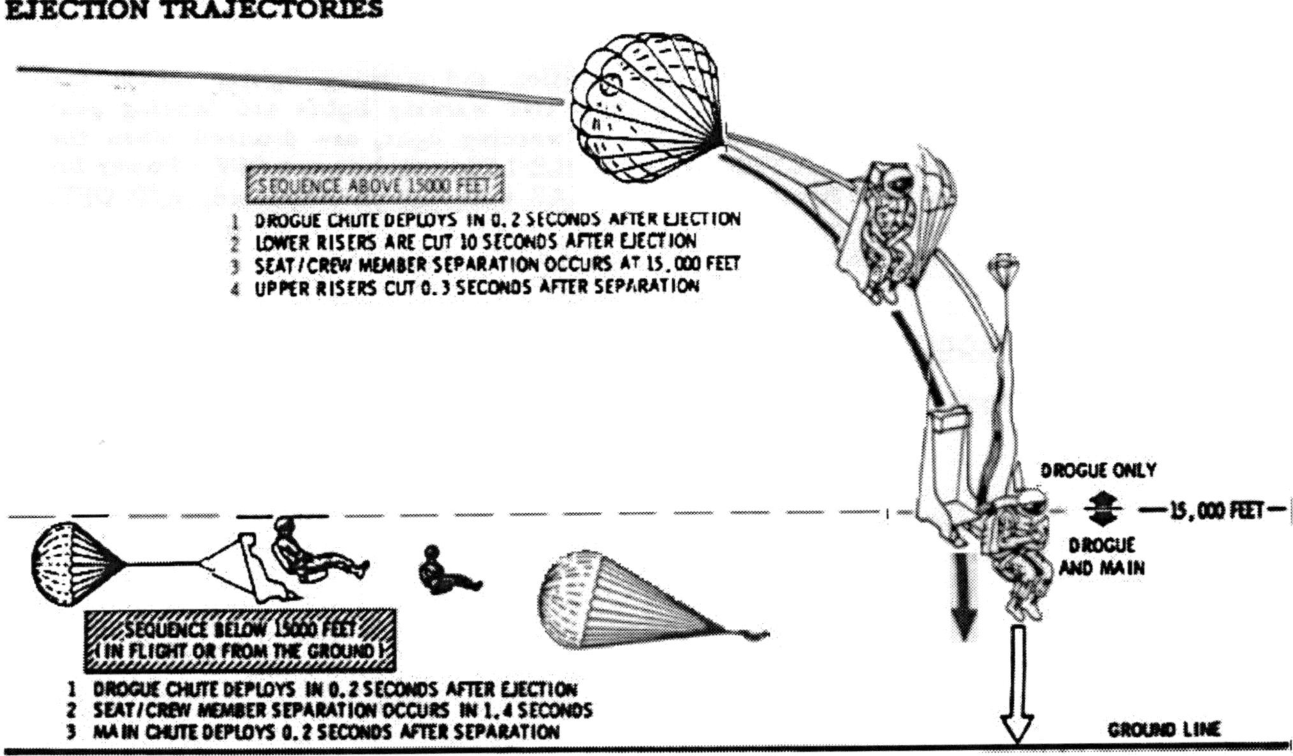
SEQUENCE ABOVE 15000 FEET
1 DROGUE CHUTE DEPLOYS IN 0.2 SECONDS AFTER EJECTION
2 LOWER RISERS ARE CUT 10 SECONDS AFTER EJECTION
3 SEAT/CREW MEMBER SEPARATION OCCURS AT 15,000 FEET
4 UPPER RISERS CUT 0.3 SECONDS AFTER SEPARATION
DROGUE ONLY
15,000 FEET
DROGUE AND MAIN
SEQUENCE BELOW 15000 FEET (IN FLIGHT OR FROM THE GROUND)
1 DROGUE CHUTE DEPLOYS IN 0.2 SECONDS AFTER EJECTION
2 SEAT/CREW MEMBER SEPARATION OCCURS IN 1.4 SECONDS
3 MAIN CHUTE DEPLOYS 0.2 SECONDS AFTER SEPARATION
GROUND LINE

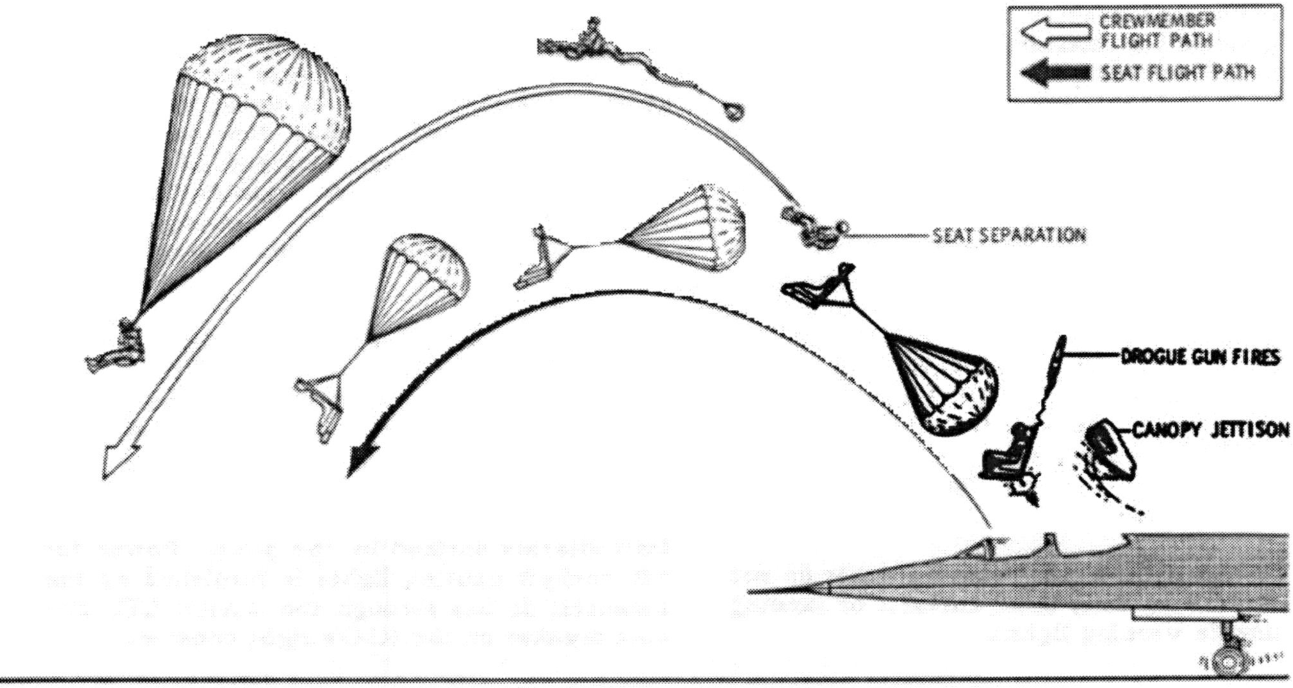
CREWMEMBER FLIGHT PATH
SEAT FLIGHT PATH
SEAT SEPARATION
DROGUE GUN FIRES
CANOPY JETTISON

F303-96(b)

Figure 1-85

track. When a seat ejects, the respective light illuminates in the opposite cockpit. The aft cockpit BAILOUT light illuminates automatically when the PILOT EJECTED light illuminates. Power for the lights is provided directly from the battery through a BAILOUT LT circuit breaker in the E-bay.

EMERGENCY WARNING EQUIPMENT

MASTER WARNING SYSTEM

Forward Cockpit

An annunciator panel, located below the instrument panel, contains amber caution and red warning lights, with engraved legends, to indicate abnormalities of certain systems.

An amber master CAUTION light and a red master WARNING light are located at the top of the instrument panel. Annunciator lights flash when initially activated. A flashing amber or red annunciator light will cause the corresponding master CAUTION or master WARNING light to illuminate steady. Depressing the master light (amber or red) extinguishes the master light and causes the associated annunciator light to illuminate steady. Any subsequent annunciator light that illuminates will flash until the associated master light is pressed.

The annunciator lights do not give an indication of multiple failures such as could occur in the A or B HYD system. For example: The loss of A system hydraulic fluid illuminates the A HYD light and the master CAUTION light. The A HYD light will flash until the master CAUTION Light is reset, then will remain on steady. If the A HYD pressure subsequently fails, the A HYD light will remain on steady and the master CAUTION will remain off.

The master caution and warning lights do not illuminate with fire, inlet unstart, or landing gear unsafe warning lights.

All caution and warning lights, except the nacelle fire warning lights and landing gear handle warning light, are dimmed when the CONSOLE LTS switch is not OFF. Power for the KEAS warning, RSO ejected, A/P OFF, both IGV, and both derich lights is furnished by the essential dc bus through the WARN 2 circuit breaker on the pilot's left console. Power for all annunciator caution and warning lights and all other caution and warning lights except the nacelle fire warning lights, the inlet unstart lights, the landing gear handle warning light and lights on the WARN 2 circuit breaker is furnished by the essential dc bus through the WARN 1 circuit breaker on the pilot's left console.

NOTE

The L and R OIL TEMP annunciator lights are not functional. The OIL TEMP lights only illuminate when the IND & LT TEST switch is pressed.

Aft Cockpit

An annunciator panel and an amber master CAUTION light are located on the instrument panel. The annunciator lights and CAUTION light function the same as those in the forward cockpit except that the annunciator lights illuminate steady instead of flashing. The annunciator lights and master CAUTION light cannot be dimmed. There are no red annunciator lights or master warning light in the aft cockpit.

An amber PILOTS CAUTION light illuminates when the pilot's master WARNING and/or master CAUTION light illuminates. The PILOTS CAUTION light remains illuminated until the master light(s) in the forward cockpit are pressed off. This permits the RSO to alert the pilot if the forward lights are not immediately noticed by the pilot. Power for aft cockpit caution lights is furnished by the essential dc bus through the WARN LTS circuit breaker on the RSO's right console.

Indicators and Warning Lights Test Button

Forward Cockpit

A push-button indicator and warning lights test switch, labeled IND & LT TEST, is located on the left instrument side panel. When the IND & LT TEST switch is depressed:

The following indicators move to their full counter-clockwise positions:

- Liquid nitrogen quantity
- Liquid oxygen quantity
- Fuel quantity
- CIP
- CG (forward and aft cockpit)
- Spike and forward bypass

The following lights illuminate:

- Annunciator panel caution and warning
- Master CAUTION and master WARNING
- L & R UNST
- Air refueling READY and DISC
- Fuel derich
- IGV
- KEAS
- SHAKER
- Landing gear indication (3 green lights)
- Nacelle fire warning. Illumination of each fire warning light indicates that the respective fire warning loop is functioning.
- Landing gear handle warning. (landing gear warning tone also heard in both headsets if one or both throttles are in idle or OFF).

All annunciator lights illuminate steady 1-1/2 seconds after the IND & LT TEST switch is depressed. Depressing the switch also illuminates the following aft cockpit lights: CG, master CAUTION, and PILOTS CAUTION.

Aft Cockpit

An indicator and warning light push-button test switch, labeled LAMP TEST, is located on the PWR & SENSOR control panel. Power for the switch is furnished by the essential dc bus. When the LAMP TEST switch is depressed:

Oxygen quantity indicator needle moves toward zero.

Fuel quantity needle moves toward zero.

The following lights illuminate:

- Annunciator panel
- UHF TRANS
- Master CAUTION
- ANS control panel
- PWR & SENSOR control
- Radar control, test and display panel
- DEF control and DEF warning panel

NACELLE FIRE WARNING SYSTEM

A fire warning system indicates the presence of a fire or hot spot in the engine nacelles. A high temperature, anywhere along the length of the detection circuits, illuminates the warning light for the corresponding nacelle.

There are two pairs of fire warning loops in each nacelle. One forward and aft pair is on the outboard side of the nacelle and the other forward and aft pair is on the inboard side. Both wires of a pair must sense a hot condition before a warning is given. Pressing the pilot's IND & LT TEST switch illuminates the FIRE LEFT NAC and FIRE RIGHT NAC warning lights if the fire warning loops are intact.

Nacelle Fire Warning Lights

FIRE LEFT NAC and FIRE RIGHT NAC warning lights, on the pilot's instrument panel, illuminate when nacelle temperature near the turbine or afterburner exceeds 1200° (± 50°) F. Covers are provided for reducing the brightness of the lights at night. Power for the lights is furnished by the emergency ac bus through the L and R FIRE WARN circuit breakers on the pilot's right console.

Fire Warning Loop Test Switches

Two switches, one for each engine, are installed above and aft of the left console in the forward cockpit. The switches, labeled L LOOP SEL SW and R LOOP SEL SW, are used by maintenance personnel to checkout the fire warning system.

MISCELLANEOUS EQUIPMENT

Dinghy Stabber

A dinghy stabber, on the pilot's glareshield and on the RSO's right console, is provided to deflate the dinghy if it accidentally inflates in the cockpit.

WARNING, CAUTION AND CONDITION LIGHTS (Forward Cockpit)

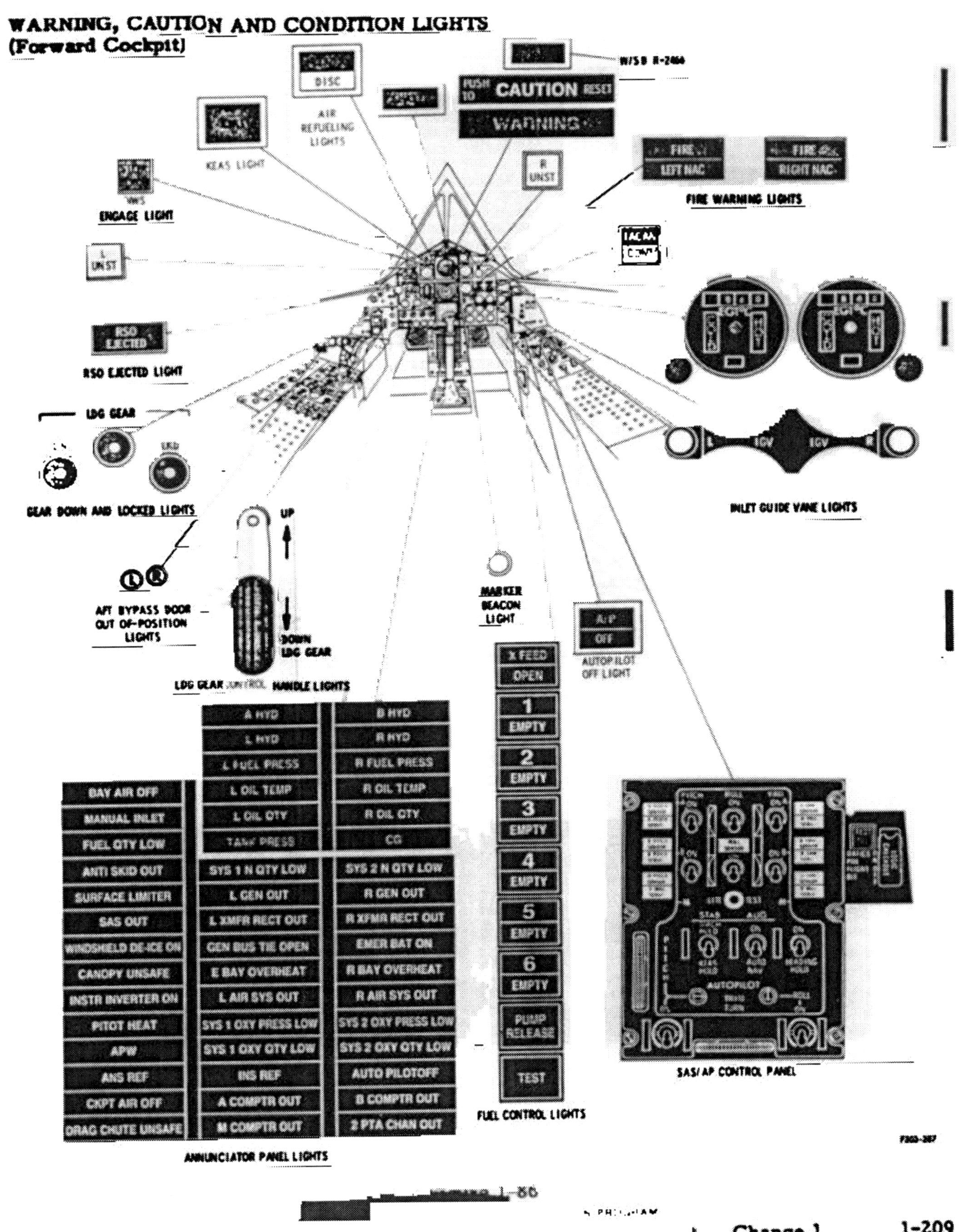

WARNING, CAUTION AND CONDITION LIGHTS (Aft Cockpit Instrument Panel)

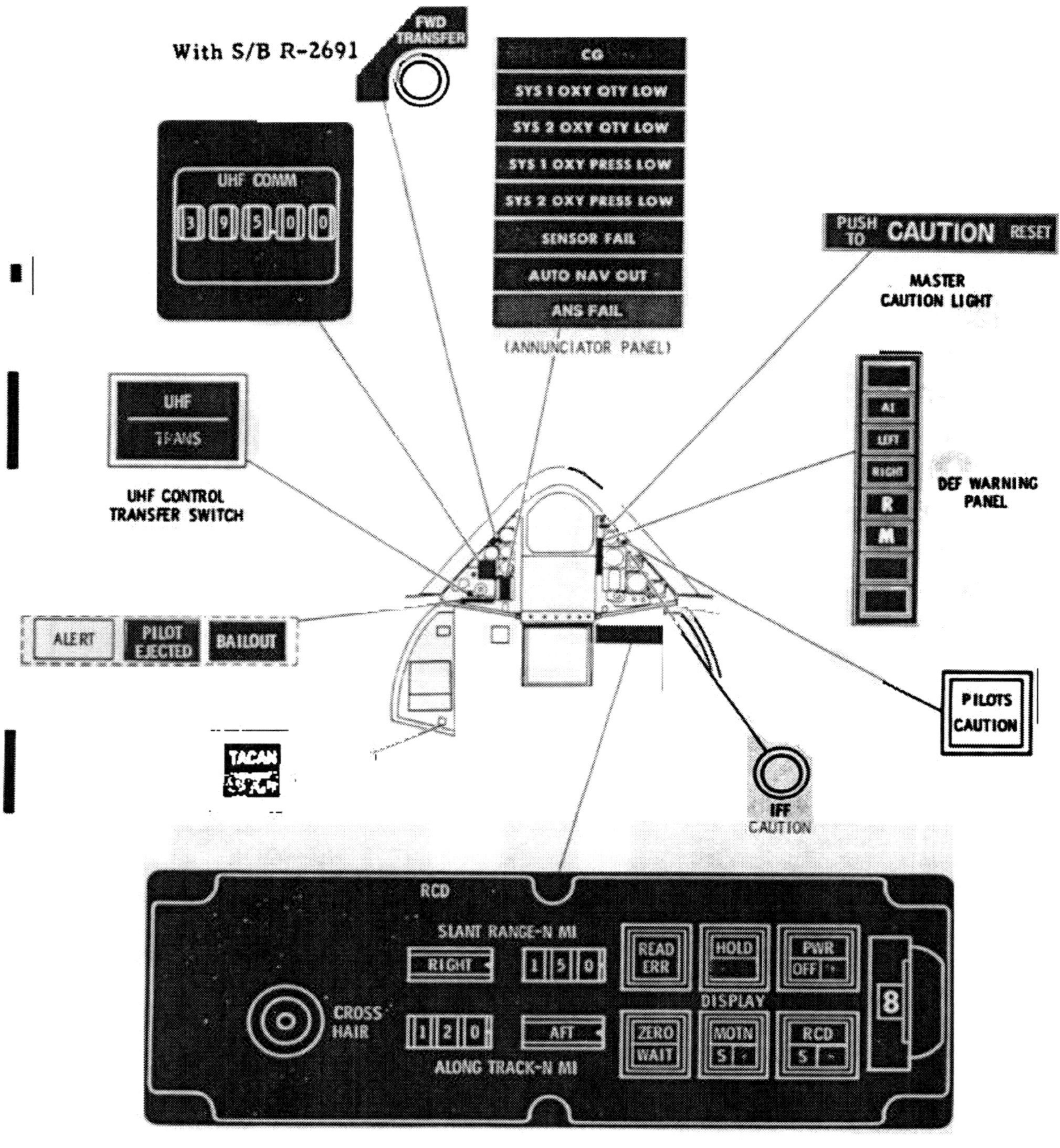

F203-167(c)

Figure 1-87

WARNING, C... (Aft Cockpit Consoles)

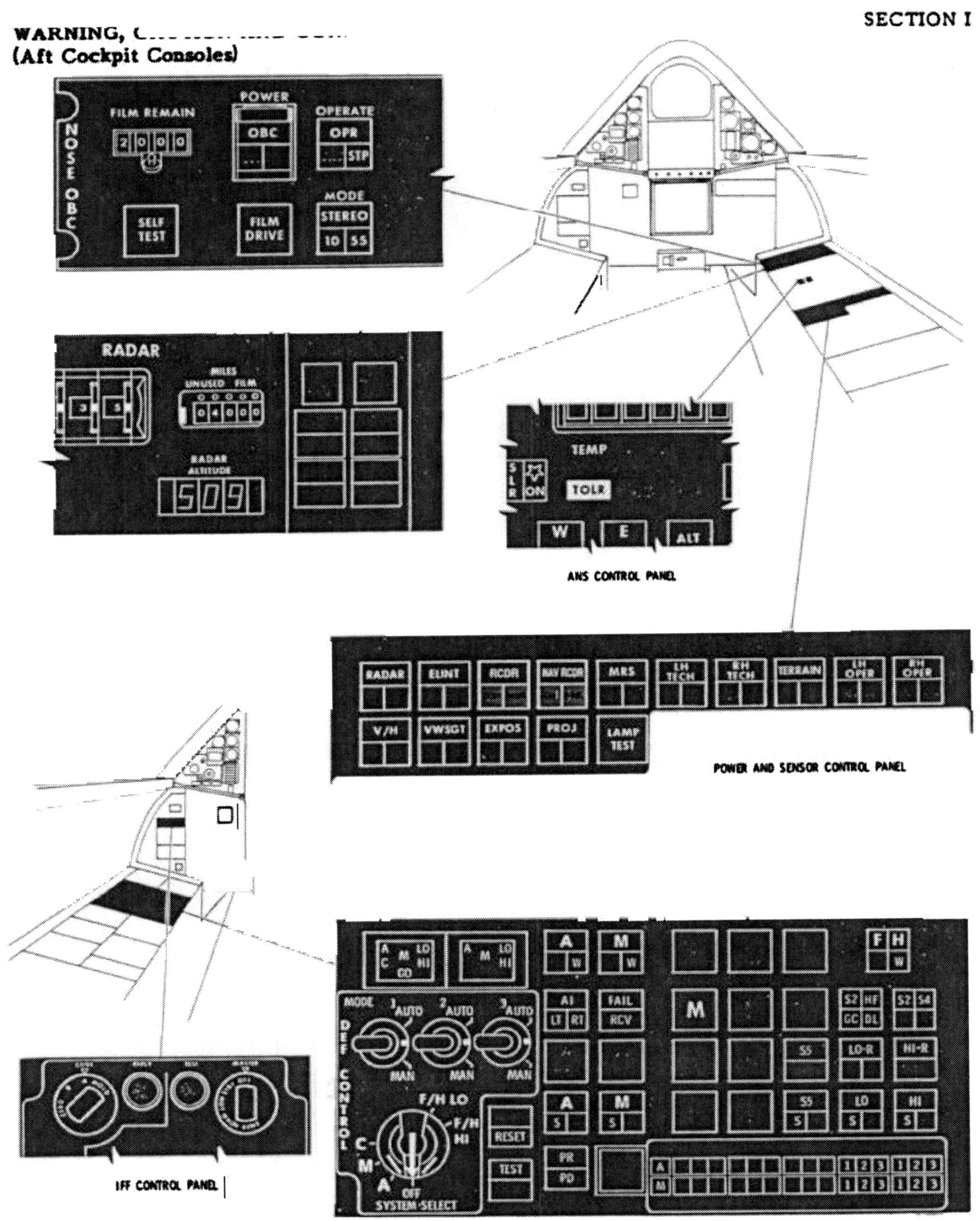

Figure 1-88

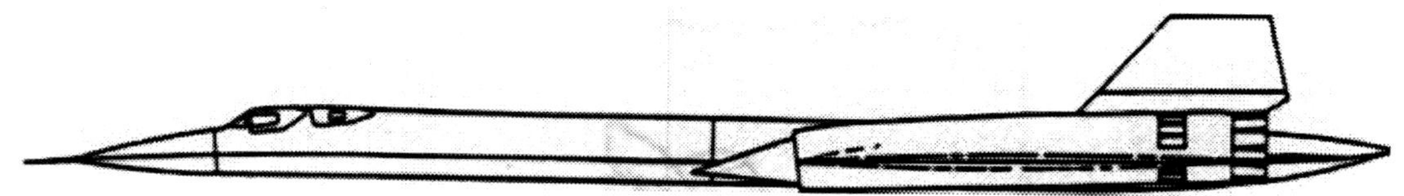

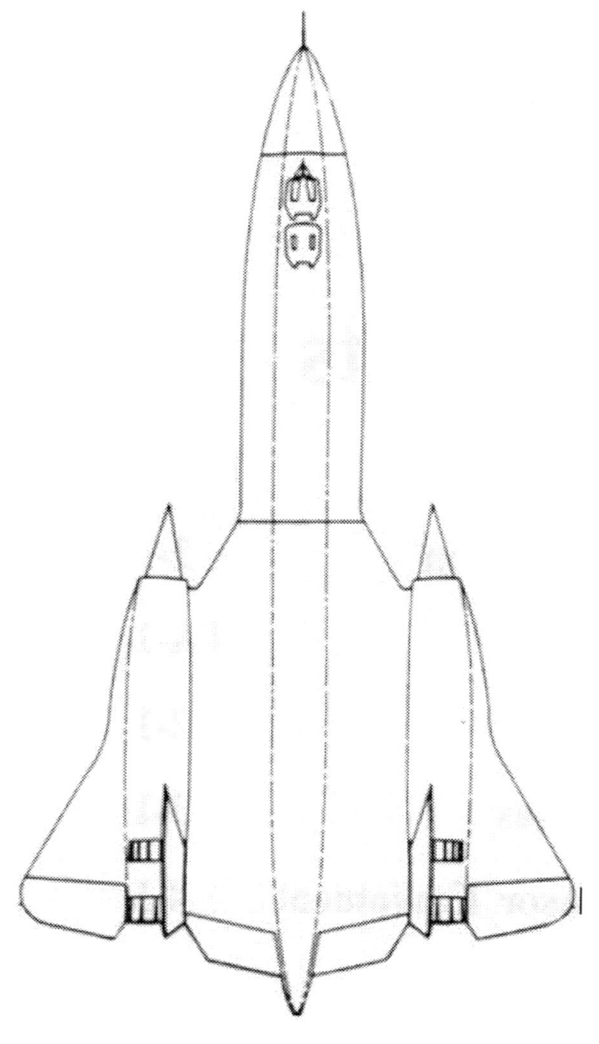

THIS PAGE INTENTIONALLY LEFT BLANK OR STILL CLASSIFIED.

Section 1A

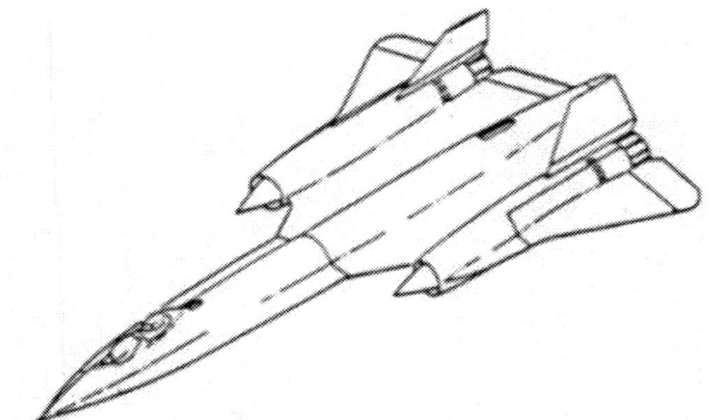

Description And Operation

TABLE OF CONTENTS

SR-71 TRAINER

INTRODUCTION

This subsection of Section I describes the SR-71B trainer aircraft. The normal and emergency operating procedures, limitations, and performance data for this aircraft are included in the appropriate sections of the SR-71A-1 Flight Manual. The aircraft systems and controls are identical with the like systems and controls in the SR-71A aircraft except as indicated in the following paragraphs. The controls and indicators in the aft cockpit are identical with the like controls in the forward cockpit except as indicated.

THE SR-71B AIRCRAFT

The SR-71B is a trainer aircraft with a full set of engine and flight controls in the aft cockpit. For pilot training purposes, the student pilot (S/P) occupies the forward cockpit and the instructor pilot (I/P) occupies the aft cockpit. For other than training flights the pilot will normally occupy the forward cockpit.

The forward cockpit of the SR-71B appears identical to the forward cockpit of an SR-71A except for the addition of a control transfer panel, an emergency intercom system switch, a panel to indicate autopilot modes selected by the other cockpit, ALERT AND BAILOUT warning lights, and the lack of a TACAN CONT switch and a system 3 liquid nitrogen gage. The aft cockpit duplicates most of the functions in the forward cockpit and retains basic navigational capabilities. The trainer does not have camera equipment, viewsight, radar, or electromagnetic radiation gear. The trainer external configuration is distinguished by a stabilizing fin extending downward along the bottom centerline of each engine nacelle, and the aft cockpit canopy is higher than the forward cockpit canopy.

AIRCRAFT GROSS WEIGHT

The gross weight of the SR-71B trainer aircraft with two pilots and full fuel load is approximately 139,200 pounds. Zero fuel weight is approximately 59,000 pounds. Refer to weight and balance handbook.

ENGINE AND AFTERBURNER

Two throttles are installed in each cockpit. The throttles are interconnected by a cable system which provides each cockpit with the capability of controlling engines from idle cutoff to maximum afterburner.

TEB Remaining Counters

The TEB counters are installed only on the throttle quadrant in the forward cockpit.

EGT Gages

There are two EGT gages, one for each engine, in each cockpit. The two gages in the aft cockpit repeat the forward cockpit temperature indications but do not affect the operation of the fuel derich system. The fuel derich system is actuated only by the forward cockpit gage indications. Because of gage tolerances, the rear cockpit gages can indicate as much as 16°C different from the forward cockpit gages. HOT and COLD flag indicators in each cockpit independently display signals from the automatic EGT control system. The overtemperature warning light in each gage operates independently.

CAUTION

The fuel derich system is activated only by the forward cockpit gage indications. Failure of a forward cockpit EGT gage results in loss of derich protection for the respective engine.

INSTRUMENT PANEL - FORWARD COCKPIT

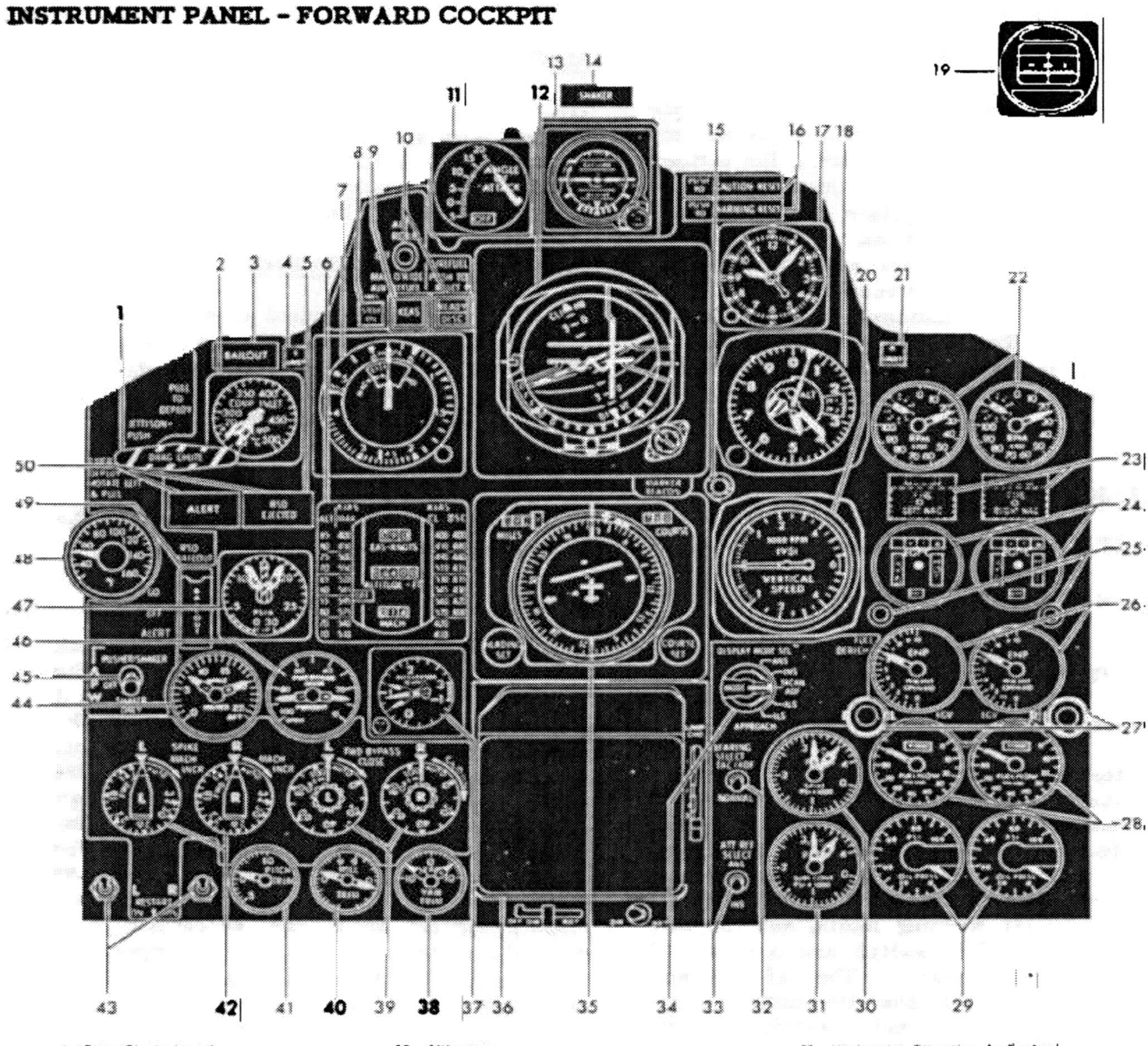

1 Drag Chute Handle
2 Compressor Inlet Temp Gage
3 Egress Light
4 Left Unstart Indicator Light
5 RSO Ejected Light
6 Triple Display Indicator
7 Airspeed-Machmeter
8 Nosewheel Steering Engaged Light
9 KEAS Warning Light
10 Air Refuel Switches
11 Angle of Attack Indicator
12 Attitude Director Indicator
13 Standby Attitude Indicator
14 Shaker Indicator Light
15 Marker Beacon Light
16 Master Caution and Warning Light
17 Elapsed Time Clock
18 Altimeter
19 Standby Compass (In Canopy)
20 Inertial-Lead Vertical Speed Indicator
21 Right Unstart Indicator Light
22 Tachometer
23 Fire Warning Lights
24 Exhaust Gas Temperature Indicators
25 Fuel Derich Lights
26 Exhaust Nozzle Position Indicators
27 IGV Lights
28 Fuel Flow Indicators
29 Oil Pressure Indicators
30 L and R Hydraulic Systems Pressure Gage
31 A and B Hydraulic Systems Pressure Gage
32 Bearing Select Switch
33 Attitude Reference Selector Switch
34 Display Mode Selector Switch
35 Horizontal Situation Indicator
36 Navigation Map Projector
37 Accelerometer
38 Yaw Trim Indicator
39 Forward Bypass Control Knobs
40 Roll Trim Indicator
41 Pitch Trim Indicator
42 Spike Control Knobs
43 Inlet Restart Switches
44 Spike Position Indicator
45 APW Switch
46 Forward Bypass Position Indicator
47 Compressor Inlet Pressure Gage
48 Temperature Indicator
49 RSO Bailout Switch
50 Egress Lights

F203-141(h)

Figure 1A-1

INSTRUMENT SIDE PANELS - FORWARD COCKPIT

1 Manifold Temperature Switch
2 Landing and Taxi Light Switch
3 Suit Heat Rheostat
4 Cockpit Temperature Control and O'Ride
5 Face Heat Rheostat
6 Cockpit Temperature Control
7 Defog Switch
8 L Refrigeration Switch
9 Temperature Indicator Selector Switch
10 R Refrigeration Switch
11 Center of Gravity Indicator
12 Fuel Quantity Indicator
13 Fuel Crossfeed Switch
14 Liquid Nitrogen Quantity Indicator
15 Forward Transfer Switch
16 Emergency Fuel Shutoff Switches
17 Battery Switch
18 Emergency AC Bus Switch
19 Instrument Inverter Switch
20 Generator Bus Tie Switch
21 L and R Generator Switches
22 Fuel Dump Switch
23 Fuel Quantity Indicator Selector Switch
24 Fuel Boost Pump Light Test Switch
25 Pump Release Switch
26 Manual Aft Transfer Switch
27 Igniter-Purge Switch
28 Fuel Boost Pump Switches
29 Fuel Tank Pressure Indicator
30 Wet-Dry Switch
31 Brake Switch
32 Indicators and Light Test Switch
33 Fuel Derich Switch
34 Gear Signal Release Switch
35 Landing Gear Lever
36 Landing Gear Indicator Lights
37 Cabin Altimeter
38 Liquid Oxygen Quantity Indicator
39 Bay Air Switch
40 Cockpit Pressure Dump Switch

F203-142(F)

Figure 1A-2

INSTRUMENT PANEL - AFT COCKPIT

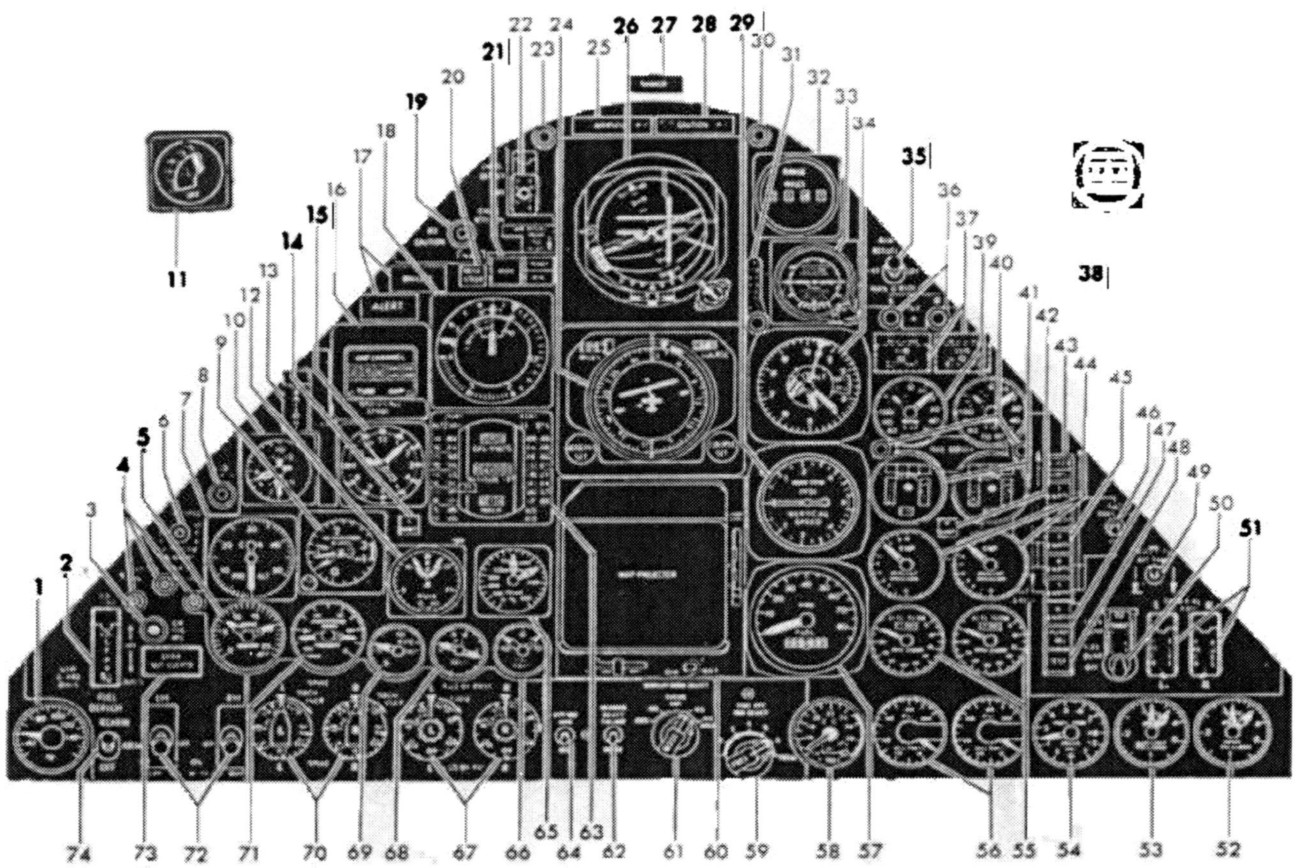

1 Temperature Indicator
2 Landing Gear Switch
3 Gear Signal Release Switch
4 Landing Gear Indicator Lights
5 Spike Position Indicator
6 Brake Switch
7 Cabin Altimeter
8 Indicators and Light Test Switch
9 Accelerometer
10 Liquid Oxygen Quantity Indicator
11 Angle of Attack Indicator (Mounted in Glare Shield)
12 Compressor Inlet Pressure Gage
13 S/P Bailout Switch
14 Left Unstart Indicator Light
15 Elapsed Time Clock
16 UHF Frequency Indicator
17 Egress Lights
18 Airspeed - Mach Meter
19 IFF Caution Light
20 Nose Wheel Steering Engaged Light
21 KEAS Warning Light
22 Drag Chute Switch
23 Supplementary Master Warning Light
24 Horizontal Situation Indicator
25 Master Warning Light
26 Attitude Director Indicator
27 Shaker Indicator Light
28 Master Caution Light
29 Inertial-lead Vertical Speed Ind.
30 Supplementary Master Caution Light
31 Marker Beacon Light
32 Distance Indicator
33 Standby Attitude Indicator
34 Altimeter
35 Air Refuel Switch
36 Inlet Guide Vane Lights
37 Engine Fire Warning Lights
38 Standby Compass (Mounted in Canopy)
39 Tachometer
40 Fuel Derich Lights
41 Exhaust Gas Temperature Indicators
42 Right Unstart Indicator Light
43 Fuel Crossfeed Switch
44 Exhaust Nozzle Position Indicators
45 Fuel Boost Pump Switches
46 Fuel Forward Transfer Switch
47 Pump Release Switch
48 Fuel Boost Pump Lights Test Switch
49 Manual Aft Transfer Switch
50 Fuel Dump Switch
51 Emergency Fuel Shutoff Switches
52 Hydraulic Pressure-Spike Dual Indicator
53 Hydraulic Pressure-Surface Control Dual Indicator
54 Fuel Tank Pressure Indicator
55 Fuel Flow Indicators
56 Engine Oil Pressure Indicators
57 Fuel Quantity Indicator
58 Center of Gravity Indicator
59 Fuel Quantity Indicator Select Switch
60 Navigation Map Projector
61 Display Mode Selector Switch
62 Bearing Select Switch
63 Triple Display Indicator
64 Attitude Reference Selector Switch
65 Compressor Inlet Temperature Indicator
66 Yaw Trim Indicator
67 Forward Bypass Switches
68 Roll Trim Indicator
69 Pitch Trim Indicator
70 Spike Switches
71 Forward Bypass Position Indicator
72 Restart Switches
73 Gear Unsafe Warning Light
74 Fuel Derich Switch

F203-144(F)

Figure 1A-3

LEFT AND RIGHT CONSOLES - FOR

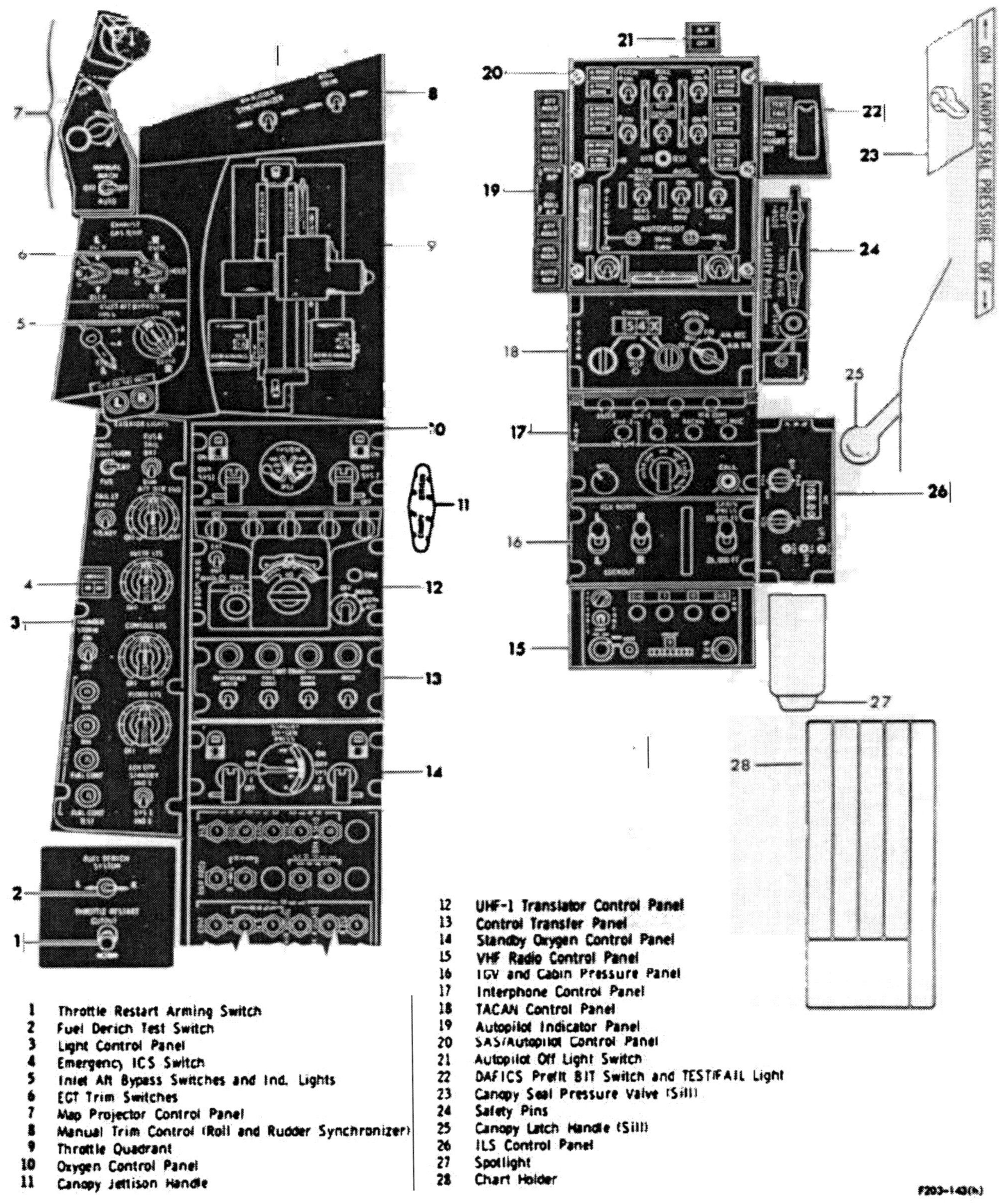

1 Throttle Restart Arming Switch
2 Fuel Derich Test Switch
3 Light Control Panel
4 Emergency ICS Switch
5 Inlet Aft Bypass Switches and Ind. Lights
6 EGT Trim Switches
7 Map Projector Control Panel
8 Manual Trim Control (Roll and Rudder Synchronizer)
9 Throttle Quadrant
10 Oxygen Control Panel
11 Canopy Jettison Handle
12 UHF-1 Translator Control Panel
13 Control Transfer Panel
14 Standby Oxygen Control Panel
15 VHF Radio Control Panel
16 IGV and Cabin Pressure Panel
17 Interphone Control Panel
18 TACAN Control Panel
19 Autopilot Indicator Panel
20 SAS/Autopilot Control Panel
21 Autopilot Off Light Switch
22 DAFICS Preflt BIT Switch and TEST/FAIL Light
23 Canopy Seal Pressure Valve (Sill)
24 Safety Pins
25 Canopy Latch Handle (Sill)
26 ILS Control Panel
27 Spotlight
28 Chart Holder

F203-143(h)

Figure 1A-4

LEFT AND RIGHT CONSOLES - AFT COCKPIT

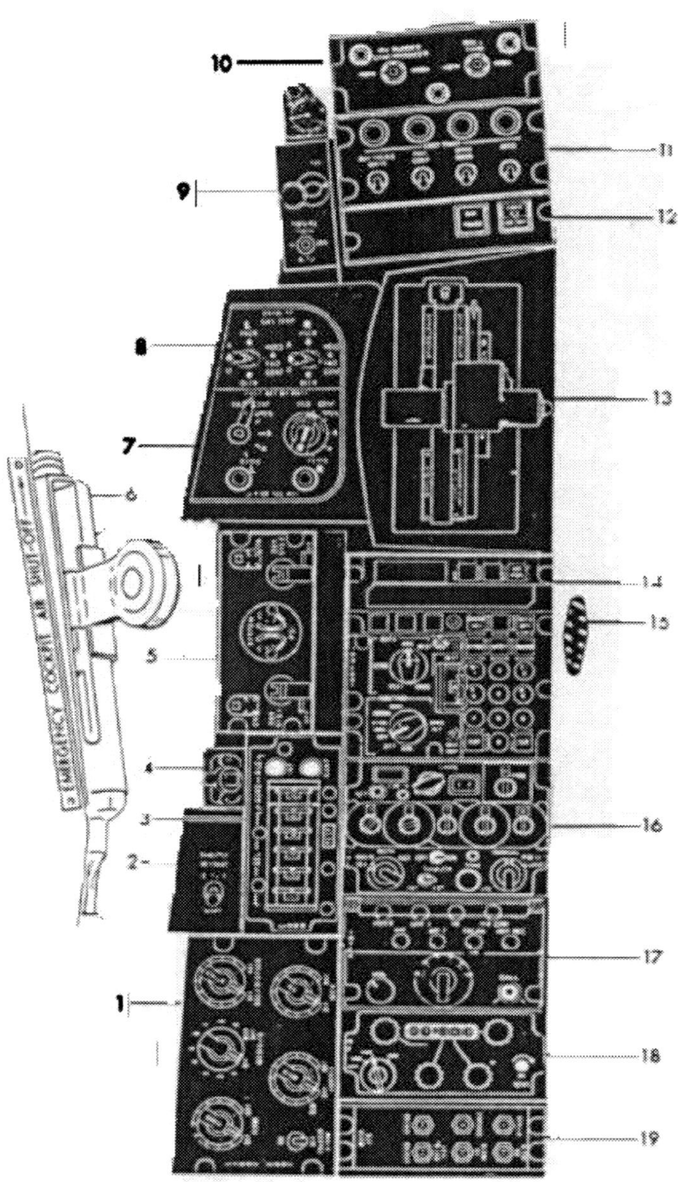

1. Light Control Panel
2. Throttle Restart Arming Switch
3. UHF Modem Control Panel
4. Spotlight Receptacle
5. Oxygen Control Panel
6. Cockpit Air Shutoff Control
7. Inlet Aft Bypass Switches and Indicator Lights
8. EGT Trim Switches
9. Map Projector Control Panel
10. Manual Trim Control (Roll and Rudder Synchronizer)
11. Control Transfer Panel
12. Emergency ICS Switch and UHF Transfer Control Switch
13. Throttle Quadrant
14. INS Control Panel
15. Emergency Canopy Jettison Handle
16. UHF-2 Radio Control Panel
17. Interphone Control Panel
18. HF Radio Control Panel
19. Circuit Breaker Panel
20. IFF Control Panel
21. Navigation Control and Display Panel
22. Tacan Control Panel
23. SAS/Autopilot Control Panel
24. Autopilot Indicator Panel
25. Autopilot OFF Light Switch
26. Canopy Seal Pressure Valve
27. DAFICS Preflt BIT Switch and TEST/FAIL Light
28. Temperature Indicator Selector Switch
29. Cabin Pressure Switch
30. Face Heat Switch
31. Canopy Latch Handle
32. MRS Power Switch
33. ILS Control Panel
34. INS Heading Slew Control
35. Circuit Breaker Panel
36. Seat and Canopy Safety Pins

F203-1450(a)

Figure 1A-5

ANNUNCIATOR PANEL - FORWARD COCKPIT

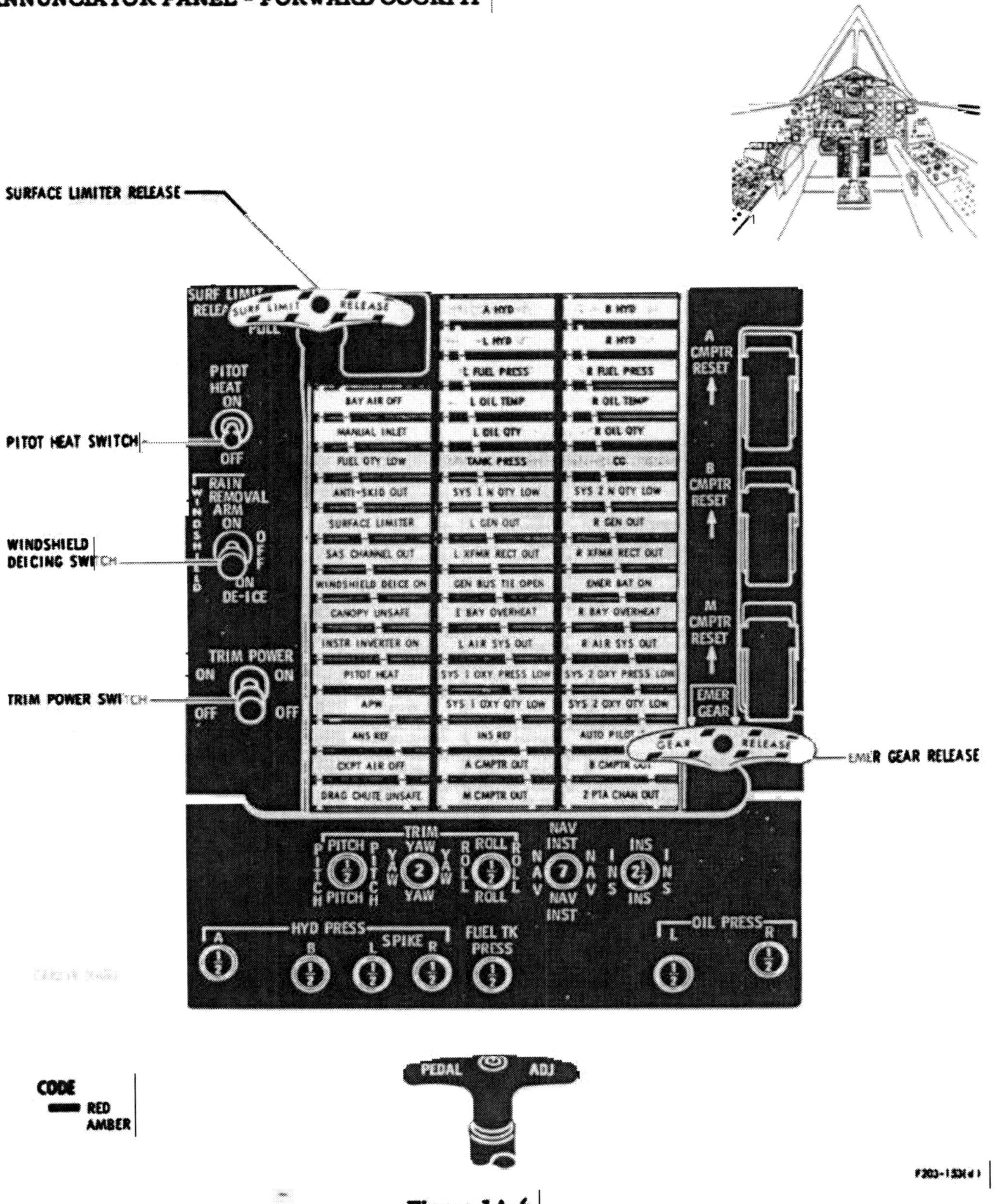

SURFACE LIMITER RELEASE
PITOT HEAT SWITCH
WINDSHIELD DEICING SWITCH
TRIM POWER SWITCH
EMER GEAR RELEASE

SURF LIMIT RELEASE PULL
PITOT HEAT ON OFF
WINDSHIELD RAIN REMOVAL ARM ON OFF ON DE-ICE
TRIM POWER ON ON OFF OFF

A HYD
B HYD
L HYD
R HYD
L FUEL PRESS
R FUEL PRESS
BAY AIR OFF
L OIL TEMP
R OIL TEMP
MANUAL INLET
L OIL QTY
R OIL QTY
FUEL QTY LOW
TANK PRESS
CG
ANTI-SKID OUT
SYS 1 N QTY LOW
SYS 2 N QTY LOW
SURFACE LIMITER
L GEN OUT
R GEN OUT
SAS CHANNEL OUT
L XFMR RECT OUT
R XFMR RECT OUT
WINDSHIELD DEICE ON
GEN BUS TIE OPEN
EMER BAT ON
CANOPY UNSAFE
E BAY OVERHEAT
R BAY OVERHEAT
INSTR INVERTER ON
L AIR SYS OUT
R AIR SYS OUT
PITOT HEAT
SYS 1 OXY PRESS LOW
SYS 2 OXY PRESS LOW
APW
SYS 1 OXY QTY LOW
SYS 2 OXY QTY LOW
ANS REF
INS REF
AUTO PILOT
CKPT AIR OFF
A CMPTR OUT
B CMPTR OUT
DRAG CHUTE UNSAFE
M CMPTR OUT
2 PTA CHAN OUT

A CMPTR RESET
B CMPTR RESET
M CMPTR RESET
EMER GEAR
GEAR RELEASE

TRIM PITCH YAW ROLL
NAV INST
INS
HYD PRESS A B
SPIKE L R
FUEL TK PRESS
OIL PRESS L R

PEDAL ADJ

CODE
RED
AMBER

F303-153(d)

Figure 1A-6

ANNUNCIATOR PANEL - AFT COCKPIT

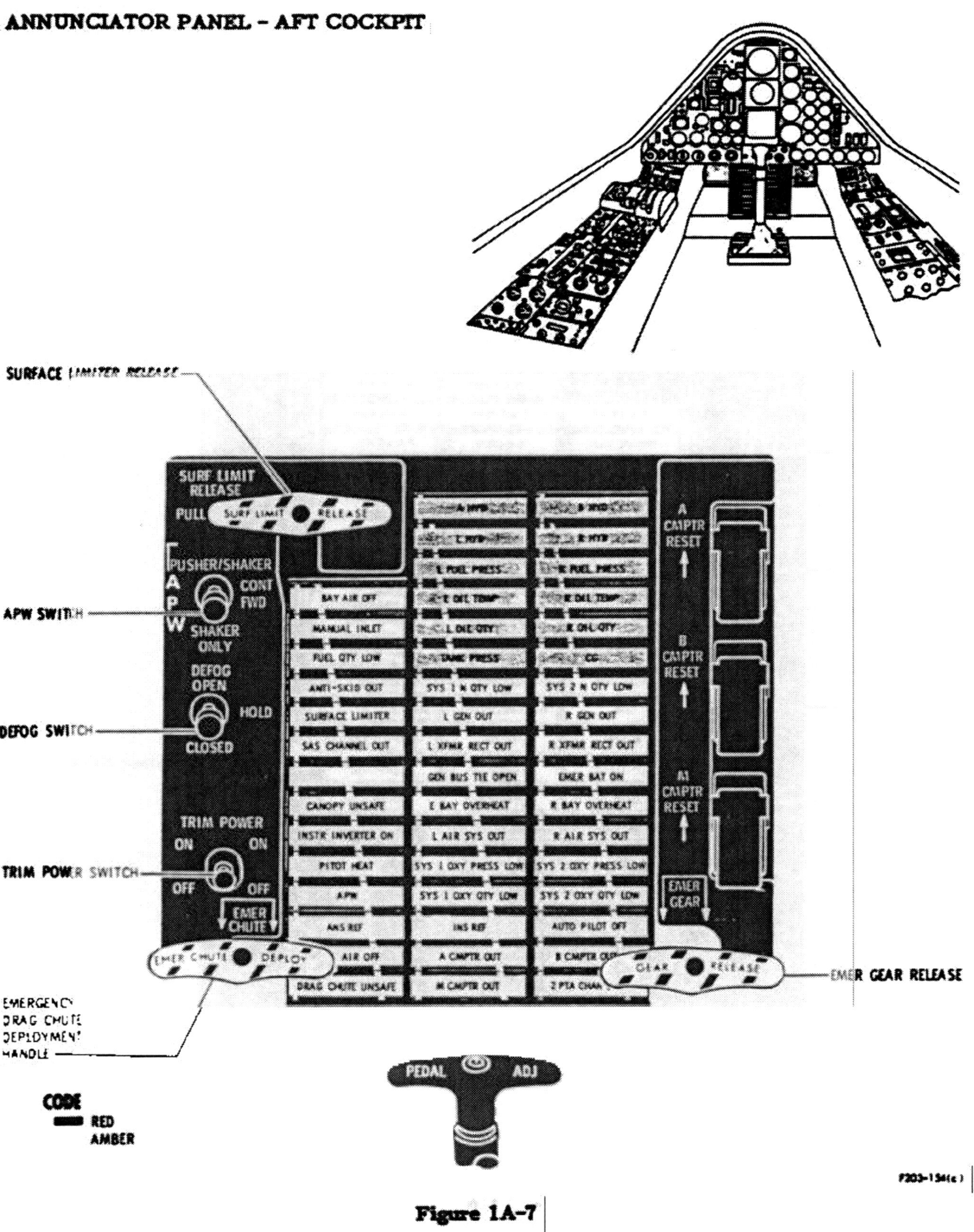

Figure 1A-7

EGT Trim Switches

EGT trim switches are installed in both cockpits. Operation of the aft cockpit switches out of the center position (marked HOLD & FWD CONT) overrides the switch positions in the forward cockpit.

Fuel Derich Arming Switch

A fuel derich arming switch is installed in each cockpit. Only the switch in the cockpit with the FUEL CONT transfer light illuminated on the control transfer panel is functional.

Fuel Derich System Test Switch

A fuel derich test switch is installed in the forward cockpit only. During the derich system test the aft cockpit EGT gages indicate only 200°C above the nominal EGT indication.

Fuel Derich Warning Light

Two fuel derich warning lights, one for each engine, are installed in each cockpit. The aft cockpit lights repeat the forward cockpit lights and are not affected by aft cockpit EGT indications.

IGV Lockout Switches

IGV lockout switches are installed in the forward cockpit only. IGV postion lights are installed in each cockpit.

Igniter Purge Switch

The igniter purge switch is installed in the forward cockpit only.

AIR INLET SYSTEM

Inlet Aft Bypass Switches and Indicators

The aft bypass control switches and indicator lights are installed in both cockpits. The aft cockpit switches have an extra position, labeled FWD CONTROL, which allows the forward cockpit to control the aft bypass doors. Operation of the switches in the aft cockpit out of the FWD CONTROL position overrides the switch positions in the forward cockpit.

Spike and Forward Bypass Control Knobs

The four rotary knobs for spike and forward bypass control are only functional in the cockpit with the SPIKE DOOR transfer light illuminated on the control transfer panel.

Inlet Restart Swtiches

The forward cockpit restart switches are only functional when the forward cockpit has the SPIKE DOOR transfer light illuminated on the control transfer panel. The RESTART ON position of either aft cockpit restart switch: puts the corresponding inlet in restart; overrides the SPIKE DOOR transfer switches to put the aft cockpit in control of the spikes and forward bypass doors of both inlets, regardless of which cockpit previously had SPIKE DOOR control; and illuminates the MANUAL INLET caution lights in both cockpits and the SPIKE DOOR transfer light on the aft cockpit control transfer panel. When both aft cockpit restart switches are returned to the OFF position, spike and forward bypass control as well as SPIKE DOOR transfer light illumination reverts to the cockpit selected on the control transfer panel.

Throttle Restart Switch

The forward cockpit throttle restart switch is only functional when the forward cockpit has SPIKE DOOR control. The aft cockpit throttle restart switch, when armed, is always functional. In addition, positioning the aft cockpit throttle restart switch out of the OFF position overrides the SPIKE DOOR transfer switches to put the aft cockpit in control of the spikes as well as the forward bypass doors of both inlets regardless of which cockpit previously had SPIKE DOOR control, and illuminates the MANUAL INLET caution lights in both cockpits and the SPIKE

DOOR transfer light on the aft cockpit control transfer panel. When the aft cockpit throttle restart switch is returned to the OFF position, spike and forward bypass control as well as SPIKE DOOR transfer light illumination reverts to the cockpit selected on the control transfer panel.

Throttle Restart Arming Switch

A throttle restart arming switch is installed in each cockpit. Each restart arming switch operates independently from the other.

Manual Inlet Indicator Light

The MANUAL INLET lights on the annunciator panels in both cockpits illuminate simultaneously when: the crewmember with SPIKE DOOR control moves one or more of the four rotary spike and/or forward bypass control switches out of the AUTO position; the crewmember with SPIKE DOOR control moves any inlet restart switch out of the OFF position regardless of which cockpit previously had SPIKE DOOR control.

FUEL SUPPLY SYSTEM

FUEL SYSTEM CONTROLS, INSTRUMENTS AND INDICATOR LIGHTS

Crossfeed Switch

The cockpits have independent operation of the crossfeed switch, but actual control of the crossfeed valves (and the X-FEED and OPEN light indications in both cockpits) is retained by the cockpit that has the FUEL CONT transfer light illuminated on the control transfer panel. When control of the fuel system is transferred, the crossfeed valve assumes the position commanded in the cockpit taking control.

Fuel Boost Pump Switches and Indicator Lights

Identical square fuel boost pump switches are installed in each cockpit. Manual boost pump control and the pump release switch are only functional in the cockpit that has the FUEL CONT transfer light illuminated on the control transfer panel. Numerals showing boost pump relays actuated and tank EMPTY lights appear simultaneously in both cockpits, thus providing monitoring of fuel sequencing to both cockpits. When control of the fuel system is transferred, fuel sequencing reverts to automatic, even though it may have been manually supplemented previously. If manual sequencing is desired by the cockpit taking control, boost pump switches must again be manually depressed.

Tank Lights Test Switch

The square pushbutton tank lights TEST switch below the pump release switch in each cockpit tests only the lights in that cockpit. The switch works regardless of which cockpit has the FUEL CONT transfer light illuminated on the control transfer panel.

Forward Fuel Transfer Switch

A two-position forward transfer switch, is installed in each cockpit.

NOTE

> Either cockpit can initiate forward fuel transfer regardless of which cockpit has the FUEL CONT light illuminated on the transfer control panel. Both switches must be off to terminate transfer.

Aft Transfer Swtich

An aft transfer switch is installed in each cockpit. Both switches must be off to terminate transfer.

Fuel Dump Swtich

A fuel dump switch is installed in each cockpit.

NOTE

Operating either fuel dump switch to the FUEL DUMP position will open the fuel dump valves regardless of which cockpit has the FUEL CONT Light illuminated on the transfer control panel. Operating either switch to the EMER DUMP position will dump all remaining fuel. Both switches must be off to terminate dumping.

Emergency Fuel Shutoff Switch

Emergency Fuel Shutoff switches are installed in each cockpit. Operating either switch to off (up) shuts off the fuel to the respective engine.

Fuel Quantity Selector Switch

A fuel quantity selector switch is installed on the right console in the forward cockpit and on the lower portion of the instrument panel in the aft cockpit. Operation of a fuel quantity selector switch and its respective quantity indicator is independent of the selector switch and indicator in the other cockpit.

Liquid Nitrogen Quantity Indicator

A liquid nitorgen quantity indicator is installed in the forward cockpit only. System 3 liquid nitrogen system and indicator are not installed.

Air Refuel Ready Switch

An air refuel ready switch is installed in each cockpit. Positioning this aft cockpit air refuel switch out of the OFF position overrides the forward cockpit switch positions.

Air Refuel Reset Switch and Indicator Light

A square, air refuel mechanism resetting pushbutton, is located on each instrument panel. The switches operate in parallel, so that depressing either switch will reset the signal amplifier and recycle the refueling receptacle locking mechanism.

Disconnect Trigger Switch

A disconnect trigger switch, is installed on each control stick grip. Depressing either trigger switch will initiate a refueling disconnect. Depressing either trigger opens the receptacle latches during air refueling using manual override.

ELECTRICAL SYSTEM

The electrical supply system of the SR-71B aircraft is basically identical with the system in the SR-71A aircraft except for some of the circuit breakers, (see Figure 1A-8). Other than circuit breakers, there are no electrical system contols in the aft cockpit.

NOTE

The following controls are only provided in the forward cockpit.

- L and R generator switches
- Generator bus tie switch
- Battery switch
- Instrument inverter switch
- Emergency ac bus switch

LANDING GEAR SYSTEM

The landing gear is controlled from the forward cockpit by a landing gear lever, and from the aft cockpit by a switch.

Landing Gear Lever

A landing gear lever is located in the forward cockpit only.

Landing Gear Switch

A guarded, lock-wired, three-position toggle switch for operating the landing gear is located on the lower left side of the instrument panel in the aft cockpit. The UP (up) and DOWN (down) positions of the switch override the position of the landing gear lever in the forward cockpit. After use, if the switch is returned to OFF (center) and the landing gear lever in the forward cockpit is not in agreement with the actual position of the landing gear, the gear position does not change but hydraulic pressure and

CIRCUIT BREAKER FUNCTION TABLE – TRAINER

CIRCUIT BREAKER	EFFECT OF POWER INTERRUPTION
	Essential DC BUS (Forward Cockpit)
WARN	Disabled: All forward cockpit warning and caution lights except nacelle fire and landing gear warning lights, and 2 PTA CHAN OUT, SAS OUT, and A, B & M COMPTR OUT caution lights.
	Essential DC BUS (Aft Cockpit)
AFCS CONT XFR	Disabled: AFCS transfer control and light. AFCS control reverts to forward cockpit.
TEMP IND	Disabled: Aft cockpit temperature indication for cockpit, R-Bay and E-Bay.
SPK-DR IND	Disabled: Aft cockpit left and right spike and door position indicators.
MAP PROJ	Disabled: Aft cockpit map projector.
WARN LTS	Disabled: All aft cockpit warning and caution lights except nacelle fire and landing gear warning lights, and 2 PTA CHAN OUT, SAS OUT, and A, B & M CMPTR OUT caution lights.
TRANS CONT	Disabled: NAV/TAC/ILS INSTR, FUEL CONT, and SPIKE DOOR transfer control and lights. NAV/TAC/ILS INSTR, FUEL CONT, and SPIKE DOOR control reverts to the forward cockpit.
AFCS A LT	Disabled: A PITCH and A YAW SENSOR/SERVO, A ROLL SERVO, ROLL SENSOR, A CMPTR OUT, 2 PTA CHAN OUT, SAS OUT and BIT TEST lights. (Both cockpits)
AFCS B & M LTS	Disabled: B PITCH and B YAW SENSOR/SERVO, M PITCH and M YAW SENSOR, B ROLL SERVO, B CMPTR OUT, M CMPTR OUT, and BIT FAIL lights. (Both cockpits)
	Essential AC BUS (Aft Cockpit)
L EGT IND (AØ)	Disabled: Aft cockpit left EGT indicator.
R EGT IND (AØ)	Disabled: Aft cockpit right EGT indicator.
L ENP (AØ)	Disabled: Aft cockpit left nozzle position indicator.
R ENP (AØ)	Disabled: Aft cockpit right nozzle position indicator.
L CIT IND (AØ)	Disabled: Aft cockpit left compressor inlet temperature indication.
R CIT IND (AØ)	Disabled: Aft cockpit right compressor inlet temperature indication.
PANEL LTS (AØ)	Disabled: LEGEND, LEGEND BRT & TEST and CONSOLE circuit breakers on left console.
MAP PROJ (BØ)	Disabled: Aft cockpit map projector speed control.
INSTR LTS (BØ)	Disabled: LH INSTR, RH INSTR and ATTACK circuit breakers on left console.
CIP (CØ)	Disabled: Aft cockpit left and right compressor inlet pressure indicators. Barber pole continues to function.
LEGEND (AØ, Left Console)	Disabled: Aft cockpit switch legends when in the dimming range.
LEGEND BRT & TEST (AØ, Left Console)	Disabled: Aft cockpit switch legends when in bright and in Warning lights test.
CONSOLE (AØ, Left Console)	Disabled: Aft cockpit console lights
LH INSTR (BØ, Left Console)	Disabled: Aft cockpit ADI, HSI and left hand instrument panel lighting.
RH INSTR (BØ, Left Console)	Disabled: Aft cockpit right hand instrument panel lighting.
ATTACK (BØ, Left Console)	Disabled: Aft cockpit angle of attack indicator light.
	26 VOLT EMERGENCY AC BUS (Forward Cockpit)
HSI CRS & HDG	Disabled: Manual course and heading inputs to HSI in cockpit with NAV/TAC/ILS INSTR control are not repeated in the other cockpit.

Figure 1A-8

CIRCUIT BREAKER PANELS

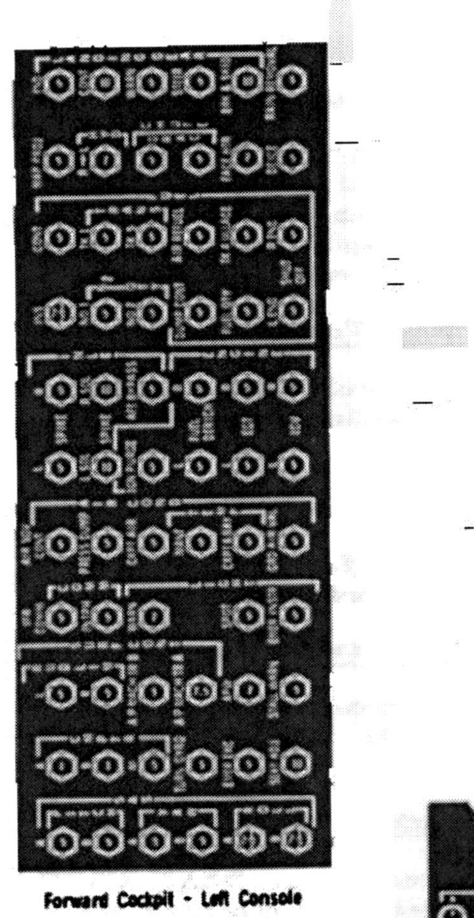

Forward Cockpit - Left Console

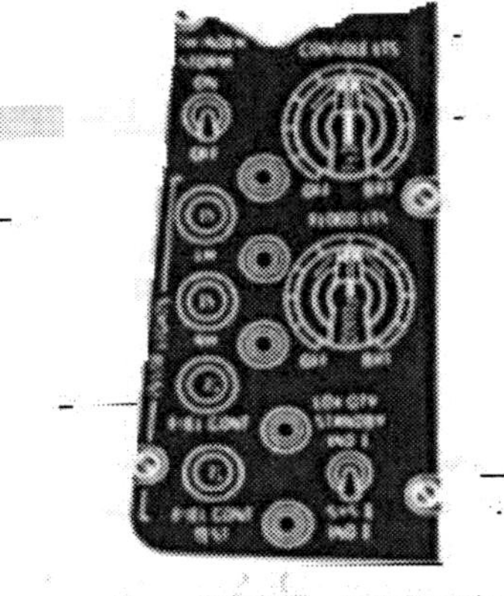

Forward Cockpit - Left Console

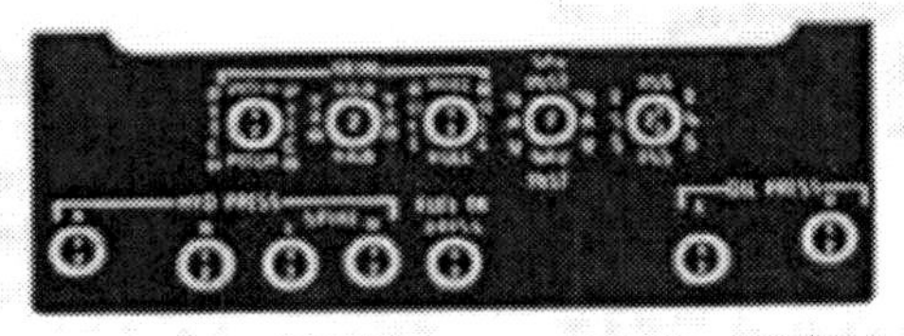

Forward Cockpit - Center Stand

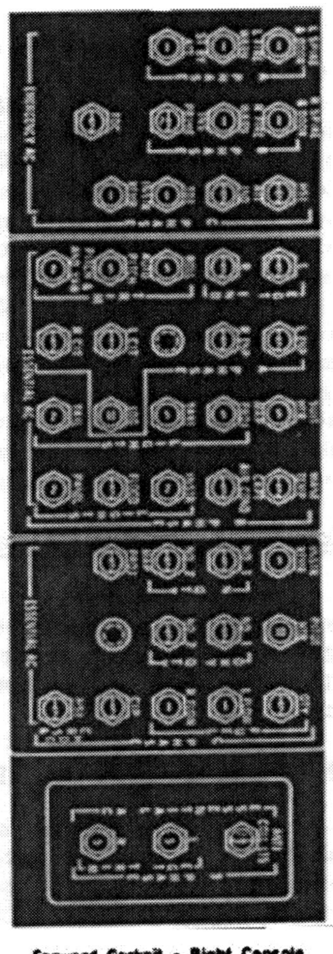

Forward Cockpit - Right Console

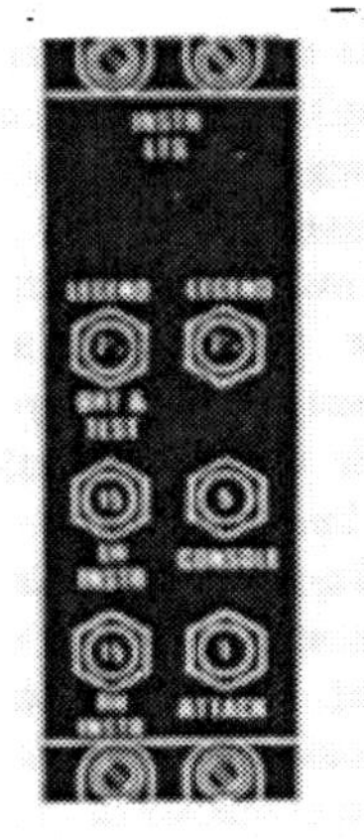

Aft Cockpit - Left Console

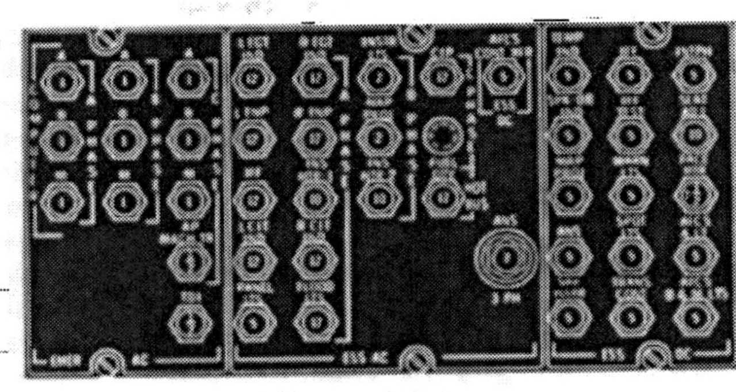

Aft Cockpit - Right Console

Figure 1A-9

electrical command of gear position is removed from the landing gear system. If the switch is returned to OFF and the landing gear lever in the forward cockpit is in the position of the landing gear or is moved to the position of the landing gear, the landing gear lever and system again become operative. The switch is normally lock-wired in the OFF position and must be in that position for normal gear operation from the forward cockpit. Location of the switch in its center position can be verified by noting the appearance of a yellow dot in the end of the switch which should be visible through an aperture in the guard when it is lowered.

Manual Landing Gear Release Handle

A manual landing gear release handle is installed in each cockpit; the two handles are interconnected mechanically and pulling either handle will release the landing gear latches.

Landing Gear Warning Cutout Button

A landing gear warning cutout pushbutton is installed in each cockpit. Depressing the button in either cockpit will disable the warning circuit in both cockpits.

Gear Unsafe Warning Light

The gear unsafe warning light, labeled GEAR NOT LOCKED, is located on the lower left portion of the instrument panel in the aft cockpit only. The light functions similarly to the light in the gear handle in the forward cockpit, lighting red whenever the gear is not in the position called for by the gear handle position or the aft cockpit landing gear switch.

NOSEWHEEL STEERING SYSTEM

Nosewheel Steering Button

A nosewheel steering pushbutton is provided on the control stick grip in each cockpit. Either pushbutton may be depressed to engage or release nosewheel steering. A nosewheel steering engaged light is installed in each cockpit.

WHEEL BRAKE SYSTEM

Brake Switch

A brake switch is located in both cockpits. Positioning the switch in the aft cockpit out of the OFF position overrides the forward cockpit switch positions. To control the wheel brake system from the forward cockpit, the aft cockpit brake switch must be in the OFF (center) position.

Antiskid Disconnect Trigger Switch

With S/B R-2695, antiskid system operation is interrupted while either trigger switch is held depressed.

DRY-WET Switch

The dry-wet switch for braking selection is installed in the forward cockpit only.

DRAG CHUTE SYSTEM

Normal and emergency deployment of the drag chute can be initiated from either cockpit.

Drag Chute Switch

A guarded three-position DRAG CHUTE switch is provided in the upper left side of the instrument panel in the aft cockpit. (See Figure 1A-10.) Its guarded OFF (center) position corresponds to the stowed position of the drag chute handle in the forward cockpit. The guard must be raised to move the switch to the CHUTE DEPLOY (up) position, or to the CHUTE JETTISON (down) position. The switch is automatically reset to the OFF position when the guard is lowered. Location of the switch in its center position can be verified by noting the appearance of a yellow dot in the end of the switch which should be visible through an aperture in the guard when it is lowered. The control in the aft cockpit can always be used to operate the drag chute mechanism. The aft cockpit switch must be placed in its guarded OFF position to transfer control of the drag chute mechanism to the forward cockpit crewmember for normal deployment or jettisoning.

NOTE

If the aft cockpit deploys the drag chute and the forward cockpit handle remains stowed, returning the aft cockpit drag chute switch to OFF will jettison the drag chute.

INSTRUMENT PANEL - Aft Cockpit SR-71B

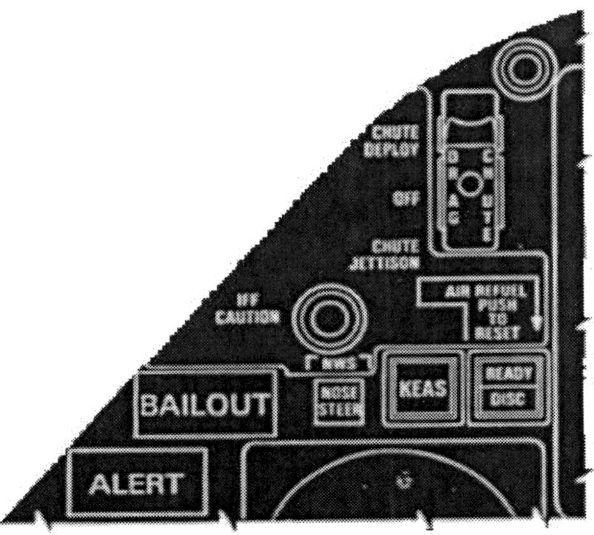

Figure 1A-10

Drag Chute Emergency Deployment

In the aft cockpit, an EMER CHUTE DEPLOY Tee-handle is provided at the lower left edge of the annunciator panel. This handle is attached to the same cable and mechanism which is used to deploy the chute manually from the forward cockpit, and approximately the same pull force and handle motion are required.

FLIGHT CONTROL SYSTEM

The forward cockpit and aft cockpit flight controls are connected in tandem and work together.

Surface Limiter Control Handle

A surface limiter control handle is installed in each cockpit. The two handles are interconnected so that they move together; consequently, either pilot may operate the surface limiter system.

MANUAL TRIM SYSTEM

Trim Power Switch

A trim power switch is installed in each cockpit. The switches operate in series so that both must be ON to apply power to the trim motors. Operating either cockpit switch to OFF will turn off trim power.

Pitch and Yaw Trim Switch

A pitch and yaw trim switch is located on the control stick grip in each cockpit. The aft cockpit switch overrides the forward cockpit switch if they are operated simultaneously. Operation of either control stick trigger switch disables the control stick trim switch in both cockpits.

Roll Trim Switch

A roll trim switch is installed in each cockpit. Operating the aft cockpit roll trim switch overrides the forward cockpit switch if they are operated simultaneously.

Right Hand Rudder Synchronization Switch

A right hand rudder synchronization switch is installed in each cockpit. The switches are in parallel so that operating either switch will actuate the trim motor. Both must be in center (off) position to stop the trim motor.

DIGITAL AUTOMATIC FLIGHT AND INLET CONTROL SYSTEM (DAFICS)

DAFICS PRELIGHT BUILT IN TEST (BIT)

The DAFICS Preflight BIT switch only operates in the cockpit with the AFCS control transfer light illuminated on the control transfer panel. DAFICS Preflight BIT TEST/FAIL light indications (including ANR failure) appear simultaneously in both cockpits. The DAFICS Preflight BIT tests for switch input faults only in: SAS and autopilot switch inputs from the cockpit with AFCS control; inlet control switch inputs from the cockpit with SPIKE DOOR control; and one APW switch input (aft cockpit unless the aft cockpit APW switch is in CONT FWD).

Computer Reset Switches

Individual computer reset switches are installed in both cockpits. The computers can be restarted manually at any time with the computer RESET switches in either cockpit.

SAS/AUTOPILOT FUNCTION SELECTOR

Identical DAFICS function selector panels are installed in both cockpits. The cockpit with the AFCS transfer light illuminated on the control transfer panel has control of the SAS and autopilot.

SAS CONTROLS AND INDICATORS

SAS will not disengage when AFCS control is transferred provided the channel engage switches are on in the cockpit assuming control. The SAS channel engage switches and the ROLL SENSOR/SERVO recycle functions only operate in the cockpit with AFCS control. PITCH and YAW pushbutton SENSOR/SERVO recycle functions operate in either cockpit regardless of which cockpit has AFCS control. Identical SENSOR/SERVO light indications appear simultaneously in both cockpits. The SAS LITE TEST switch on each SAS control panel operates independently but illuminates the Preflight Bit TEST/FAIL lights in both cockpits.

WARNING

Only the positions of the SAS channel engage switches in the cockpit that has AFCS control effect the SENSOR/SERVO and SAS OUT caution lights. No warning is displayed if the SAS channel engage switches are not ON in the cockpit that does not have AFCS control. If the SAS engage switch(es) are OFF in the cockpit that does not have AFCS control, transferring AFCS control to that cockpit results in loss of SAS until the switch(es) are engaged.

AUTOPILOT SYSTEM

The cockpit with the AFCS transfer indicator light illuminated on the control transfer panel has control of the SAS and autopilot. The autopilot will disconnect if AFCS control is transferred to the opposite cockpit. Autopilot mode engage switches, trim wheels, and alignment indices only function on the SAS control panel in the cockpit with AFCS control.

Autopilot Indicator Panel

An autopilot indicator panel is located to the left of the SAS control panel in each cockpit. Autopilot mode lights only illuminate in the cockpit not in control of the AFCS to allow monitoring of the autopilot modes selected in the other cockpit.

Control Stick Command Switch

A control stick command (CSC) switch is installed in each cockpit. The switch is only functional in the cockpit with AFCS control.

Autopilot Disconnect Switch

An autopilot trigger disconnect switch is installed in each cockpit. Depressing either switch disengages the autopilot regardless of which cockpit has AFCS control.

A/P OFF Light

An A/P OFF light is installed in each cockpit. Depressing the light in either cockpit extinguishes the AUTOPILOT OFF annunciator light in both cockpits.

APW AND HIGH ANGLE OF ATTACK WARNING SYSTEMS

APW System Stick Warning Switch

A three-position APW system stick warning control switch is provided on the left side of the annunciator panel in the rear cockpit. The functions of the PUSHER/SHAKER (up) and SHAKER ONLY (down) positions are the same as for the APW control switch in the forward cockpit. Selection of either switch position overrides the position of the forward cockpit switch. Selection of the CONT FWD (center) position transfers control of the APW stick warning system to the forward cockpit.

PITOT-STATIC SYSTEMS

Pitot Heat Switch

The pitot heat switch is installed in the forward cockpit only.

FLIGHT INSTRUMENTS

The front cockpit flight instruments are duplicated in the aft cockpit except as indicated. (See Figures 1A-1 and 1A-3.)

Attitude Reference Selector Switch

Each cockpit can independently select the attitude reference source for the ADI in that cockpit. DAFICS autopilot and analytical redundancy inputs are determined by the position of the attitude reference switch in the cockpit with the AFCS transfer light illuminated on the control transfer panel.

HORIZONTAL SITUATION INDICATOR

The HSI displays and the corresponding ADI steering indications selected by the cockpit with the NAV/TAC/ILS INSTR transfer light illuminated on the control transfer panel are repeated in the other cockpit.

COMMUNICATION & AVIONIC EQUIPMENT

Microphone Switches

The forward cockpit microphone switches are identical with the like switches in the SR-71A aircraft forward cockpit. The aft cockpit has four microphone switches, one on the control stick, one on the inboard throttle, and one on each side of the floorboard near the scuff plates. The floor-mounted switch on the right side is only for operation of the interphone.

EMERGENCY INTERCOMMUNICATIONS SYSTEM

A separate, press to talk emergency system is provided for communicating between the two cockpits when ac power is lost. There is no hot microphone capability when the emergency ICS is activated. The system is controlled by identical square, self-illuminated pushbutton switches located on the light control panel in the forward cockpit and on a separate panel on the left console in the aft cockpit. The top half of the pushbuttons are labeled EMERG ICS. The EMERG ICS is put into operation by depressing either or both control pushbutton(s), which (1) illuminates a green ON legend on the pushbutton face in both cockpits, (2) connects the intercom directly to the battery bus and (3) isolates all other sources of audio from the headsets of the pilots.

Depressing the same pushbutton switch again deselects the EMERG ICS. If neither cockpit has EMERG ICS selected the amber OFF light in both cockpit pushbutton switches illuminates and the isolated audio sources are reconnected.

Power is furnished to the emergency ICS by the 28v dc battery through the EMER INTPH circuit breaker located in the E bay.

UHF COMMUNICATIONS AND NAVIGATION SYSTEM

UHF Control Transfer Switch

The UHF control transfer switch is located on the left console in the aft cockpit.

UHF Remote Frequency Indicator

The UHF remote frequency indicator is located on the aft cockpit instrument panel.

UHF Modulator/Demodulator Control Panel

The UHF modulator/demodulator (MODEM) control panel is located on the left console in the aft cockpit.

UHF Distance Indicator

The UHF distance indicator is located on the upper right of the instrument panel in the aft cockpit.

HF RADIO

The HF radio is located on the left console in the aft cockpit.

INSTRUMENT LANDING SYSTEM

An ILS control panel is installed on the right console in both cockpits. Only the ILS control panel in the cockpit with the NAV/TAC/ILS INSTR transfer light illuminated on the control transfer panel is operative. A marker beacon indicator is installed in each cockpit.

IFF

The IFF control panel is located on the right console in the aft cockpit.

TACAN

A TACAN control panel is installed on the right console in each cockpit.

WINDSHIELD

The aft cockpit windshield is similar to the forward cockpit but does not have hot-air deicing provisions or the liquid rain-removal system.

Defog Switch

Defog air is controlled independently in each cockpit. The defog switch in the aft cockpit is located on the left side of the annunciator panel.

Windshield Deice On Caution Light

The forward cockpit WINDSHIELD DEICE ON caution light is the only annunciator caution light installed on only one annunciator panel. Only the master caution light in the forward cockpit illuminates when the WINDSHIELD DEICE ON light illuminates.

CANOPIES

Canopy Unsafe Warning Light

The CANOPY UNSAFE warning light in both cockpits illuminates when either one or both of the canopies is not latched down and/or properly sealed.

REAR VIEW PERISCOPE

A rear view periscope is installed on each canopy. The field of view of the front periscope is partially blocked by the aft cockpit.

LIGHTING EQUIPMENT

EXTERIOR LIGHTS

All exterior lights including the landing, taxi, anti-collision, fuselage, and the tail light are controlled by switches located in the front cockpit only.

AFT COCKPIT INTERIOR LIGHTING

The instrument panel lights, console panel lights and floodlights are controlled by rheostat switches located on the lighting panel on the left console. A two position thunderstorm light switch is located on the lighting panel. Separate rheostat switches are provided on the aft cockpit lighting panel for the segment (alpha-numeric) lights on the INS control panel and the angle of attack gage.

The green landing gear indicator lights, the GEAR NOT LOCKED, NOSE STEER, KEAS, air refuel READY/DISC, SHAKER, IGV and DERICH lights in the aft cockpit are dimmed when the forward cockpit console lights switch is out of the OFF position. The remaining aft cockpit warning and caution lights are dimmed when the aft cockpit console lights switch is out of the OFF position.

ENVIRONMENTAL CONTROL SYSTEM

The following controls are only provided in the forward cockpit:

- Refrigeration switches
- Cockpit temperature control switch
- Cockpit air temperature control and override switch
- Manifold temperature control switch
- Bay air switch
- Cockpit pressure dump switch

A cabin pressure selector switch, cabin altimeter, temperature indicator, and temperature indicator selector switch are installed in each cockpit. The cockpit air handle is in the aft cockpit.

Temperature Indicator and Selector Switch

In the aft cockpit, the temperature indicator is located on the lower left side of the instrument panel and the temperature selector switch is on the right console. The temperature indicator and temperature selector switch for each cockpit operates independently.

LIFE SUPPORT SYSTEMS

Suit Heat Rheostat

The pressure suit heat control is installed only in the forward cockpit.

Visor Heat Rheostat

Visor heat is controlled independently in each cockpit. The visor heat rheostat in the aft cockpit is located on the right console.

CONTROL TRANSFER PANEL - SR-71B

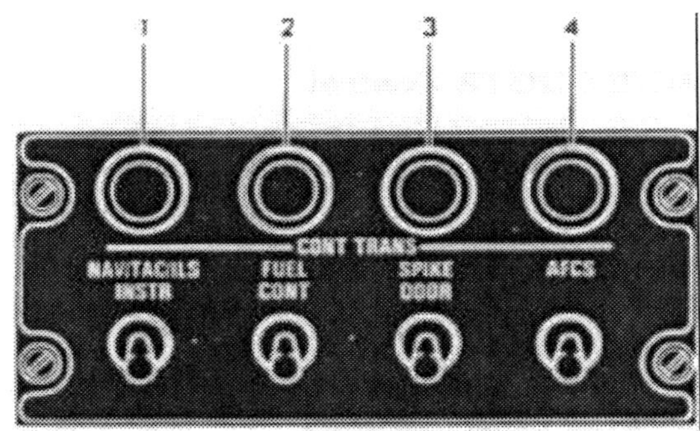

1 NAV/TAC/ILS control transfer switch and indicator light
2 FUEL CONT control transfer switch and indicator light
3 SPIKE DOOR control transfer switch and indicator light
4 AFCS control transfer switch and indicator light

F260-301

Figure 1A-11

CONTROL TRANSFER PANELS

A control transfer panel is located on the left console in both cockpits. Each panel contains four two-position toggle switches and four associated control transfer indicator lights. The switches allow either crewmember to give or take control of the aircraft system(s) associated with each switch. (See Figure 1A-11.) Changing the position of a control transfer switch once (fore or aft) on either control transfer panel transfers control from whichever cockpit previously had control to the other cockpit. Control of a system is indicated in the cockpit with control by illumination of the associated transfer indicator light. Control transfer and transfer indicator lights are powered by the Essential DC bus through the TRANS CONT and AFCS CONT XFR 28v dc circuit breakers.

NOTE

In the event of control transfer relay malfunction or loss of power in one of the four transfer circuits, control of the associated system reverts to the forward cockpit.

NAV/TAC/ILS INSTR Control

The cockpit with the NAV/TAC/ILS INSTR transfer light illuminated has control of the Display Mode Select switch, Bearing Select switch, HSI Course and Heading Set knobs, and the TACAN and ILS control panels. The ADI steering commands and HSI displays selected by the controlling cockpit are repeated in the other cockpit.

Fuel System Control

The cockpit with the FUEL CONT transfer light illuminated has control of those components within the fuel system that are operated manually: derich, crossfeed, boost pumps and pump release.

NOTE

When control of the fuel system is transferred, fuel sequencing reverts to automatic, even though it may have been manually supplemented previously. If manual sequencing is desired by the cockpit taking control, boost pump switches must again be manually depressed.

The crossfeed valve assumes the position commanded in the cockpit taking control.

Inlet Spike and Forward Bypass Control

The cockpit with the SPIKE DOOR transfer light illuminated has control of the inlet spike and forward bypass door control knobs. The forward cockpit restart switches and throttle restart switch are only functional when the forward cockpit has SPIKE DOOR control.

NOTE

When out of the OFF position, either aft cockpit restart switch or the aft cockpit throttle restart switch overrides the SPIKE DOOR transfer switch to put the aft cockpit in control of the spikes and forward bypass doors of both inlets and illuminates the MANUAL INLET caution lights in both cockpits and the SPIKE DOOR transfer light on the aft cockpit control transfer panel. When all aft restart switches are returned to the OFF position, SPIKE DOOR control as well as SPIKE DOOR transfer light illumination reverts to the cockpit selected on the control transfer panel.

Automatic Flight Control System Control

The cockpit with the AFCS transfer light illuminated has control of the stability augmentation system, autopilot controls, and the DAFICS Preflight BIT switch.

INTERTIAL NAVIGATION SYSTEM (INS)

Inertial Control Panel

The INS control panel is located on the left console in the aft cockpit.

Heading Slew Knob

The heading slew knob is located on the outboard aft portion of the right console in the aft cockpit.

INS Segment Lights Control

The INS segment (alpha-numeric) lights on the inertial control panel are controlled by the INS SEG LTS rheostat on the lighting control panel.

ASTROINERTIAL NAVIGATION SYSTEM

The Astroinertial Navigation System (ANS) control panel is located on the right console in the aft cockpit. Navigational accuracy may be degraded to the extent that position accuracy cannot be updated by reference to viewsight and radar system data.

Star tracking may be degraded because the field of view of the astrotracker is reduced by the position of the astrotracker window behind the aft canopy.

SENSOR EQUIPMENT

There is no sensor equipment installed in the trainer aircraft.

MISSION RECORDER

The mission recorder is controlled by a square, self-illuminated, pushbutton switch labeled MRS on the top half of the pushbutton and located on the right console in the aft cockpit. Depressing the pushbutton turns on the mission recorder, illuminating the lower left quarter of the pushbutton ON in green: depressing a lighted pushbutton turns off the recorder. In case a failure occurs in the 28-VDC or 400-cycle power supply to the recorder, the bottom right quarter of the pushbutton face lights FAIL in red. If power is subsequently restored the recorder will resume operation; however, the FAIL light will not go out, having to be reset on the ground.

EGRESS COORDINATION SYSTEM

The bailout switch located on the left side of the instrument panel in the forward cockpit is labeled RSO (rear seat occupant) BAILOUT; the bailout switch in the corresponding location in the aft cockpit is labeled S/P (student pilot) BAILOUT. Each switch is covered by a red guard and has three positions, OFF (center), ALERT (down), and GO (up). An ALERT and BAILOUT light are located on the left side of the instrument panel in both cockpits. Operating either bailout switch to the GO position causes the BAILOUT light in the other cockpit to illuminate. Operating either switch to the ALERT position illuminates the ALERT light in the other cockpit. The bailout switch is normally in the OFF position, with the guard down. When the aft cockpit seat ejects, RSO EJECTED illuminates in the forward cockpit, signifying that the forward cockpit pilot may eject safely; if the forward cockpit seat ejects first, the rear seat occupant must rely on sound and vision to determine when the forward seat is gone.

EMERGENCY WARNING EQUIPMENT

MASTER WARNING SYSTEM

The annunciator panel lights in the aft cockpit duplicate all the warnings and cautions displayed on the forward cockpit annunciator panel except one. The WINDSHIELD DEICE ON annunciator caution light is only installed in the forward cockpit.

Master Warning and Master Caution Lights

Depressing the master warning or master caution light extinguishes the master caution or warning light and causes the associated annunciator light to illuminate steady in that cockpit only.

A supplementary post-type warning and master caution light is located at the top of the instrument panel in the aft cockpit. These lights are fitted with a plastic lighttube so that they are visible when the canopy is down and the seat is in an elevated position.

Indicators and Warning Lights Test Button

The indicator and warning lights test button, labeled IND & LT TEST, is installed in each cockpit. Depressing the button in either cockpit will test the indicators and lights in both cockpits.

MISCELLANEOUS EQUIPMENT

Dinghy Stabber

A dinghy stabber is located on the glareshield in the forward cockpit and on the right side of the canopy in the aft cockpit.

Section II

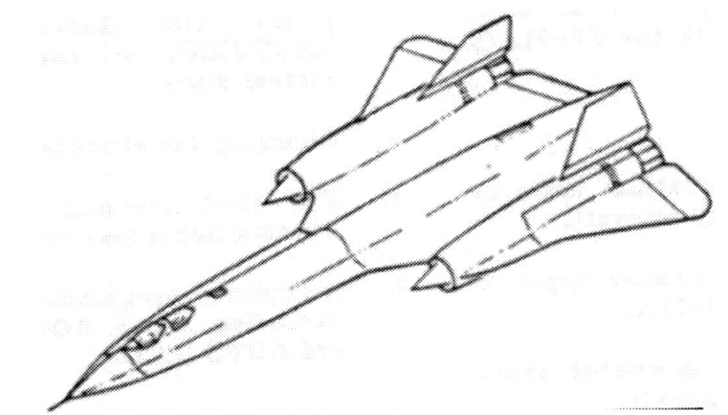

Normal Procedures

TABLE OF CONTENTS

INTRODUCTION

The following procedures provide an amplified listing which applies to the SR-71A/B aircraft.

Symbol Coding

1. Steps without special notations apply to the forward cockpit of all aircraft.

(2.) Steps with an enclosed number apply to the aft cockpit of the SR-71A.

▲3. Steps preceded by the ▲ symbol apply to both cockpits of all aircraft.

T 4. Steps preceded by a T apply to the forward cockpit of all aircraft as well as the aft cockpit of the SR-71B.

(T5.) Steps with an enclosed T and step number apply to the aft cockpit of SR-71A/B.

(T) 6. Steps preceded by an enclosed T apply only to the aft cockpit of the SR-71B.

The same system is used for abbreviated checklists which are provided separately for the Pilot and RSO. Interior Preflight checklists are provided for each crew position. From Starting Engines on, checklists are common.

CREW COORDINATION

Crew coordination is paramount to mission success and safety of flight. Communication between crewmembers should be continuous when accomplishing checklists. Verbal coordination between crewmembers is required prior to:

a. Going off interphone.

b. Going off aircraft oxygen system or opening faceplate.

c. Changing the programmed mission or steering reference points, changing the pilot's ANS distance display mode (DP/TURN), or changing navigational system mode.

d. Changing the attitude reference.

e. The pilot pressing the indicator and warning lights test button.

f. Autopilot engagement or disengagement (including KEAS HOLD, MACH HOLD, and AUTO NAV).

g. Change of fuel panel settings or fuel crossfeed or transfer operation.

The RSO must monitor aircraft attitude, altitude, and airspeed and advise the pilot if a potentially dangerous situation exists. This is particularly important during critical phases of flight involving substantial changes in aircraft attitude, altitude, and speed.

It may be advantageous to use the interphone system HOT MIC feature for crew communication during some flight phases.

PREPARATION FOR FLIGHT

FLIGHT RESTRICTIONS

Refer to Section V for operating restrictions and limitations.

FLIGHT PLANNING

Refer to Appendix I.

TAKEOFF AND LANDING DATA

Refer to Appendix I.

WEIGHT AND BALANCE

For detailed loading information, refer to Handbook of Weight and Balance data.

PERSONAL EQUIPMENT HOOKUP - Pressure Suit

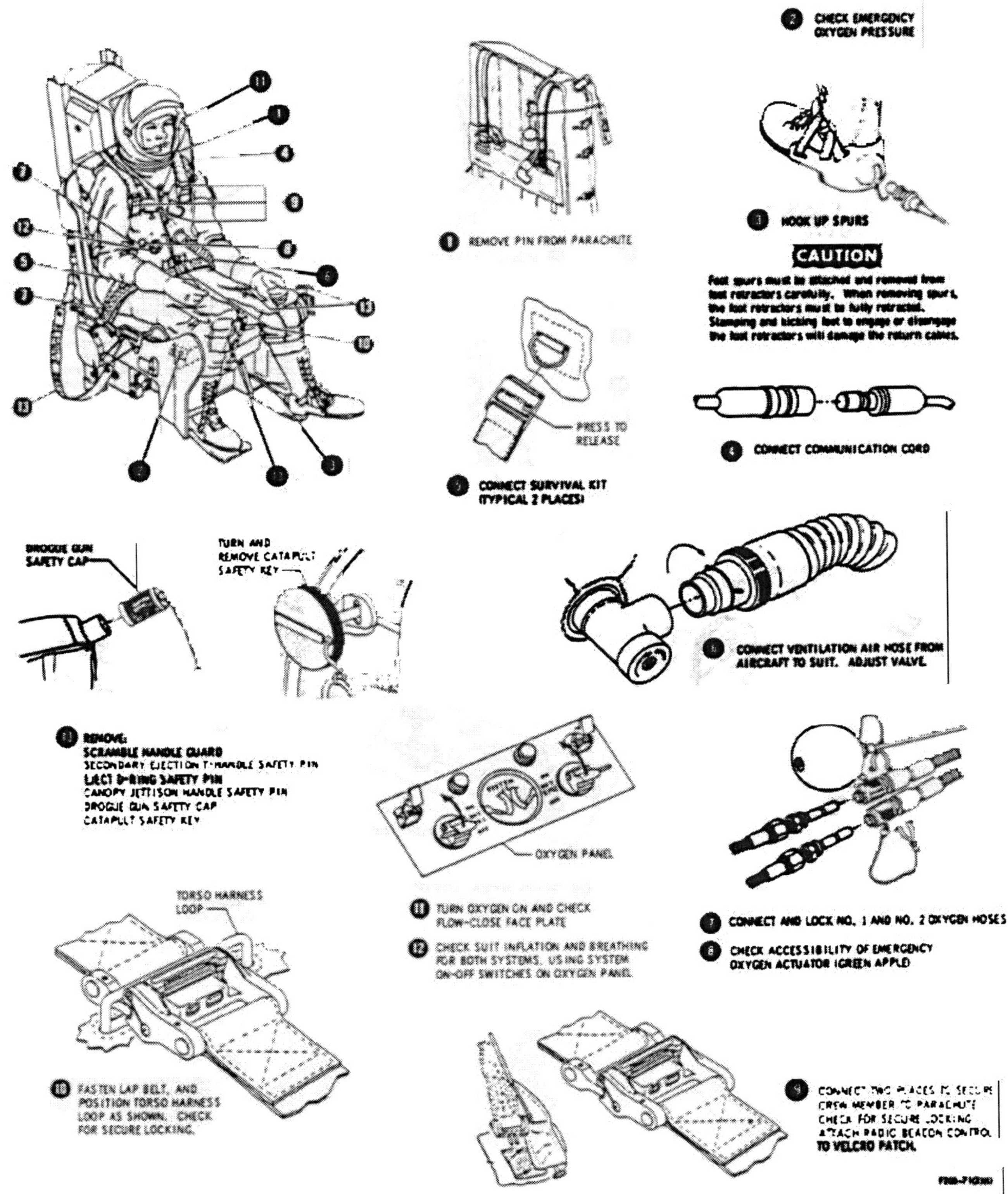

Figure 2-1 (Sheet 1 of 2)

PERSONAL EQUIPMENT HOOKUP - Shirt

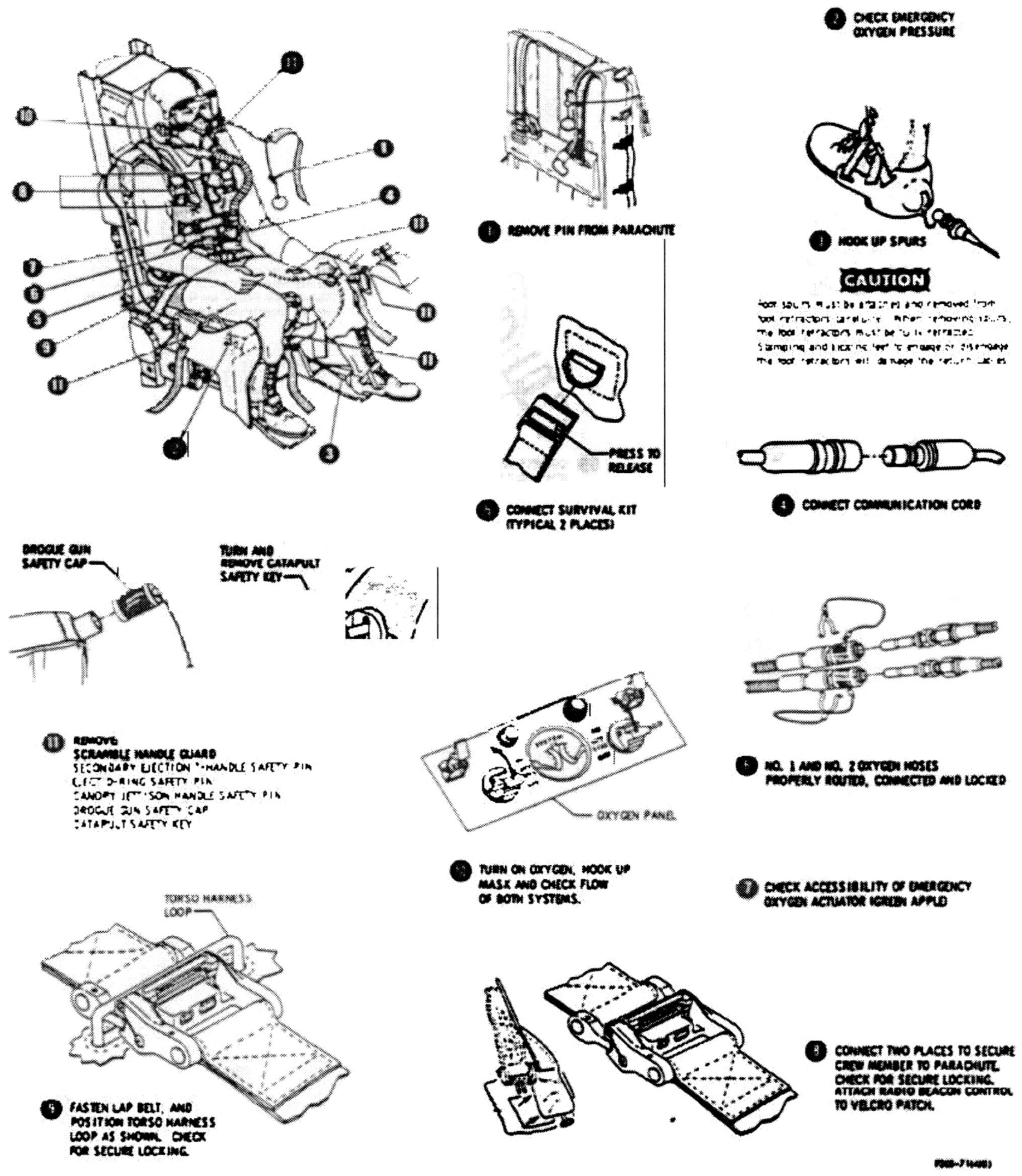

Figure 2-1 (Sheet 2 of 2)

Before each flight, check takeoff and anticipated landing gross weights and weight-and-balance clearance (Form 365F or local substitute). Note weight and moment values programmed for CG mode selector box.

NOTE

Recommended weight and/or c.g. limits can be exceeded by seemingly normal loading arrangements. Check loading documents carefully.

AIRCRAFT STATUS

Refer to AF Form 781 for engineering, servicing, and equipment status.

PREFLIGHT CHECK - SR-71A/B

EXTERIOR INSPECTION

Because it is not practical for the flight crew to perform an exterior inspection while wearing pressure suits, the exterior inspection should be accomplished by other qualified personnel.

BEFORE ENTERING COCKPIT - SR-71A/B

▲1. Ejection seat and canopy pins -Installed

▲2. Circuit breakers - Checked in (set).

If any DAFICS computer circuit breakers are open, those in the aft cockpit should be reset approximately one minute prior to resetting those in the front cockpit.

▲3. Canopy handles - Checked.

Fwd cockpit - Locked forward.
Aft cockpit - Aft position.

4. Mode selector reference moment setting - Checked.

▲5. Publications - Checked.

FRONT COCKPIT INTERIOR CHECK - SR-71A/B

Check personal equipment hookup. (See Figure 2-1). Hookup will be performed by personal equipment personnel.

Left Console - Pilot

1. Throttle restart arming switch - NORM.
2. Liquid oxygen quantity indicators -Check SYS 1, SYS 2, and STANDBY.
   a. LOX QTY selector switch - SYS 1, IND 1.
   b. LOX quantity gage - Check both full.
   c. LOX QTY selector switch - STANDBY, IND. 1.
   d. LOX quantity gage - Check No. 1 full.
   e. LOX QTY selector switch - SYS 1, IND 1.
3. Light rheostat switches - Checked.
4. Thunderstorm lights switch - As desired.

   At night, use of these lights can facilitate the P.E. Hookup.
5. Emergency ICS switch OFF. (Trainer only)
6. Standby oxygen system switches -OFF.
7. Control transfer panel lights - All on. (Trainer only)

Cycle control transfer switches if needed to obtain control in the forward cockpit and illuminate all four transfer lights.

8. UHF radio - ON and set.

   a. Mode - INT.

   b. VOL - Nearly full clockwise.

   c. PWR - Set.

   d. Frequency - Set.

   e. Function select - Set.

9. Oxygen control panel - Set.

10. Aft bypass position lights - Checked.

    Press to test.

11. L and R aft bypass switches -CLOSE.

12. EGT trim switches - AUTO.

13. Map projector controls - Set.

14. Throttles - OFF.

15. Throttle friction lever - Set.

16. Throttle restart switch - Cycle to OFF.

    Slide the switch to the forward bypass open and then to the restart position. Check that the MANUAL INLET and CAUTION lights illuminate. Return the switch to OFF.

17. TEB counters - 16.

Instrument Panel - Pilot

1. Cockpit pressure dump switch - OFF.

2. Bay Air switch - ON.

3. Manifold temperature switch - AUTO.

4. Landing/Taxi light switch - OFF.

5. Suit heat rheostat - OFF.

6. Face heat rheostat - Set.

   Use face heat at all times. Adjust for comfort.

CAUTION

- Do not use the HIGH face heat position when equipped with the PPG (glass) visor except for emergency heating. Continuous use of the HIGH position may delaminate the visor.

- The face heat switch should not be set above 5 with the visor raised, or the faceplate may be damaged.

7. Cockpit temperature control rheostat - 12 o'clock position.

   Cockpit temperature control may have to be adjusted for varying ambient conditions.

8. Temperature indicator selector switch - R BAY.

9. L and R refrigeration switches - OFF.

10. Cockpit temperature mode selector and override switch - AUTO.

11. Defog switch - CLOSED.

12. Brake switches - Set.

    a. Set ANTI-SKID ON, or ALT STEER & BRAKES, respectively, depending on whether the left or right engine is to be started first.

    b. Set WET/DRY switch DRY.

13. Indicators and warning lights test button - Press.

    a. Spike and forward bypass position indicators full counterclockwise, (0 inches and 100% open).

b. LN_2 and LOX quantity indicators decrease to zero.

c. All cockpit caution and warning lights illuminate.

d. Gear warning tone sounds in headset.

e. All CIP indicator needles decrease to zero.

f. Fuel quantity indicator needle moves to zero. The c.g. indicator indicates 14%.

g. The annunciator panel C.G. warning light remains illuminated until c.g. indicator needle is above 17%.

14. Fuel derich switch - ARM.

15. Landing gear lever - DOWN.

16. Cabin altimeter - Field elevation.

17. Standby attitude indicator - Erecting.

If required, pull the cage knob to erect the instrument, then release the knob. The instrument will erect and then seek 7° nose-down and 0° roll (if the aircraft is level).

NOTE

A jitter of ± 1/2° in the pitch axis is acceptable and may occur at any pitch angle.

18. Angle of attack indicator - Checked.

Check OFF flag out of view and AOA indicates zero.

19. Drag chute control Checked in, light off.

Verify that the drag chute handle is in the full forward detent JETTISON position and that the DRAG CHUTE UNSAFE annunciator light is not illuminated.

WARNING

The red marking on the drag chute handle shaft must not be visible.

20. Compressor inlet temperature (CIT) gage - Checked.

Check needles together and ambient temperature indicated.

21. Airspeed/Mach Meter - Checked.

a. Limit hand setting - 460 KIAS.

b. Airspeed indication - 60 knots or less.

c. Mach number indication - Right half of window blanked. Disregard Mach reading in left half of window.

22. RSO EJECTED light - Press to test.

23. Compressor inlet pressure (CIP) gage - Checked.

L and R needles and reference pointer together and indicating barometric pressure.

24. APW switch - PUSHER/SHAKER.

With ANS, INS, or TACAN selected on the Display Mode Selector switch, the ADI glide slope pointer should deflect to the lowest dot on the glide slope displacement scale. The pointer may fluctuate if there is fuselage motion in the pitch axis.

25. Spike and forward bypass position indicators - Checked.

    a. Spikes - 0 in. aft.

    b. Forward bypass - Open 100%.

26. Accelerometer - Reset.

27. L and R spike and forward bypass controls - Cycle, then AUTO.

    Check knobs for security. Check that the MANUAL INLET and CAUTION lights are on when not in AUTO.

28. L and R restart switches - Cycle to RESTART ON, then off, individually.

    Check operation of the MANUAL INLET light when in RESTART ON.

29. Projector - Checked.

    a. Verify proper loading.

    b. Check controls and lights.

    c. Illumination as desired.

30. Surface limiter release handle - Pulled, and SURFACE LIMITER caution light off.

31. Pitot heat switch - OFF.

    Check that the PITOT HEAT caution light is on.

32. Windshield rain removal and de-ice switch - OFF.

33. Trim power switch - ON.

34. A, B and M CMPTR RESET switches - Normal (Guard down).

35. Clock - Set.

36. Altimeter - Set.

NOTE

It is possible to rotate the barometric set knob through full travel so that the 10,000-foot pointer is 10,000 feet in error. Check that the 10,000-foot pointer is reading correctly.

37. Vertical velocity indicator - Checked.

    Check for zero indication.

38. TACAN control transfer switch - CONT illuminated (SR-71A).

    Press to obtain control in the forward cockpit.

39. Engine instruments - Checked.

40. Igniter purge switch - Off.

41. Liquid nitrogen quantity gages -Checked.

42. Forward transfer switch - OFF.

43. Emergency fuel shutoff switches - Fuel on (guards safety wired down).

44. Fuel dump switch - OFF (guard down).

45. Battery - BAT.

46. Emergency ac bus switch - NORM.

47. Generators - OFF.

48. Instrument inverter switch - NORM.

    Place switch to TEST and check that INST INVERTER ON light illuminates, then set to NORM.

Right Console - Pilot

1. PVD - OFF.

2. ILS power switch - ON.

3. SAS - OFF.

4. **SAS lights - Test.**

   All SAS panel warning lights should illuminate when the test switch is depressed including the DAFICS BIT TEST and FAIL lights.

5. **Autopilot - OFF.**

   Press the right console A/P OFF switch and check that the A/P OFF light is on.

6. **TACAN mode selector - T/R.**
7. **Interphone control panel - Set.**
8. **IGV Lockout switches - NORM.**
9. **Cockpit pressure selector switch - Set.**

   Select either the 10,000 or 26,000 foot setting. The 26,000 foot setting is normally desired.

10. **VHF radio - TR and set.**
    a. **Mode select switch - As desired.**
    b. **Frequency control/Emergency select switch - PRE or MAN.**
    c. **Frequency - Set.**
    d. **Volume control - Nearly full clockwise.**
11. **Canopy seal - OFF.**

TRAINER AFT COCKPIT INTERIOR CHECK

Left Console - Instructor Pilot

1. **Thunderstorm lights switch - As desired.**

   At night, use of these lights can facilitate the P.E. Hookup.

2. Light rheostat switches - Checked.
3. Throttle restart arming switch - NORM.
4. UHF modulator/demodulator (Modem) control - Set.
   a. Code selector switches - Set.
   b. Range address switch - Set.
5. Oxygen control panel - Set.
6. HF radio - OFF and set.
7. Interphone control panel - Set.
8. UHF radio - ON and set.
9. INS - Check aligning.
   a. Check present position.
   b. Function switch - NORM or STOR HDG.

NOTE

INS must be in NAV to obtain a valid mag heading.

   c. **Check/enter desired DP's.**
   d. **Adjust INS segment lights as desired.**
10. **Aft bypass position lights - Checked.**

    Press to test.

11. **L and** R **aft bypass switches FWD CONT.**
12. EGT **trim switches - HOLD & FWD** CONT.

13. Map projector controls - Set.

14. Throttles - OFF.

15. Throttle friction lever - Set.

16. Throttle restart switch - Cycle to OFF.

    Check that the MANUAL INLET and CAUTION lights are on and the SPIKE DOOR transfer light on the control transfer panel is on when the throttle restart switch is not in the OFF position.

17. Emergency ICS switch - OFF.

18. UHF TRANS switch - Set.

19. Control transfer panel - Cycle to forward control.

    Verify lights illuminate when aft cockpit has control and then transfer to forward control.

20. Cockpit air handle - Off (forward).

Instrument Panel - Instructor Pilot

1. Indicators and warning lights test button - Press.

    a. Spike and forward bypass position indicators full counterclockwise (0 inches and 100% open).

    b. LOX quantity indicators decrease to zero.

    c. All cockpit caution and warning lights illuminate.

    d. Gear warning tone sounds in both headsets.

    e. All CIP indicator needles decrease to zero.

    f. Fuel quantity indicator needle moves to zero. The c.g. indicator indicates 14%.

2. Brake switch - OFF.

3. Landing gear switch - OFF.

4. Fuel derich switch - ARM.

5. Drag chute switch - OFF, light off.

    Check that the yellow dot in the end of the switch is visible with the guard down and the DRAG CHUTE UNSAFE annunciator light is not illuminated.

6. Airspeed/Mach Meter - Checked.

    a. Limit hand setting - 460 KIAS.

    b. Airspeed indication - 60 knots or less.

    c. Mach number indication - Right half of window blanked. Disregard Mach reading in left half of window.

7. Liquid oxygen quantity indicators -Check SYS 1, SYS 2 and STANDBY.

    Coordinate with forward cockpit to check System 1 and Standby system.

8. Clock - Set.

9. Cabin altimeter - Field elevation.

10. Accelerometer - Reset.

11. Compressor inlet pressure (CIP) gage Checked.

    L and R needles and reference pointer together and indicating barometric pressure.

12. Compressor inlet temperature (CIT) gage - Checked.

    Check needles together and ambient temperature indicated.

13. Spike and forward bypass position indicators - Checked.

    a. Spikes - 0 in. aft.

b. Forward bypass - Open 100%.

14. L and R spike and forward bypass controls - Cycle, then AUTO.

Check knobs for security. MANUAL INLET and CAUTION lights will not illuminate when knobs are not in AUTO unless aft cockpit has SPIKE DOOR transfer light illuminated on control transfer panel.

15. L and R restart switches, - Cycle to RESTART ON, then off, individually.

Check that the MANUAL INLET and CAUTION lights are on and the SPIKE DOOR transfer light on the control transfer panel is on when in RESTART ON.

16. Projector - Checked.

a. Verify proper loading.

b. Check controls and lights.

c. Illumination as desired.

17. Surface limiter release handle - Pulled, and SURFACE LIMITER caution light off.

18. APW switch - CONT FWD.

19. Defog switch - CLOSED.

20. Trim power switch - ON.

21. Drag chute emergency deploy switch Stowed and safetied.

22. A, B, and M CMPTR RESET switches - Normal (Guard down).

23. Standby attitude indicator - Erecting.

If required, pull the cage knob to erect the instrument, then release the knob. The instrument will erect and then seek 7° nose down and 0° roll (if the aircraft is level).

NOTE

A jitter of ± 1/2° in the pitch axis is acceptable and may occur at any pitch angle.

24. Air refuel switch - OFF.

25. Altimeter - Set.

NOTE

It is possible to rotate the barometric set knob through full travel so that the 10,000-foot pointer is 10,000 feet in error. Check that the 10,000-foot pointer is reading correctly.

26. Vertical velocity indicator - Checked.

Check for zero indication.

27. Engine instruments - Checked.

28. Forward transfer switch - OFF.

29. Fuel dump switch - OFF (guard down).

30. Emergency fuel shutoff switches - Fuel on (guards safety wired down).

Right Console - Instructor Pilot

1. SAS - OFF.

2. SAS lights - Test.

All SAS panel warning lights should illuminate when the test switch is depressed including the DAFICS BIT TEST and FAIL lights.

3. Autopilot - OFF.

Press the right console A/P OFF switch and check that the A/P OFF light is on.

4. TACAN mode selector - T/R.

5. Temperature indicator selector switch R BAY.

6. Cockpit pressure selector switch - Set.

   Select either the 10,000 or 26,000 foot setting. The 26,000 foot setting is normally desired.

7. Face heat rheostat - Set.

   Use face heat at all times. Adjust for comfort.

   CAUTION

   - Do not use the HIGH face heat position when equipped with the PPG (glass) visor except for emergency heating. Continuous use of the HIGH position may delaminate the visor.
   - The face heat switch should not be set above 5 with the visor raised, or the faceplate may be damaged.

8. ANS - Checked, MAG set.

   a. DATA Switch - TEST

      Press DISPLAY push-button switch to display data.

   b. Check mission tape number and Star Catalog number.

   c. DATA Switch - As required.

      Check Control and Display Panel readouts. Check mission modifications as required.

   d. MAG/GRID push-button switch - MAG.

9. ANS DATA Switch - NORMAL

   Press DISPLAY push-button switch to display selected data.

10. IFF - Set.

11. ILS power switch - ON.

12. MRS power switch - ON.

    Check green ON illuminated, red FAIL not illuminated.

13. Canopy seal - OFF.

AFT COCKPIT INTERIOR CHECK - SR-71A

Left Console - Aft Cockpit

1. Cockpit air handle - Off (forward).

2. Light rheostat switches - Set.

3. HF radio - OFF and set.

4. UHF radio - On and set.

   a. Mode - INT.

   b. Vol - Nearly full clockwise.

   c. PWR - Set.

   d. Frequency - Set.

   e. Function select - Set.

   f. UHF TRANS switch - Set.

5. Interphone control panel - Set.

6. INS - Check aligning.

   a. Check present position.

   b. Function switch - NORM or STOR HDG.

      NOTE

      INS must be in NAV to obtain a valid mag heading on BDHI and HSI.

   c. Check/enter desired DP's.

d. Adjust INS segment lights as desired.

7. DEF systems - Off.

System A: System A power ON legend not illuminated.
System H: Warmup and standby lights extinguished and the mode switch MAN and AUTO legends off.

WARNING

Assure that the System H Mode Indicator lights for the H LO and H HI bands are extinguished. System H transmitter radiation while on the ground is hazardous to personnel if antenna hoods are not installed.

System M: System M power ON legend not illuminated.

8. UHF modulator/demodulator (Modem) control - Set.

   a. Code selector switches - Set.

   b. Range address switch - Set.

9. Oxygen control panel - Set.

10. DEF gating generator switch - Guard down.

This switch is nonfunctional.

Instrument Panel - Aft Cockpit

1. TACAN CONT transfer switch light -Off

2. TACAN mode selector - T/R.

3. IFF - Set.

4. G-Band Beacon switch - OFF.

5. RCD display brightness control Full clockwise.

6. Egress lights - Press to test.

Press to test the ALERT, PILOT EJECTED and BAILOUT lights.

7. Cockpit pressure selector switch - Set.

Select either the 10,000 or 26,000-foot position. The 26,000-foot position is normally set.

8. Face heat rheostat - Set.

Use face heat at all times. Adjust for comfort.

CAUTION

- Do not use the HIGH face heat position when equipped with the PPG (glass) visor except for emergency heating. Continuous use of the HIGH position may delaminate the visor.

- The face heat switch should not be set above 5 with the visor raised, or the faceplate may be damaged.

9. Camera exposure control - Checked and set.

   a. Rotate the exposure dial full clockwise to align the 90° index with the first high reflectivity dot.

   b. Set the briefed sun angle value.

NOTE

If the 90° index does not align with the first high reflectivity dot in the full clockwise position, the dial is not correctly installed. The corresponding electrical value on the sun dial will be incorrect.

10. Attitude indicator - Checked and set.

   a. Check indicator movement and set zero pitch angle.

   b. Attitude Reference Selector - ANS.

11. V/H indicator M pointer - Set.

12. Liquid oxygen quantity indicators Check.

13. Clock - Set.

14. BDHI No. 1 needle select switch - ADF.

15. BDHI heading select switch - INS.

NOTE

INS must be in NAV to obtain a valid mag heading on BDHI and HSI.

Viewsight Control Panel - Aft Cockpit

1. V/H select switch - BUS.
2. Map drive switch - Set.
3. Map rate control - Set.
4. MAP/DATA film select switch -Set.

Right Console - Aft Cockpit

1. Canopy seal - OFF.
2. OBC Power switch - Off.
3. ANS - Checked, MAG set.
   a. Check mission tape number.
   b. Normal Display - Check C&D panel readouts.
   c. Mission modifications Check as required.
   d. MAG/GRID push-button switch - MAG.
4. Sensor power switches - STP then OFF (ON extinguished).
   a. RADAR
   b. ELINT
   c. RCDR
   d. LH and RH TECH
   e. TERRAIN
5. NAV RCDR power switch - ON.
6. MRS power switch - ON.
7. V/H power switch - ON.
8. VWSGT power switch - ON.
9. EXPOS power switch - ON.
10. Map projector - Checked.
    a. Verify proper loading.
    b. Check controls and lights.
    c. Illumination as desired.
11. LAMP TEST - Press to test.

    Check all instrument panel and console lights.
12. Left and right technical camera CONT switches - A (Auto).
13. FMC switches - V/R.
14. V/H SOURCE - NAV.

AFT COCKPIT CHECK (SOLO FLIGHT) - SR-71A/B

NOTE

Abbreviated checklists are not supplied for this procedure.

Before flight, check the following items in the rear cockpit. The trainer aircraft shall be flown solo only from the front cockpit.

1. Lap belt, shoulder harness and all personal leads - Secured.
2. All circuit breakers - In.

WARNING

For the SR-71B:

- The following controls in the aft cockpit can override the forward cockpit:

  Aft bypass switches
  EGT Trim switches
  Throttle restart switch
  Brake switch
  Landing gear switch
  Drag chute switch
  Restart switches
  APW switch
  Air refuel switch
  Trim switches

- Trim power must be ON in both cockpits to enable the trim system.

- Fuel forward transfer switches, fuel dump switches, and emergency fuel shutoff switches must be off in both cockpits to turn the respective systems off.

Left Console - SR-71A/B (Solo)

1. Cockpit air handle - On (aft).
2. Panel, instrument, and thunderstorm light switches - OFF.
3. (T) Throttle restart arming switch - CUT-OUT.
4. UHF modulator/demodulator (Modem) control - Set.
5. Oxygen control panel - Sys 1 and 2 ON.
6. HF radio control panel - Set.
7. Interphone control panel - Set.
8. UHF radio - BOTH, frequency set.
9. INS - Checked, set to NAV.
   a. Check present position.
   b. Check alignment complete (NAV RDY light flashing).
   c. Check/enter desired DP.
   d. Set FUNCTION switch to NAV.

NOTE

INS must be in NAV to obtain a valid mag heading.

SR-71A:

10. DEF systems - Off.
11. (T) L and R aft bypass switches - FWD CONTROL.
12. (T) EGT trim switches - HOLD & FWD CONT.
13. (T) Throttle friction - OFF.
14. (T) Emergency ICS panel - OFF.
15. (T) UHF TRANS switch - OFF.

    Check UHF TRANS switch off to provide UHF-1 with ADF and external mode operating capability.
16. (T) Control transfer panel - Set (lights OFF).

Instrument Panel - SR-71A (Solo)

1. TACAN - T/R, frequency set.
2. IFF - NORMAL, modes and codes set.
3. Cockpit pressure switch - 26,000 FT.
4. Face heat switch - OFF.
5. UHF TRANS switch - OFF.

   Check UHF TRANS switch off to provide UHF-1 with ADF and external mode operating capability.

Instrument Panel - SR-71B (Solo)

1. Brake switch - OFF.
2. Landing gear switch - OFF; guard safety wired.
3. Drag chute switch - OFF.
4. S/P BAILOUT switch - OFF.
5. L and R spike and forward bypass controls - AUTO.
6. L and R restart switches - OFF.
7. APW shaker - CONT FWD.
8. Defog switch - CLOSED.
9. Trim power switch - ON.
10. Emergency chute deployment handle - Stowed and safetied.
11. A, B, and M CMPTR RESET switches - Normal (Guard down).
12. Bearing select switch - TAC/ADF.
13. Display mode select switch - ILS APCH.
14. Air refuel switch - OFF.
15. Forward transfer switch - OFF.
16. Fuel dump switch - OFF (guard down).
17. Emergency fuel shutoff switches - Fuel on (guards safety wired down).

Right Console - SR-71A/B (Solo)

Ⓣ 1. SAS - ON.

Ⓣ 2. Autopilot - OFF.

Ⓣ 3. TACAN - T/R, frequency set.

Ⓣ 4. Cockpit pressure switch - 26,000 FT.

Ⓣ 5. Face heat switch - OFF.

6. ANS - Checked and set.

SR-71A:

7. Sensor power - OFF.

Ⓣ 8. IFF - NORMAL, modes and codes set.

Ⓣ 9. ILS panel - ON, frequency set.

10. MRS power switch - ON.

    Check green ON illuminated, red FAIL not illuminated.

11. Canopy seal - ON.

    After the engines are started, the canopy seal will inflate and remain inflated until engine shutdown.

    Close rear cockpit canopy and lock externally immediately prior to engine start.

CAUTION

Leave the rear cockpit canopy open until just before engine start to maintain adequate cooling in the equipment bays. Close the front cockpit canopy followed by the rear cockpit canopy immediately prior to engine start.

DANGER AREAS - Engine Operation

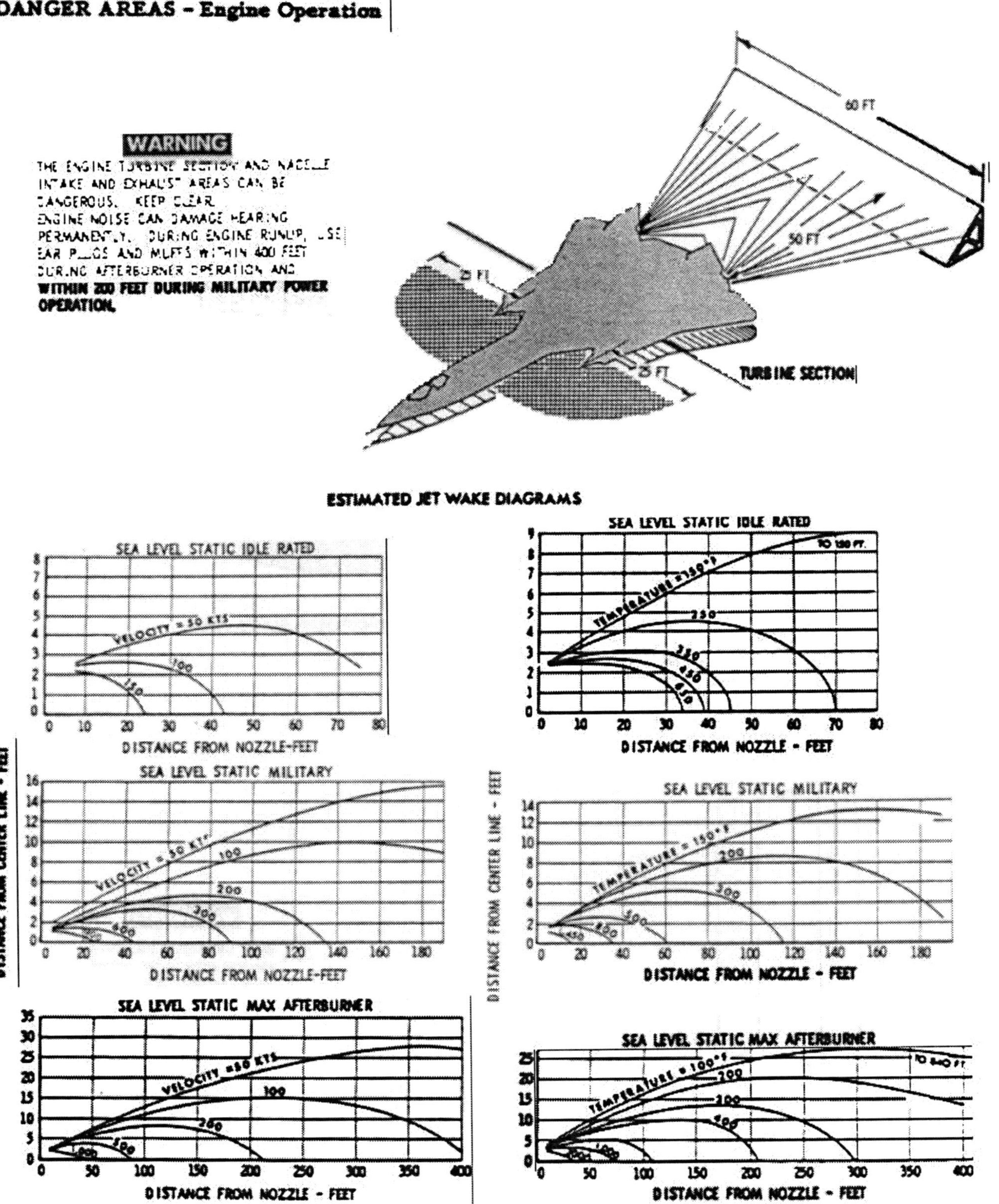

Figure 2-2 (Sheet 1 of 2)

DANGER AREAS - EMF Radiation

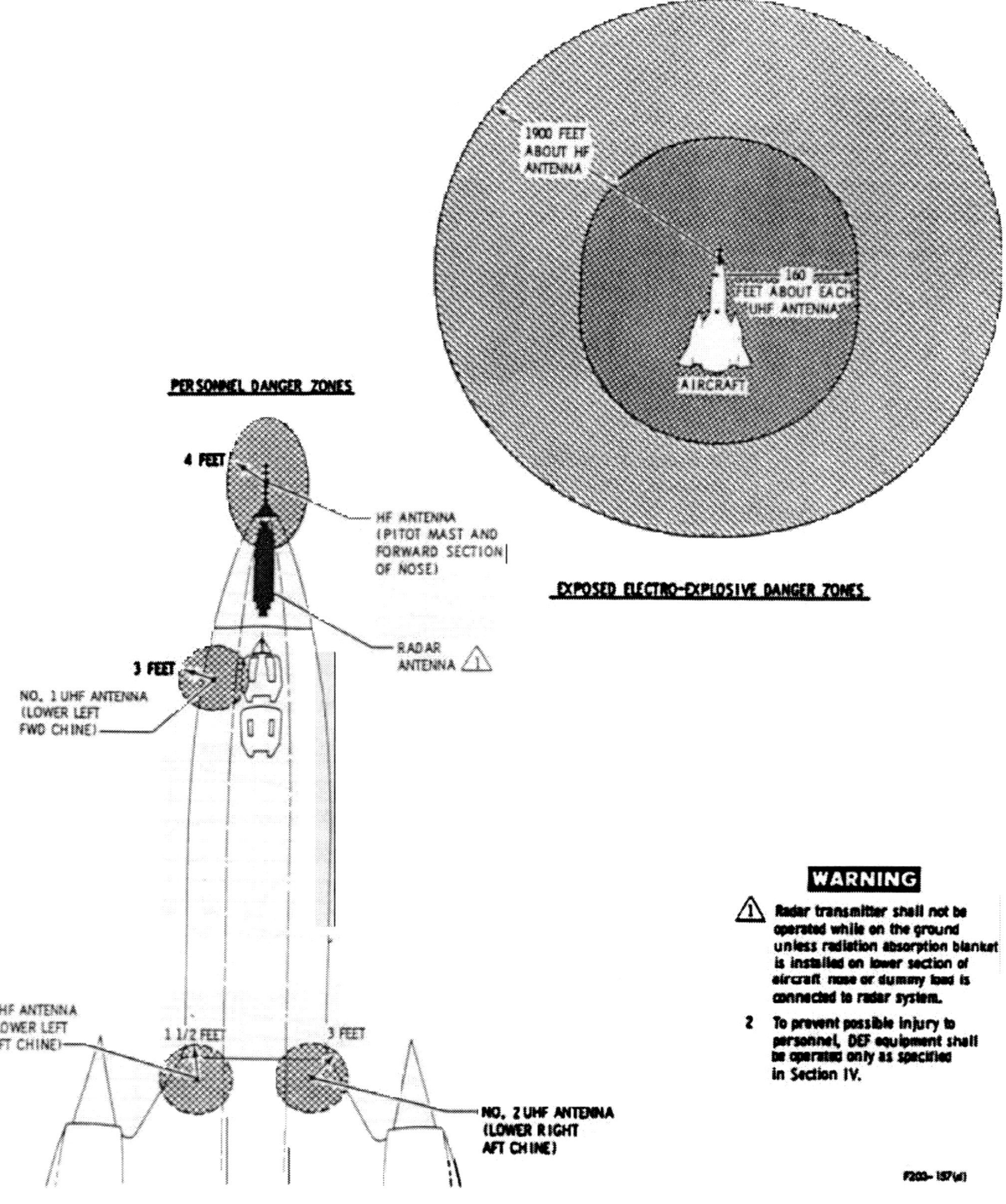

Figure 2-2 (Sheet 2 of 2)

All subsequent checklists apply to both cockpits of the SR-71 A/B.

NOTE

Pilot and RSO (or IP) coordination is required. RSO (or IP) reads -Pilot responds. Alphabetized items need not be read.

STARTING ENGINES

▲1. Interphone - Checked.

Check CALL, HOT MIC, and normal functions (and emergency ICS in SR-71B).

▲2. BAILOUT light - Checked.

Coordinate ALERT and BAILOUT light illumination with switch position. Return switch to OFF (guard down).

▲3. Triple display indicator - Check.

a. Altitude - Within ±200 feet of pressure altimeter indication when altimeter is set at 29.92 inches Hg. Maximum difference between TDI's, fwd and aft cockpits, 100 feet.

b. Airspeed - 75 to 110 KEAS.

c. Mach - 0.11 to 0.2 normal.

▲4. Fuel quantity indicating system - Checked.

a. Individual tank quantities - Check. (within 550 lb between cockpits)

b. Sum of individual tank quantities - Check. (within 780 lb of TOTAL)

c. TOTAL fuel quantity - Check. (within 850 lb between cockpits)

▲5. Indicated and corrected computed c.g. - Within 0.5% MAC.

Check indicated c.g. reading between cockpits and compute c.g. with manual computer.

NOTE

While on the ground, c.g. computed using the manual c.g. computer should be corrected as follows to allow for the effect of level rather than flight attitude (with normal fuel distribution per T.O. 1-1B-40).

Total Fuel	Correction to MAC for computed c.g.*
Full tanks	-0.3%
70,000	-0.4%
65,000	-0.55%
60,000	-0.7%
55,000	-0.85%
55,000 (tank 6 empty)	-0.5%
50,000	-1.0%
45,000	-1.2%
45,000 (tank 6 empty)	-0.8%
40,000	-1.4%
35,000	-1.2%
30,000 (tank 6 empty)	-1.0%
25,000 (tank 6 empty)	-1.0%
20,000 (tank 6 empty)	-1.0%

*In level attitude, computed c.g. is aft of actual c.g. (The c.g. gage should read actual c.g.).

NOTE

When tank 6 is not full, use tank 6A and 6B scale. Fuel distribution must be obtained from mission loading form.

One minute before starting engines:

▲6. No. 1 and No. 2 oxygen systems - ON and checked.

Sys 1 and Sys 2 oxygen supply levers both latched ON and pressure checked.

▲7. **Baylor bar - Latched and locked.**

8. **Exterior light switches - ON.**

   a. **FUS & TAIL switch - BRT.**

   b. **TAIL LT switch - STEADY.**

   c. **ANTI - COLLISION switch - ANTI-COLLISION.**

9. **Brake switch - Setting checked.**

   Set ALT STEER & BRAKES if the right engine is to be started first.

10. **First engine - Start.**

   Although either engine can be started first, it is recommended that the left engine be started first (and shut down first) for odd numbered flights. Start the right engine first for even numbered flights. This enables a flight control system check to be made on alternate single-hydraulic systems immediately after starting.

   Ground personnel using interphone equipment will observe exhaust nozzle and nacelle inspection panels during start.

   WARNING

   Determine intake and exhaust areas are clear of personnel and ground equipment. Check fire guard(s) standing by.

   CAUTION

   - Before starting an engine, assure wheels are chocked. There is no parking brake.
   - Do not move control stick until at least 1500 psi can be maintained on the A or B hydraulic system.

   NOTE

   The crewchief will call the pilot when the starting unit is connected, and the pilot will instruct the crewchief to turn the unit on after verbally confirming that the engine combustion chamber drain valves are open and fuel is draining from each engine.

   a. **Pilot - Signal for engine rotation.**

   b. **Throttle - IDLE at first indication of rpm increase.**

      When necessary, an alternate technique of advancing the throttle at 1000 rpm may be used.

   c. **Fuel flow - Checked for increase.**

   d. **IGV lights - Off.**

   NOTE

   With pressurization of the engine fuel hydraulic system during start, the IGV position light must be extinguished (IGV cambered); if not, discontinue the start and determine the cause.

   e. **Ignition - Verify within 15 seconds when using gas engine cart or within 20 seconds when using 3AG1100 air turbine starter. If no ignition indicated by an rpm increase and a rise in EGT within the allowable time, move throttle to off and continue cranking engine for 30 seconds at 1000 rpm.**

CAUTION

o In case of a false start, use Clearing Engine procedure, this section.

o When using the 3AG1100 air turbine starter, do not exceed 5 seconds steady-state cranking operation between 1370 and 1470 rpm. Resonant frequency of the air turbine is in this range. No problems are encountered accelerating through this range, providing the transition period is less than 5 seconds.

f. Ground starting unit - Signal for disconnect at 3200 rpm.

g. If 565°C is exceeded, move throttle to OFF. If 649°C is exceeded, do not attempt to restart the engine.

NOTE

If the engine does not accelerate smoothly to idle rpm, but appears to "hang" in the 2600 to 2800 rpm range, retard the throttle to OFF and then quickly return it to IDLE. This "double clutching" procedure momentarily leans the fuel/air mixture and positions the flame front correctly in the burner cans so the engine can accelerate normally to idle rpm.

h. Idle rpm - Checked.

Engine idle speed is 3975 ± 50 rpm below 60C (140°F).

CAUTION

o When using the MA-1A carts (or equivalent) abort start if Idle rpm is not obtained after 90 seconds.

o When using the shelter airstart system, abort start if Idle rpm is not obtained after 120 seconds.

i. Engine and hydraulic instruments Check normal indications.

(1) EGT - 350° to 565°C (start limit).

(2) Fuel flow - 4600 to 6300 lb/hr. A lower indication is evidence of heat sink system malfunction. If this occurs, shutdown and request investigation of circulating systems.

(3) Oil pressure - 35 psi minimum.

CAUTION

Discontinue start if oil pressure rise is not observed by the time IDLE rpm is obtained.

(4) Hydraulic system pressures Checked.

(5) CIP should decrease to slightly below ambient.

Two minutes after engine start:

11. Flight controls - Steady neutral position.

Confirm with maintenance that the control surfaces arrive at a steady neutral position.

WARNING

Abort flight if maintenance detects surface movement without stick or rudder inputs.

12. Flight control system - Checked.

With nosewheel steering disengaged, individually check each axis for full deflection and freedom of travel in both directions. Confirm correct deflection and normal response by ground crew observation using the sequence: nose up, nose down, left roll, right roll, nose left, nose right.

If the right engine was started first and the elevon up travel is restricted, the pusher piston may be extended. Inform maintenance of the restriction, wait until the mixer access panel is removed, and then overpower the restriction. Check the flight control system after the left engine start.

If a restriction to rudder travel is felt and a force of approximately 10 pounds overpowers the restriction, the cause may be an extended rudder servo limiter piston due to the nonoperating engine. The rudders must be checked after the second engine is started to insure that the rudders are not restricted.

If nosewheel steering will not disengage, rudder control will be severely restricted in-flight with the gear down.

NOTE

Rapid control surface deflection while near idle rpm may result in temporary illumination of an A or B HYD warning light. The light should extinguish when flow demands diminish and normal pressure is restored.

13. Second engine - Start.

Use the same sequence as for items a thru i of step 11.

If the right engine was started first and the elevon up travel was restricted, check the flight control system after left engine start.

If a restriction to rudder travel occurred after the first engine start and a force of approximately 10 pounds overpowered the restriction, check the rudders again to insure that the rudders are not restricted. If a restriction to rudder travel occurs again, a flight control system problem exists.

TEB counters - Checked.

15. Generators - On (NORM), and lights off.

Check R and L GEN OUT lights extinguish.

NOTE

With transfer of electrical power while on the ground, the DAFICS will undergo ground re-initialization indicated by momentary illumination of the A, B, and M CMPTR OUT caution lights, OFF flags in both TDIs, and TDI resynchronization to 55,000 ft., Mach 2.0, and 300 KEAS. If DAFICS indications are abnormal, notify maintenance.

16. Generator Bus Tie light - Off.

17. External power -Disconnected.

Signal ground crew to disconnect.

T 18. Fuel system - Checked.

a. Check all pump, tank empty, crossfeed and pump release lights are on when TEST is pressed.

b. Press the crossfeed switch to obtain OPEN.

Illumination of the OPEN portion of the switch confirms crossfeed is on.

c. Press pump switches 1 through 6 ON in sequence.

d. Press ON an additional tank containing fuel.

e. Press pump release switch and check that the manually selected tank is released.

f. Press crossfeed switch OFF.

g. Tanks 1, 3, & 6 (or 5) boost pump lights on.

19. Left and right forward bypass - Both confirmed open.

    Ground crew confirms doors open.

T 20. Spike and forward bypass position indicators - Check.

    a. Spikes - 0 in. aft.
    b. Forward bypass - 100% open.

T 21. Brakes - Normal & alternate systems checked, set ANTI SKID ON.

    Pump brakes and check normal feel while crew chief visually confirms brake actuation on both trucks. Normal feel does not necessarily indicate braking action. Perform the check both in ANTI SKID ON and ALT STEER & BRAKE. While applying moderate brake pressure, cycle the brake switch and check for a slight pedal movement (thump) and small position change when shifting between hydraulic systems. The absence of the thump indicates only one braking system available.

    Pause slightly while passing through the ANTI SKID OFF position and observe the ANTI-SKID OUT light illuminated. If the light does not illuminate, there may be an electrical/switch failure and only one braking system may be available.

    With S/B R-2695, check the antiskid disconnect feature of the trigger switch. With the brake switch in ANTI SKID ON and/or ALT STEER & BRAKE, check the ANTI-SKID OUT annunciator caution light illuminates while the trigger is depressed and extinguishes when the trigger is released.

    Set ANTI SKID ON at the conclusion of the check.

NOTE

If both engines must be shut down temporarily after start and it is necessary to retain ANS alignment, insure that ground air and power are connected and on, and turn generators off before shutting down the second engine.

CLEARING ENGINE

Cool the engine and remove trapped fuel and vapor as follows:

1. Throttle - OFF.

CAUTION

Allow a minimum of 1 minute for fuel drainage and coast down before motoring engine.

2. Starter - Engage and motor engine for at least 30 seconds and until EGT is below 150°C.

   Signal ground crew to motor engine at 1000 rpm. Crew chief will advise pilot when engine is clear and ready for start.

CAUTION

Do not motor the engine with the fuel shut off switch in the fuel off position except in an emergency. Damage to the engine may result with the engine fuel-hydraulic system off.

NOTE

If an electrical power interruption has occurred, cycle the MRS power switch off (light extinguished) then ON to assure reestablishment of MRS operation. Fuel boost pump circuit breakers should also be checked after electrical power interruption.

AFT TOWING - ENGINE OPERATING

Aft towing of the aircraft with engines running is permitted with:

a. Engines at idle.

b. 120,000 pounds gross weight or less.

c. Interphone communications maintained between the pilot and tow operation observer.

d. All braking accomplished by the tow tractor.

WARNING

The pilot shall not use aircraft braking except in an emergency.

e. Aircraft steering accomplished by a ground crewmember, using a nose wheel tow bar.

WARNING

Do not move the rudder pedals during hookup of the nose steering linkage at completion of towing.

BEFORE TAXIING

(T1) IFF - STBY.

(T2) HF - On.

Refer to Danger Areas, Figure 2-2, for extent of danger to personnel and exposed electro-explosive devices.

WARNING

Do not transmit on ground until safe to do so.

(T3) INS - Checked, set to NAV

Check alignment complete (NAV RDY light flashing)

NOTE

INS must be in NAV to obtain a valid mag heading on BDHI and HSI.

T 4. DAFICS Preflight BIT - Check.

a. SAS channel engage switches - ON.

b. SENSOR/SERVO lights - Checked off.

c. Cycle controls in pitch, roll and yaw and check for abnormal control surface oscillation or vibration.

d. Autopilot pitch and roll engage switches - ON.

e. Control stick trigger switch - Depress.

Check autopilot disengagement.

(T) f. Aft cockpit SAS channel engage switches - ON.

(T) g. AFCS control - Transfer to aft cockpit.

Check AFCS transfer light illuminated in aft cockpit.

(T) h. SAS Lights - Off.

Check SENSOR/SERVO lights remain off.

(T) i. Cycle aft cockpit controls in pitch, roll and yaw and check for abnormal control surface oscillation or vibration.

(T) j. Aft cockpit autopilot pitch and roll engage switches -ON.

Check autopilot disengagement.

(T) l. AFCS control - Transfer to forward cockpit.

m. Autopilot pitch and roll engage switches - ON.

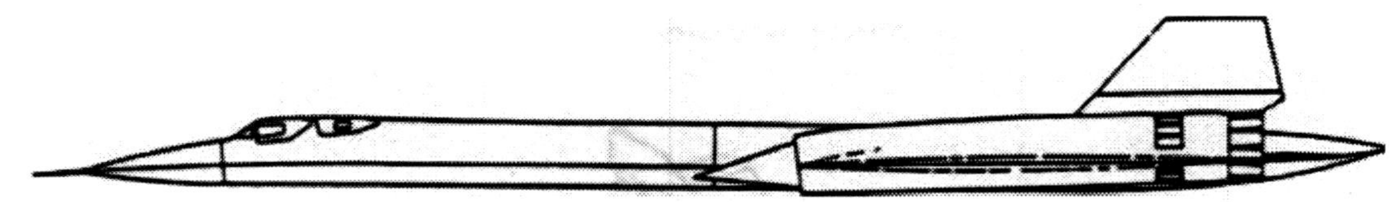

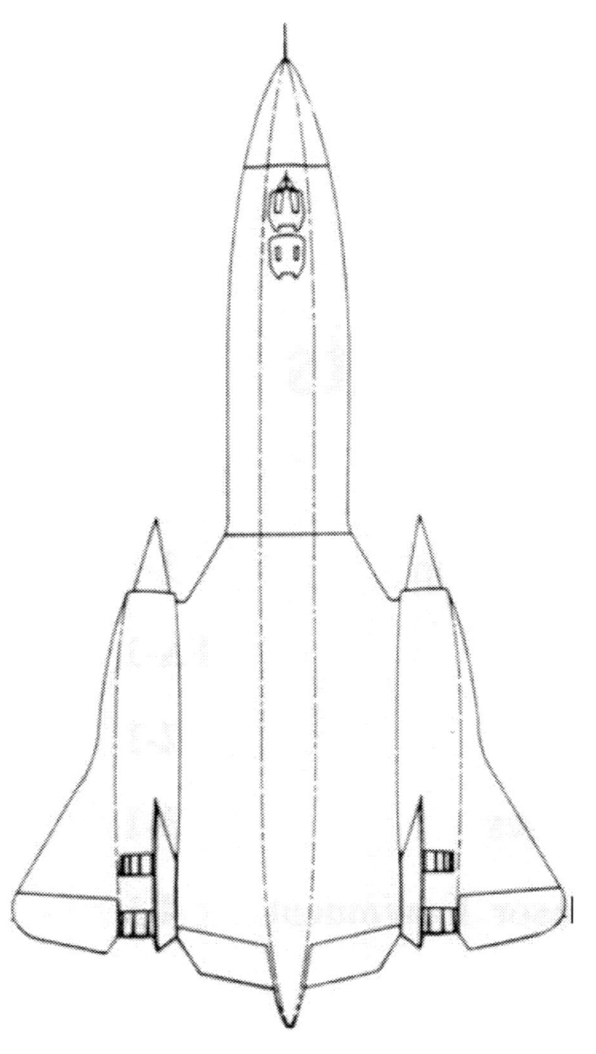

THIS PAGE INTENTIONALLY LEFT BLANK OR STILL CLASSIFIED.

(T) k. Aft cockpit control stick trigger switch - Depress.

Check autopilot disengagement.

(T) l. AFCS control - Transfer to forward cockpit.

m. Autopilot pitch and roll engage switches - ON.

n. Forward cockpit switch positions for DAFICS PREFLIGHT BIT - Set.

- ATT REF SELECT switch - INS
- KEAS HOLD switch - ON
- HEADING HOLD switch - ON

o. DAFICS PREFLIGHT BIT switch - ON.

The BIT TEST light illuminates steady green while the test is running. The BIT TEST light also illuminates when the function selector on the maintenance analyzer panel is not in the OFF position.

The PREFLIGHT BIT check can be terminated manually (once it is initiated) by stopping any DAFICS computer.

Pressure from A hydraulic system is required to engage the DAFICS PREFLIGHT BIT. Low pressure or flow from A, B, L or R hydraulic system will cause the DAFICS preflight BIT to fail.

If the DAFICS PREFLIGHT BIT switch will not engage, recheck:

1) CSC/NWS switch - Released.
2) ATT REF SELECT switch - INS
3) APW switch PUSHER/ SHAKER
4) SPIKES & FWD BYPASS doors - AUTO
5) RESTART switches - Off
6) Throttle Restart switch -Off
7) SAS channel engage switches ON
8) AUTOPILOT PITCH & ROLL engage switches - ON
9) KEAS HOLD switch - ON
10) HEADING HOLD switch -ON

NOTE

If at BIT completion the FAIL light, any SENSOR light, any SERVO light, or any CMPTR OUT light illuminates, notify maintenance.

After one minute:

p. Check BIT TEST light flashing green, sensor and servo lights extinguished, BIT FAIL light extinguished, and OFF Flags in both TDI's. The CIP barber pole reads zero.

q. Check autopilot pitch and roll engage switches, KEAS HOLD switch, and HEADING HOLD switch-Off. AUTOPILOT OFF and SAS OUT lights illuminated.

The flashing BIT TEST light and SAS OUT light indicates that the SAS is still in the ground test mode.

r. Check DAFICS PREFLIGHT BIT switch - OFF (guard down).

s. SENSOR/SERVO recycle switches - Press one of the six.

Pressing one of the six SENSOR/ SERVO recycle switches resets the DAFICS system to the flight mode. Check SENSOR/SERVO lights, BIT TEST light, and SAS OUT lights are out. Check both spikes have returned to the full forward position and the CIP barber pole has returned to normal. Both TDI's will initiate resynchronization and run up to 55,000 ft, Mach 2.0, and 300 KEAS. AOA will indicate 10°. AOA will return to 0° in approximately 1 min 15 sec and TDI indications will return to normal in approximately 2 min 15 sec after the DAFICS system has been reset to the flight mode. The A, B, and M CMPTR OUT annunciator panel lights will flash momentarily when the DAFICS system is reset.

WARNING

The SAS is non-functional while in the ground test mode. DAFICS will not operate normally until the system is reset. Failure to press a SENSOR/SERVO recycle switch after the DAFICS Preflight BIT is complete will cause the DAFICS to remain in the ground test mode.

T 5. Flight instruments and navigation equipment - Checked.

a. Turn display mode selector switch to:

(1) INS - Check that bearing pointer and DME display TACAN/ADF data with Bearing Select switch set to TAC/ADF. Check that bearing pointer and DME display INS data with Bearing Select switch in NORMAL. Adjust Course Set knob to check for proper CDI indications.

(2) TACAN/ADF - Tune and identify TACAN station. Check for bearing pointer and DME indication. Adjust Course Set knob to check proper CDI and To-From indications.

(3) ILS - Tune and identify ILS station. Set Course Set knob to final approach course and check CDI and glide slope indications in relation to present position.

(4) ILS/APPROACH - Adjust Course Set knob to align the course arrow with the top index. Depress ILS test buttons on ILS control panel and check for proper indications on steering bars, glide slope, and CDI.

T6. UHF-1 and UHF-2 radios - Checked.

Check external and internal operation. For external operation, using power level 4 or less, depress CONT and INT lights to confirm normal operation. The interrogate light should remain on for three to four seconds.

7. VHF radio - Checked.

T 8. SAS channel engage switches - OFF.

T 9. Trim - Checked.

Check pitch (full travel), roll, and yaw trim and set to zero. Confirm that direction of movement corresponds with indication in the sequence: nose up, nose down, left roll, right roll, nose left and nose right. Check RH rudder synchronizer.

T 10. Flight control system - Checked.

With nosewheel steering disengaged, check each axis for full deflection and freedom of travel in both directions, and confirm correct deflection of control surfaces in the sequence: nose up, nose down, left roll, right roll, nose left and nose right.

If nosewheel steering will not disengage, rudder control will be severely restricted in flight with the gear down.

NOTE

Rapid control surface deflection while near idle rpm may result in temporary illumination of an A and/or B HYD warning light. The light should extinguish when flow demands diminish and normal pressure is restored.

11. Shotgun cartridge - Checked.

Confirm left and right cartridges engaged.

12. Fuel derich system Both checked and rearmed.

a. Set both engines 400 rpm above idle speed.

b. Actuate the derich test switch until 860°C EGT is exceeded with LEFT and then RIGHT selected.

When the EGT indications exceed 860°C:

c. Verify that the EGT gage warning lights are on and that the Fuel Derich lights are on.

d. Note that engine speeds decrease between 50 and 400 rpm.

e. Cycle the fuel derich switch to REARM then ARM.

Verify that each engine returns to 400 rpm above idle, and EGT indications are normal.

f. Reset the throttles to IDLE.

13. Air refueling system and drag chute doors - Checked.

a. Air refuel switch - AIR REFUEL.

Check READY light on. Confirm doors open and light on, toggles unlatched.

b. Air refuel switch MAN O'RIDE. Confirm door open, and light on, toggles latched.

c. Actuate stick trigger, confirm toggles retract.

d. Air refuel switch - OFF. Confirm light off, door closed.

e. Confirm with ground crew that drag chute doors are locked. A paddle indicator on the drag chute door should be flush with the fuselage contour when viewed from alongside the cockpit at the level of the crew station.

(T14.) ANS mode - Set.

The ANS is normally placed in the INERTIAL ONLY mode.

▲15. ANS - Checked.

a. Pilot set display mode select switch to ANS. Check true heading under HSI lubber line and programmed true course in HSI course window.

b. RSO check ANS heading against INS heading. Check for proper DP code and coordinates, and crosscheck command course and distance to DP with pilot.

c. Pilot check bearing select switch in both positions for normal operation of bearing pointer and DME.

d. Pilot and RSO check for normal attitude indications in both attitude reference select switch positions.

e. Pilot check standby attitude indicator.

▲16. INS - Checked.

a. Pilot set display mode select switch to INS, bearing select switch to NORMAL.

b. RSO set BDHI SEL HEADING and NO. 1 NDL switches to INS. Display distance to DP on Inertial Control Panel (Data switch set to STRG, distance is in left display).

c. Confirm INS DP bearing and distance are the same in both cockpits.

▲17. Ejection seat and canopy pins -Removed.

▲18. Canopy - Closed and locked.

Visually check engagement of canopy hooks.

CAUTION

To prevent overheating the ANS, the RSO canopy must not be closed and locked prior to the pilot's canopy unless the cockpit air handle is off (forward).

NOTE

Severe cockpit fogging may occur if cold cockpit temperature control settings are selected unless the RSO's cockpit air handle is off.

▲19. Canopy seal switch - ON.

(T20.) Cockpit air handle - On (aft).

21. L and R refrigeration switches - ON.

Minimize the time between locking the aft canopy and activation of a ship air-conditioning system. A delay increases the possibility of overheating equipment.

T 22. CANOPY UNSAFE, L and R AIR SYS OUT and CKPT AIR OFF caution lights - Off.

The RSO should recycle the cockpit air shutoff lever if the CKPT AIR OFF caution light is illuminated.

23. Ground air - Disconnect.

Signal ground crew for disconnect. Confirm the BAY AIR OFF light extinguishes.

(24.) OBC Power switch - ON.

25. PVD - On and set.

Up to 25 seconds may be required before laser line is visible.

a. Set ROLL to index.

b. Set PITCH to index or as desired.

c. Set intensity as desired.

d. Set SCALE to NORM or as desired.

WARNING

Do not look directly into the laser beam.

(T) 26. Angle of Attack indicator -Checked.

Check OFF flag out of view and AOA indicates zero.

T 27. Periscope - Checked.

T 28. Nosewheel steering - Engaged and checked.

Nose should swing as rudder pedals are moved slightly. Nosewheel STEER ON light should illuminate.

29. Panels and gear pins - Secured and removed.

Crewchief confirms all panels and doors secured. Crewchief disconnects interphone and displays landing gear downlock pins.

30. OBC self test - Completed, OPR/STP light on.

CAUTION

If installed, the Optical Bar Camera must always be operated in the standby or operate modes while in flight; if shut down, the optical bar may be damaged.

TAXIING

Observe crewchief for signal

CAUTION

Taxi and turn at low speed to minimize side loads on the landing gear. Fast taxiing should also be avoided to prevent excessive brake and tire heating and wear.

T 1. Braking and nosewheel steering - Checked.

When clear of obstacles disengage nosewheel steering and check individual brake operation on L and R systems, and for dragging brakes. Release pedal pressure before changing hydraulic systems. Engage NWS and check steering operation.

NOTE

Rudder pedal feedback, due to nosewheel castering, indicates that nosewheel steering has not disengaged. The STEER ON light also remains on.

T 2. Turn-and-slip indicator - Checked.

Check turn needle deflection in the direction of turn and ball free in race.

T 3. SAS lights - Checked.

Check SAS control panel for PITCH, ROLL, or YAW SENSOR lights during turns or braking. Attempt to reset SENSOR lights. All SENSOR lights should be out prior to takeoff.

T4. ANS - As desired.

If the ANS is in the INERTIAL ONLY mode and NAVIGATE/ASTRO INERTIAL mode is desired for takeoff, place the ANS in NAVIGATE/ASTRO INERTIAL.

BEFORE TAKEOFF

▲1. Pilot's ANS distance display mode DP/TURN.

Ta. ANS DATA switch - TEST.

Press DISPLAY push-button switch to display data.

Tb. DP/TURN push-button switch - As desired.

RSO will coordinate the pilot's desired ANS distance display mode.

2. Flight instruments - Set for takeoff.

a. Display Mode Select switch - Set.

b. Attitude Reference Select switch - INS.

c. For instrument departure, tune and identify TACAN station.

d. HSI Course Select knob - Set.

3. Engine run - Lockout and EGT trim checked.

a. Wheels - Chocked.

b. Brakes - Apply.

c. IGV switches - LOCKOUT.

MINIMUM TURNING RADIUS

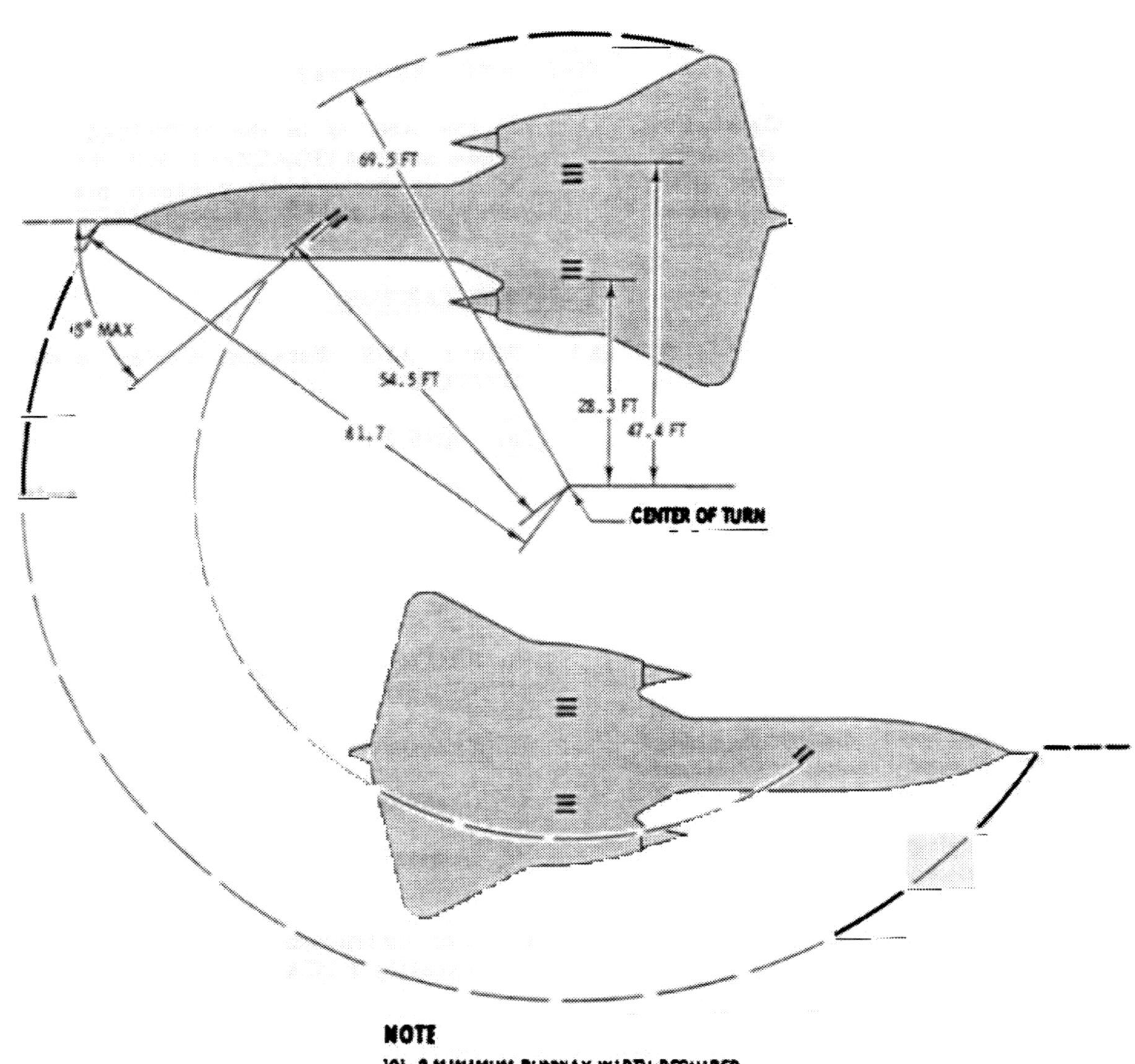

NOTE

101.9 MINIMUM RUNWAY WIDTH REQUIRED FOR 180-DEGREE TURN (MAIN GEAR WHEELS ON EDGE OF RUNWAY AT START OF TURN).

F203-54(a)

Figure 2-3

One engine at a time, with AUTO EGT selected:

d. Throttle - Military.

Move the throttle smoothly to the Military stop, observing ENP and EGT. EGT should increase and ENP indication should move toward zero. An EGT gage COLD flag will appear when the throttle reaches the Military position if EGT is below the nominal trim band.

NOTE

Automatic trimming does not occur until the throttle is positioned at or above the Military position.

e. Throttle - Retard approximately one-half inch aft of the Military position and return to Military rapidly. This removes hysteresis from the fuel control linkage. Hold the military power throttle setting for at least 30 seconds to allow MRS recording of engine parameters.

Note EGT gage COLD/HOT flag operation. If the throttle is retarded before EGT reaches the nominal trim band, disappearance of the COLD flag while the throttle is retarded confirms normal operation of the automatic EGT trim system permission circuit.

f. IGV light - Off.

The IGV position light should remain off.

g. IGV switch - NORM (as EGT approaches the nominal trim band).

h. IGV light - On.

The IGV position light should illuminate immediately when IGV NORM is selected, indicating an IGV shift to the axial position. The nozzle should open slightly at IGV shift.

NOTE

- o Do not takeoff if the IGV position light fails to illuminate.
- o An inoperative IGV lockout which is detected during the before takeoff trim check does not require aborting the flight.
- o The engine IGV light should not illuminate on rpm increase with its IGV switch in the LOCKOUT position. With IGV NORM selected, the engine IGV light should illuminate during rpm increase (approximately 300 to 800 rpm below the Military rpm schedule) and extinguish when the guide vanes reach the cambered position as the throttle is retarded to idle.

i. EGT trim - As required.

Check automatic EGT trims to within the nominal band shown by Figure 2-4.

If a HOT flag appears and EGT approaches an overtemperature condition, retard the throttle, select manual EGT control, downtrim as required, then recheck trim at Military in AUTO EGT.

If no HOT or COLD flag is observed and EGT is normal at Military power, downtrim EGT momentarily in manual control, then select AUTO EGT. At Military, the COLD flag should appear temporarily, and the engine should retrim to the AUTO EGT deadband range.

If Auto EGT is unusable, trim manually as shown in Figure 2-4.

j. RPM - Check engine speed vs the schedule shown by Figure 2-4.

k. Engine and inlet instruments - Check.

EGT Auto Trim Check Schedule

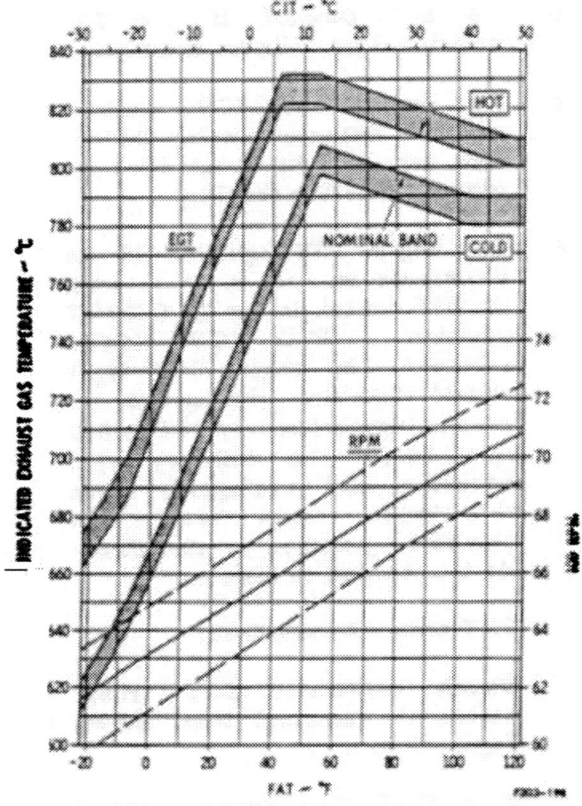

Figure 2-4

Check engine instruments for normal indications. Note ENP and fuel flow values (for engine comparison), and oil pressure values. Note L and R CIP; CIP indication should be less than at idle.

4. EGT trim switches - HOLD or AUTO

During cold temperature ground operations, engine surge (compressor stall) may occur at or above military power if EGT goes above the nominal trim band.

With EGT trim in AUTO, EGT uptrim starts when the throttle is advanced to the military position and continues until EGT increases into the deadband. Even if an engine was previously trimmed within the deadband, thermal lag results in a slight EGT overshoot when readvancing the throttle to military.

a. To avoid compressor stalls on takeoff when outside air temperature is below 15°C (59°F), position EGT trim to HOLD after stabilized in military power within the nominal EGT operating band. Select automatic trim, if desired, after takeoff.

Engine surge should not occur when ambient temperature is above 15°C (59°F).

If compressor stalls occur during engine run, retard throttle and downtrim EGT. Refer to Exhaust Gas Temperature Limits, Ground Operation, Section V.

b. Throttle - Retard smoothly to IDLE.

Check that ENP indication is normal during power reduction.

NOTE

After retarding from Military power to IDLE, do not readvance the throttle for at least ten seconds (for the engine to stablize at idle rpm). Otherwise, stall and dieout may occur. The stall may be inaudible, but dieout is indicated by decreasing rpm and, particularly, by increasing EGT. If dieout occurs, move the throttle to OFF to prevent overtemperature. The engine may be restarted as soon as a starter is available; accomplish Clearing Engine checklist. The stall and dieout occur only during ground static operation and is more likely when relatively high ambient temperatures exist.

c. IGV light - Off.

The IGV light should extinguish.

NOTE

If cockpit fog is encountered, increase cockpit temperature. Twelve-to-one o'clock auto-temperature control rheostat positions are normally sufficient.

T 5. Flight controls and trim settings -Check.

Cycle and check hydraulic pressure. Recheck trim settings zero.

▲6. Fuel sequencing - Checked.

a. Check tanks 1, 3, and 6 or 5 ON, depending on fuel load, and quantities decreasing. If less than full load, check tank 4 increasing.

NOTE

To check tank 3 pump operation, if no decrease in tank 3 fuel quantity has been noted, transfer fuel or increase left engine rpm.

▲7. CG - Checked.

Takeoff c.g. must be forward of 22%. (To check the c.g. which will occur in the flight attitude, increase the indicated c.g. value by the amount of the hand-held c.g. computer correction.) When the takeoff fuel load is above 70,000 pounds, the c.g. should be no further forward than 20%. With less than 70,000 pounds of fuel, c.g. should not be aft of 20% while level, to allow for the aft c.g. shift during takeoff.

NOTE

- A supersonic leg with less than a full fuel load may require manual control of the fuel system to achieve a desirable supersonic c.g.
- Press the Tank 5 or Tank 4 boost pumps on before transferring fuel forward. Otherwise, with crossfeed off, a reduction in fuel flow to approximately 3600 lb per hour will occur on the right side. This is less than the desired value for normal operation of the fuel heat sink system. Release the tank after completing fuel transfer.

T 8. Forward transfer - OFF.

T 9. Fuel Derich switch - ARM.

▲10. No. 1 and No. 2 oxygen systems - ON and pressure checked.

Verbally confirm oxygen latched ON with normal pressure.

▲11. Baylor bar - Latched and locked.

12. Brake switches - DRY or WET, and ANTI SKID ON.

Use the DRY position for a RCR of 21 or more. Wet runway conditions shall be assumed to exist and the WET position used if RCR is less than 21. If RCR is not available, assume a wet runway condition if moisture is visible on the runway, particularly as evidenced by glare or reflections.

▲13. Takeoff data - Review.

a. Acceleration Check.

b. Refusal speed.

c. Rotation speed.

d. Takeoff speed.

e. Single-engine speed.

NOTE

If a tire cooling period has been required, do not takeoff until ground crew signals that tire condition is satisfactory.

14. Pitot heat switch - ON and checked.

Ground crew confirms heat on.

15. Battery switch - Checked BAT.

16. Instrument inverter switch - Checked NORM.

(T17) INS altitude - Update.

Update the INS altitude to the sustained or mid-altitude expected after takeoff, i.e. enter A/R altitude, or if climbing immediately to cruise conditions enter 35,000 or 40,000 feet.

18. VHF radio antenna cover - Removed.

Ground crew displays cover to pilot.

TAKEOFF

(T1) IFF - NORMAL

Set proper mode and code.

2. SAS - Engaged, lights off.

   a. Channel engage switches - ON.

   b. SENSOR/SERVO lights - Check off.

   c. BIT TEST light - Check off.

WARNING

The SAS is non-functional while in the ground test mode. DAFICS will not operate normally until the system is reset. Failure to press a SENSOR/SERVO recycle switch after the DAFICS Preflight BIT is complete will cause the DAFICS to remain in the ground test mode.

(T) 3. Aft cockpit SAS - ON.

Channel engage switches all ON.

NOTE

For normal operations, all SAS channel engage switches should remain ON in both cockpits for entire flight regardless of which cockpit has AFCS control.

(T) 4. AFCS control - Transfer to other cockpit.

Check AFCS transfer light illuminated in cockpit not previously in control.

(T) 5. SAS Lights - Off.

Check SENSOR/SERVO lights remain off.

(T) 6. AFCS control - As desired.

WARNING

In the SR-71B, only the positions of the SAS channel engage switches in the cockpit that has the AFCS transfer light illuminated on the control transfer panel effect the SENSOR/SERVO and SAS OUT caution lights. No warning is displayed if the SAS channel engage switches are not ON in the cockpit that does not have AFCS control. If the SAS engage switch(es) are OFF in the cockpit that does not have AFCS control, transferring AFCS control to that cockpit results in loss of SAS until the SAS switch(es) are engaged.

▲7. Warning and caution lights - Checked.

Any amber caution lights (autopilot, cockpit air, etc.) which remain on must be justified by an intentional and acceptable operating situation. Do not start a takeoff if any red warning lights are on.

▲8. Circuit breakers - Checked.

9. Tank 4 boost pump switch - Press on.

▲10. Compass - Checked.

Check INS and standby compass against runway heading. Start ANS runway heading alignment when required.

11. Nosewheel steering -Engaged.

Confirm STEER ON light illuminated.

Refer to Figure 2-5 for illustration of the typical sequence of events during takeoff.

a. Brakes - Release when IGV lights illuminate (approximately 6000 rpm) as the throttles are advanced.

CAUTION

The tires may skid if the brakes are held at high thrust.

NOTE

Abort takeoff if IGV position lights fail to illuminate with Military rpm during engine acceleration.

b. At Military power - Check engine instruments for values at or approaching those observed during trim.

(1) Tachometer.
(2) EGT.
(3) Nozzle Position.
(4) Oil Pressure.

c. Throttles Advance to mid afterburner range for A/B ignition, then smoothly advance to maximum afterburner.

CAUTION

- To prevent overspeed, afterburners must not be ignited before engines reach Military rpm.
- Abort the takeoff if an afterburner fails to ignite within 3 seconds.
- Advancing the throttle will result in momentary nozzle excursion, and engine transient speed oscillation may approach 250 rpm.

d. Maximum A/B - Check engine instruments.

Exact readout of these instruments is time consuming. The readouts should be anticipated and needle position checked against a clock position. If there is any indication of deficient engine performance during throttle advancement, abort the takeoff. If possible, any abort decision should be made before the aircraft has reached high speed. Refer to takeoff performance data in the Appendix. Directional control should be maintained with nosewheel steering up to nosewheel lift-off speed.

e. Acceleration - Check.

Check KIAS against computed acceleration check speed at selected acceleration check distance. Refer to takeoff performance data in the Appendix.

NOTE

Failure of the IGV lights to remain illuminated during takeoff results in inability to develop full rated thrust; unless a more serious malfunction is indicated, the takeoff may be continued if the acceleration check speed has been reached satisfactorily.

ROTATION TECHNIQUE

In general, the tires are more vulnerable to blowouts during takeoff than at landing because of the higher groundspeeds and gross weights involved. Wing lift quickly relieves the gear load as the nose is raised. Apply smooth, constant back pressure 15 to 25 knots below computed rotation speed. Lift the nosewheel off at rotation speed, using the rotation rate required to leave the ground at computed takeoff speed. Depending on gross weight, normal takeoff attitude is 8° to 10°

TAKEOFF - Typical

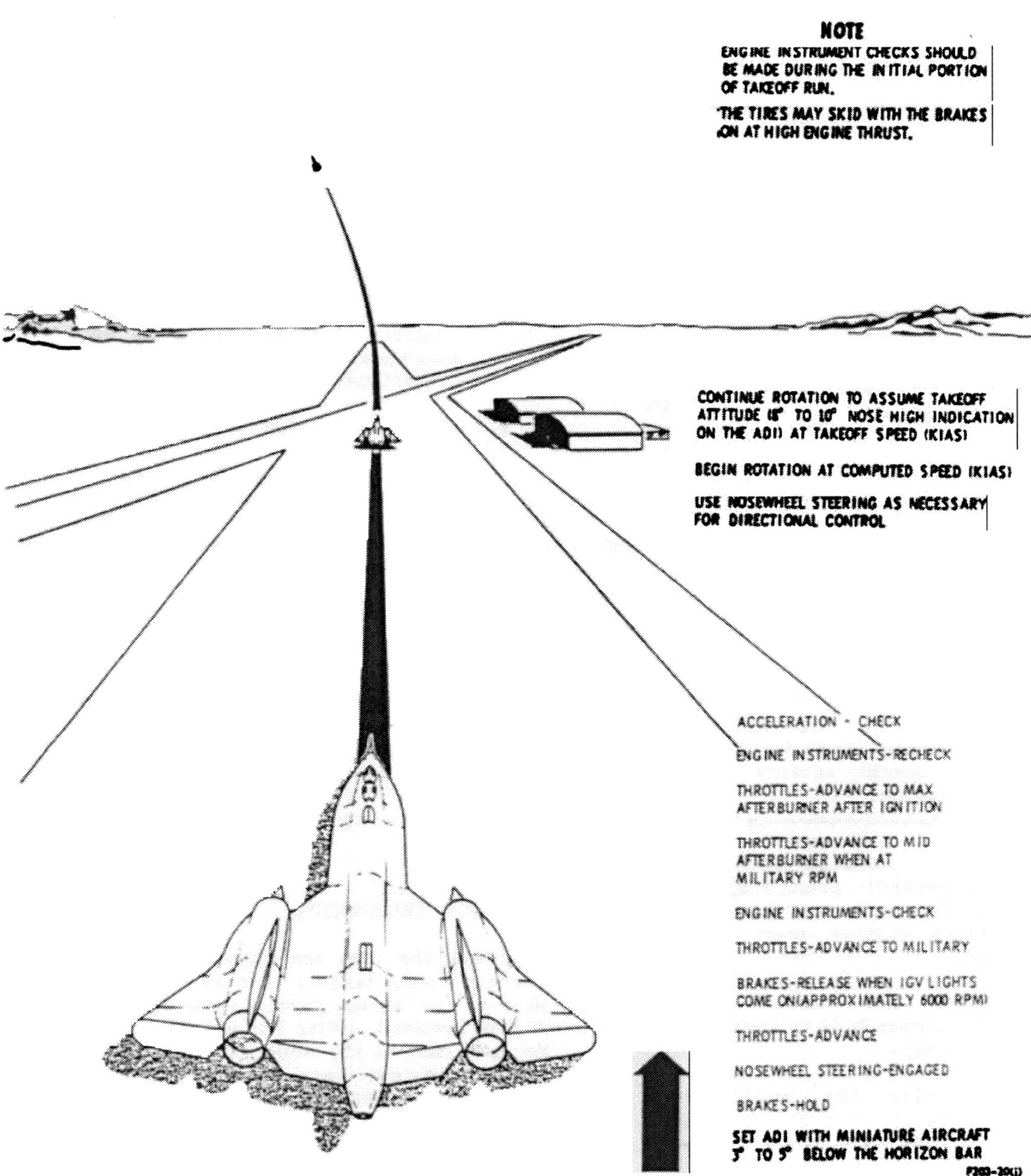

Figure 2-5

nose high indication on the ADI. The transition from start of rotation to takeoff requires approximately 5 seconds when using the normal takeoff technique. Refer to Takeoff Speed Schedule in Part II of the Appendix for rotation and takeoff speeds.

Premature nosewheel liftoff should be avoided because the unnecessary drag extends the ground run and may result in excessive tire loads.

NOTE

AOA indicates zero at airspeeds less than 100 KEAS and actual AOA at higher airspeeds.

CROSSWIND TAKEOFF

The aircraft weathervanes into the wind during crosswind takeoffs when the nosewheel lifts off and nosewheel steering is no longer available. Rudder pressure must be held to counteract the crosswind. A definite correction must be made as the aircraft breaks ground. Apply lateral control as necessary for wings-level flight. Both the directional and lateral control applications are normal and no problems should be encountered when taking off during reasonable crosswind conditions.

AFTER TAKEOFF

When definitely airborne:

1. Landing gear lever - UP.

   Gear retraction requires 12 to 16 seconds.

WARNING

- Single engine operation is critical immediately after takeoff. Increasing airspeed and decreasing angle of attack have greater benefits than gaining altitude at a maximum rate. Single engine flight capability is presented in Part II of the Appendix. With gear down, the minimum safe speed out of ground effect is approximately 30 knots greater than in ground effect.
- Immediately depress the control stick trigger switch to deactivate APW System stick pusher operation if a false stick pusher warning occurs. If the stick pusher is not deactivated by the trigger, use a pull force of 30 to 35 pounds in addition to normal stick forces to overcome the stick pusher spring. Use pitch trim to relieve stick force.

After gear retraction is completed and single engine flying speed is obtained, establish climb power as desired. A military power climb conserves fuel.

2. Engine instruments - Check.

At Mach 0.5:

3. Surface limiter - Engaged, SURFACE LIMITER light off.

   Rotate handle counterclockwise and release to engage

4. Attitude Reference - ANS (pilot).

   The RSO will crosscheck attitude sources before the pilot selects the opposite reference. This is especially critical for night or instrument flight conditions.

   NOTE

   The RSO should vigilantly monitor attitude during takeoff/climb out and crosscheck his attitude indicator references by alternately selecting ANS/INS. He will notify the pilot immediately if any abnormal attitude is suspected.

5. EGT trim switches - AUTO

CLIMB

NOTE

If the cockpit air handle was positioned OFF for takeoff, wait until a safe altitude out of the moist air before positioning the handle to on. Approximately 5000 feet above ground level should be sufficient.

Normal climb is 400 KEAS until Mach 0.90 is intercepted, then hold Mach 0.90.

WARNING

If moderate turbulence is encountered, reduce airspeed to 300 - 350 KEAS while subsonic. Climb at 400 KEAS if supersonic, or decelerate to subsonic speeds at 350 KEAS if the climb cannot be continued. Refer to Section VII, Operation in Turbulence.

NOTE

The pilot must advise the RSO of autopilot engagement or disengagement and the mode(s) affected.

1. PUMP REL switch - Press, Tank 4 released, light out.

   During climb at maximum power when the right-hand shut-off float switches in tank 1 have been actuated because of tank depletion or flight attitude, illumination of the L and/or R FUEL PRESS warning light(s) (a fuel low pressure warning) may occur if tank 4 is released. At maximum power, do not release tank 4 at low altitude.

T 2. Altimeter - Set.

   Set 29.92 as required.

(3.) Sensor power - As briefed.

   All sensor power switches except DEF may be turned on at this time. The RCD or IPD power switch ON legend should illuminate when Radar power is applied. The OBC should be in STBY before takeoff.

   Sensor warmup times are:

RADAR	-	6 min.
RCD	-	1 to 3 min.
EIP/EMR	-	2 min.
TECH	-	20 to 40 sec.

   Ensure TECHs and radar in AUTO or MANUAL modes as required.

(T)4. HF radio - Retune (618-T-only).

   Recycle HF tuning to in-flight antenna impedance by momentarily placing freq control knob off-frequency, then back to original position. Recycle antenna coupler by keying transmitter.

WARNING

Rf energy from the HF radio during tuning or transmission has caused erroneous light and instrument indications.

5. DEF system power - As briefed.

Depress DEF System A power switch to illuminate the ON legend. After two minutes warm-up, the S (standby) legend should illuminate.

Depress the DEF F/H power switch. The W (warm-up) legend illuminates immediately. In approximately five minutes, the W legend extinguishes and the S (standby) legend illuminates.

Depress the DEF M power switch to illuminate the W and ON legends. After three minutes warmup, the W legend extinguishes.

6. DEF systems - Checked.

Refer to the DEF System Procedures, Section IV.

CAUTION

To avoid DEF system damage due to overheating, do not exceed the transmission time periods scheduled for the checks while testing in the manual mode below FL 500.

NOTE

These tests can be finished after transonic acceleration if the system warm-up time requirement does not allow earlier completion.

7. G-band Beacon switch - As required.

Attitude Control

Each crew member must be aware of attitude reference operating characteristics and be alert for failures. The RSO should select INS attitude reference when the pilot has ANS selected, and vice versa. The pilot should crosscheck attitude with the standby attitude indicator. The RSO should periodically cross check attitude reference sources by alternately selecting INS and ANS.

If the ANS nav-ready signal is lost, the pilot's ANS REF annunciator caution light illuminates, and the RSO's ANS FAIL annunciator caution light illuminates. If the pilot's ATT REF select switch is in ANS: the autopilot disengages, the DAFICS ANR light flashes (flashing DAFICS Preflight BIT FAIL light), the PVD is inhibited, and the power OFF flag appears in the ADI. If the RSO's ATT IND switch is in ANS, the power OFF flag appears in the attitude indicator.

If the INS is operating in the attitude mode, the pilot's INS REF annunciator caution light illuminates. If the pilot's DISPLAY MODE SEL switch is in any position other than ANS, the course warning flag comes into view.

If the INS platform fails, the pilot's INS REF annunciator caution light illuminates. If the pilot's ATT REF SELECT switch is in INS: the autopilot disengages, the DAFICS ANR light flashes (flashing DAFICS Preflight BIT FAIL light), the PVD is inhibited, and the power OFF flag appears in the ADI. If the RSO's ATT IND switch is in INS, the power OFF flag appears in the attitude indicator.

ACCELERATION

The TDI is the primary speed and altitude reference for acceleration to, during, and for deceleration from supersonic flight. The pilot should crosscheck TDI indications with the pitot-static instruments. The RSO should monitor altitude, attitude, and speed. Cross-check navigation system ground speed against aircraft speed indications.

SUPERSONIC AIRSPEED SCHEDULES

The optimum supersonic airspeed is 450 KEAS while climbing between Mach 1.25 and

NORMAL EVENTS AFTER TAKEOFF OR DURING TRANSITION TO SUPERSONIC CRUISE

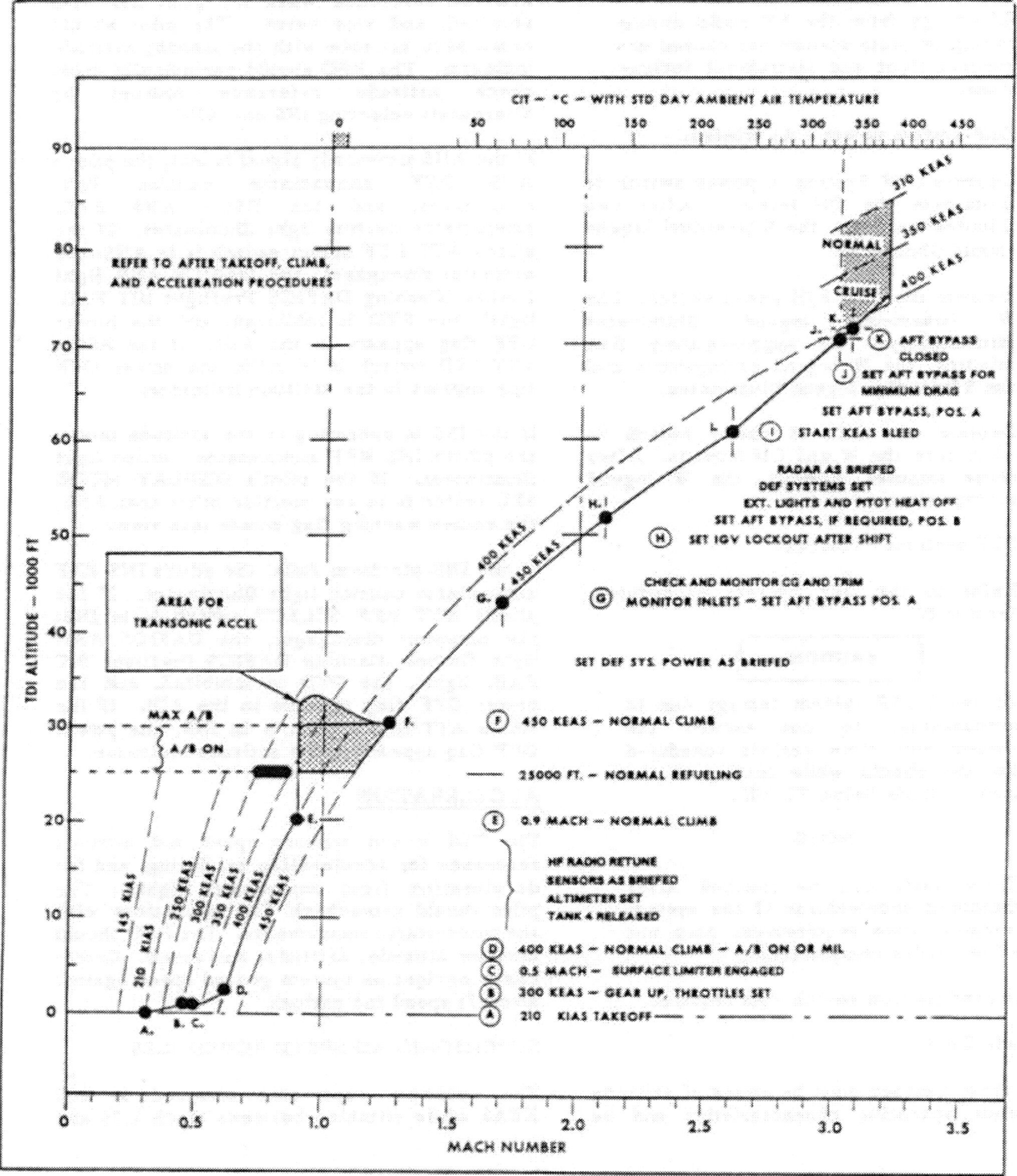

Figure 2-6

Mach 2.6. Lower airspeeds such as 400 or 375 KEAS may be used, as when turbulence is encountered, but performance is considerably degraded.

TRANSONIC ACCELERATION PROCEDURE

Transonic acceleration is accomplished at either a level altitude or during a climb-and-descent maneuver.

NOTE

> The climb-and-descent acceleration is recommended for best specific range (NM per pound of fuel used).

Level Acceleration

A level acceleration to intercept the supersonic climb speed schedule can be made at refueling altitude, normally 25,000 feet. When ambient temperatures are near or lower than standard, less time and distance are required to intercept the climb speed schedule than the climb-and-descent procedure. The total range penalty is small under these conditions.

Start the acceleration with minimum afterburner. Complete course changes while subsonic so that the additional power required for turning will not diminish the power available for transonic acceleration. Set maximum power at Mach 0.9. Gently increase pitch to climb attitude near 430 KEAS. A smooth technique is required, as 450 KEAS is only slightly more than Mach 1.1 at 25,000 feet and is still within the critical thrust/drag speed range which begins near Mach 1.05.

WARNING

> Airspeed may increase rapidly after Mach 1.1 is reached. Reduce power (below Military, if necessary) to avoid high airspeeds. Do not use excessive load factors to prevent exceeding 450 KEAS.

The procedure can be used at another altitude; however, when lower, the transition to 450 KEAS climb attitude must be made in the unfavorable speed range from Mach 1.05 to 1.10. At higher altitudes, the transition through this speed range can be completed before starting the climb, but less thrust is available. If ambient temperature increases, thrust decreases and the time, fuel and distance penalty for using the level acceleration procedure is greater.

Climb-And-Descent Acceleration

The climb-and-descent procedure requires less fuel to intercept the climb speed schedule than the level acceleration when ambient temperatures are warmer than standard.

NOTE

> The climb-and-descent procedure is recommended for best specific range (NM per pound of fuel used) at all temperatures.

WARNING

> Although angle of attack increases during the subsonic climb, pitch attitude must decrease to avoid dangerous flight conditions. Failure to monitor and control attitude, speed, and angle of attack can result in approach to pitch-up conditions.

Start the acceleration with minimum afterburner power. Intercept 0.9 Mach. Set maximum afterburner at 30,000 feet for the remainder of the acceleration, observing the 300 KEAS restriction. At 33,000 feet, increase speed to at least Mach 0.95. This speed is slightly above the start of the drag rise region. Make a smooth transition to establish a 2500 to 3000 fpm rate of descent.

NOTE

Engine stalls during the subsonic climb may indicate a potentially dangerous flight situation. Stalls can result from low CIP or high distortion in the inlet associated with aircraft operation beyond established flight limits. Refer to Subsonic Compressor Stalls, Section III.

After establishing the descent rate, maintain attitude until initiating climb. Avoid higher rates of descent since the usual result is altitude penetration below 29,000 feet and high fuel consumption.

When using the climb-and-descent procedure, it is important to exceed Mach 1.05 early in the descent, and to avoid turning until the climb is established. Begin the transition to climb near 435 KEAS so as to intercept 450 KEAS while climbing.

WARNING

- Airspeed may increase rapidly after Mach 1.1 is reached. Reduce power (below Military, if necessary) to avoid high airspeeds. Do not use excessive load factors to prevent exceeding 450 KEAS.
- In turbulence, reduce climb speed as specified in Section VII, Operation in Turbulence.

Transonic Acceleration Performance Comparison

Figure 2-7 is provided for convenience only. Refer to the appendix for detailed performance values.

TRANSONIC ACCELERATION PERFORMANCE COMPARISON
140,000 initial gross weight
Initial speed = Mach 0.75 at 25,000 feet

Temp difference from Std.		-10°C	Std	+10°C	+20°C	
Level Accel. at 25,000 Ft.	Fuel	4800	5500	6800	8800	lb
	Time	3.1	3.6	4.4	5.8	min
	Dist.	30	36	47	63	nmi
Level Accel. at 30,000 Ft.	Fuel	4900	5600	6600	8300	lb
	Time	4.0	4.6	5.4	6.7	min
	Dist.	38	44	54	69	nmi
Climb to 33,000 Ft.	Fuel	4900	5400	6100	7200	lb
	Time	4.3	4.7	5.3	6.2	min
	Dist.	39	44	51	62	nmi

Std. Temp. = -34.5°C @ 25,000 Ft., -44.4°C @ 30,000 Ft.
Min. A/B to Mach 0.9 and for subsonic climb to FL 300.
Max. A/B for level accel. above Mach 0.9 and above FL 300.

Figure 2-7

SUPERSONIC ACCELERATION PROCEDURE

1. Throttles - A/B

   Use minimum or maximum afterburner, as desired.

   Intercept Mach 0.90 and climb at that speed until starting the transition to the supersonic schedule.

   If the pitch autopilot is used during subsonic climb, the Mach Hold function must be engaged to climb at constant Mach.

   Refer to appendix Figure A1-4 for approximate differences between IAS and EAS indications. At Mach 0.9, correct airspeed indications at 30,000 feet are 324 KEAS and 347 KIAS. The TDI can be monitored by comparing TDI Mach with indicated Mach.

At start of transonic acceleration:

2. Throttles - Maximum A/B.

   Advance both throttles smoothly to maximum power and advise the RSO.

   Use either the level acceleration procedure or the climb-and-descent procedure.

   NOTE

   - o The Auto-Nav feature of the autopilot is normally used if the roll autopilot is engaged.
   - o If the autopilot KEAS hold mode is used, engage the pitch autopilot and stabilize KEAS for a few seconds before engaging KEAS hold.

After reaching speeds above Mach 1.25 when operating without the automatic EGT control engaged, trim as necessary to maintain nominal EGT as shown in Figure 2-4. EGT must not exceed 830°C below 40°C CIT and 805°C above 40°C CIT. Maintain EGT between 775 and 805°C after climb is established.

At Mach 1.7:

3. Inlet parameters - Monitor.

   Monitor spike and forward door positions and CIP.

4. Aft bypass controls - Set.

   Normally set Position A at Mach 1.7, but wait until the forward bypass doors move out of full closed to set the aft bypass.

5. C.G. and trim - Monitor.

   NOTE

   - Manual boost pump selection for the purpose of shifting c.g. is not normally necessary with a full fuel load. With reduced fuel loads, manual control of the fuel system may be required to achieve a desirable supersonic c.g. position. (See Cruise Fuel Management, this section.) The c.g. normally moves aft 1% per 5000 lb of fuel used until the right-hand shutoff switch in tank 1 operates.
   - While at maximum power with high fuel flows, press tank 5 (or tank 4) boost pumps on before using fuel forward transfer to prevent an R FUEL PRESS warning. Release the tank after completing fuel transfer.

At IGV shift (CIT limit is 150°C):

6. IGV switches - LOCKOUT.

CAUTION

Decelerate to 125°C CIT or less (approximately Mach 2.0) if the IGV lights are not extinguished upon reaching 150°C CIT (approximately Mach 2.2 with ambient temperature near standard).

The IGV lights should extinguish on completion of IGV shift at 85° to 115°C CIT (approximately Mach 1.7 to 2.3, depending on ambient temperature). The lights must be out at 150°C CIT. When the guide vanes shift, forward bypass opening will increase and fuel flow and thrust will decrease. The engines may not shift simultaneously due to tolerances in fuel control schedules.

NOTE

If the IGV lights extinguish (IGV shift to cambered) while below the normal shift range, land when practical.

7. Aft bypass controls - Position B, if required.

   a. Set Position B when the lights extinguish if forward bypass opening exceeds 20%. Drag increases noticeably if the forward bypass schedules in excess of 20% open.

   b. Monitor forward bypass door position. Shift the aft bypass from position B to A when the forward bypass approaches closed. Allowing the forward bypass doors to schedule less than 5% open with the aft bypass doors in position B reduces inlet stability and increases the probability of an unstart without appreciably increasing overall inlet efficiency.

8. Exterior lights - Off.

   a. Anticollision/fuselage lights switch - OFF.

   b. TAIL LT switch - Off (center position).

9. Pitot heat switch - OFF.

(10.) DEF systems - Set and checked.

NOTE

Set DEF system operating conditions as briefed when above FL 500.

(11.) Radar - As briefed.

At FL 600:

(T12.) IFF - Mode C OUT.

This prevents automatic altitude reporting.

At Mach 2.6:

13. Aft bypass controls - Set position A, when required.

Shift from the B to the A position when the forward bypass approaches closed.

Allowing the forward bypass doors to schedule less than 5% open with the aft bypass doors in position B reduces inlet stability and increases the probability of an unstart without appreciably increasing overall inlet efficiency.

14. KEAS bleed - Monitor.

For normal climb at 450 KEAS, decrease KEAS 10 knots per 0.1 Mach number increase above Mach 2.6.

NOTE

The minimum pitch trim indication to be expected at Mach 2.6 is +0.5°. At higher Mach, the minimum limit depends on KEAS, aircraft weight and c.g. Assure trim is at or above 0° except for the specific high Mach, high KEAS conditions at 25% c.g. depicted on Figure 6-7. Check the c.g. if less nose-up trim is indicated.

(T15) INS altitude - Update as required.

Update the INS altitude to the mid-leg altitude between level off and start of descent.

At Mach 3.0:

16. Aft bypass controls - Set.

Maintain position A or use the CLOSE setting for cruise near Mach 3.0. If the forward doors are not closed with position A selected, maintain position A for best performance.

The optimum setting for cruise near Mach 3.0 may be determined by setting the aft doors to CLOSE individually. If drag increase on the closed side is noted, cruise with aft bypass in position A.

At Mach 3.05:

17. Aft bypass controls - Set CLOSE.

Set CLOSE position for cruise above Mach 3.05.

CRUISE

Cross-check the TDI altitude, KEAS, and Mach displays with IAS, indicated Mach, and altimeter, periodically, to detect discrepancies in the TDI. If Mach number appears to be inaccurate, make appropriate adjustments so that flight limits will not be exceeded.

The checklist Mach-KEAS-Altitude Relationship chart and the Mach-Airspeed-Temperature Conversion chart can be used to check the TDI and ANS true airspeed. Indicated spike position vs. the TDI and indicated Mach provides another cross-check on Mach number, as spike position is based on DAFICS Mach output.

WARNING

In the SR-71B, no warning is displayed if the SAS channel engage switches are not ON in the cockpit that does not have AFCS control. If the SAS engage switch(es) are OFF in the cockpit that does not have AFCS control, transferring AFCS control to that cockpit results in loss of SAS until the SAS switch(es) are engaged. Verbally confirm the position of the SAS channel engage switches before transferring AFCS control.

Flight will not be extended into night or IFR conditions if the pilot's standby attitude indicator and either the ANS or INS reference for the ADI are inoperative or erroneous. Climb and penetration through overcast is permitted. If already operating in night or IFR conditions, land when practical.

Pitot Static Reference - Subsonic Operation

The pressure altimeter is the primary altitude reference for subsonic flight operation; to fly an assigned altitude, use the altimeter correction card.

IFF Mode C uses DAFICS (TDI) altitude for automatic altitude reporting.

Operation at Supersonic Speeds

Avoid sustained operation at speeds between Mach 2.5 and 2.7 when convenient. This area is normally more susceptible to inlet duct roughness than higher or lower speeds. Refer to Climb, Section VI.

Autopilot Operation, Cruise or Climb

Manual trim in the yaw and roll axes may be used while the roll autopilot is engaged in Heading Hold or Auto Nav modes to minimize track error (course hang-off). When only attitude-hold mode (pitch or roll autopilot) is engaged, use the trim wheel on the function selector panel to adjust attitude. (Manual trim inputs at this time will only cause control transients.) If Mach hold is desired, engage the pitch autopilot and stabilize the aircraft at the desired Mach before engaging the Mach hold mode.

Optimizing Trim

Trim should be optimized during cruise to minimize drag. With the autopilot on, match fuel flows then apply yaw trim as needed to center the turn-and-slip ball. The periscope may be used to visually confirm rudder trim symmetry. With the roll autopilot engaged, check for off-center displacement of the roll autopilot alignment needle. "Beep" roll trim in the direction of needle displacement until the needle is centered. It may be necessary to repeat the procedure if KEAS or Mach changes.

The max range loss factors for rudder, sideslip, or c.g. deviation (based on symmetrical power and supersonic cruise) are:

rudder: 1° - 3 nm 2° - 12 nm

sideslip: 1° - 12 nm 2° - 48 nm

c.g.: 50 nm/% c.g. fwd of 25% MAC

Turning, in Auto Nav Mode

While in the Auto Nav mode, anticipate turn entry. Use manual stick inputs, if necessary, to avoid excessive roll rates and bank angles. It is unlikely that the stick shaker warning will be encountered during the roll-in to a normal turn. If shaker warning occurs while rolling into a steep turn, manually reduce the roll rate.

Steep Turns

The pitch autopilot can be used for turns with bank angles up to 45°. Pitch and roll attitude must be monitored when making steep turns.

C.G. Crosscheck

Since center of gravity control is important for optimum cruise performance and safety, crosscheck the c.g. indication with pitch trim and, occasionally, by computation.

During climb, the c.g. normally moves aft approximately 1% per 5000 pounds of fuel used and can be expected to reach 24% to 25% when the fuel in tank 1 reaches the right-hand shutoff setting. This should occur near the level-off point in a supersonic flight profile after air refueling to full tanks. After right-hand shutoff, the automatic sequencing provides a center of gravity which will approach 25%. This minimizes elevon deflection and trim drag during supersonic operation. If c.g. should exceed 25%, transfer fuel forward (C.G. annunciator panel warning light illuminates when c.g. reaches 25.3 to 25.6%). During subsonic mission legs, successive forward transfers may be necessary to keep the c.g. forward of the subsonic aft limit (22%).

NOTE

The optimum supersonic c.g. position may not be reached automatically with a partial fuel load. Manual fuel management may be required.

Fuel Management

Maintain c.g. forward of the center of gravity limit by using forward transfer. Forward transfer should also be used for an electrical system or SAS emergency.

Select fuel crossfeed OPEN any time tanks 5 and 6 are empty with both engines operating normally, and any time the FUEL QTY LOW caution light or an L or R FUEL PRESS warning light illuminates. Even though symmetrical thrust conditions exist, the use of crossfeed during afterburner operation can result in mismatched fuel flow indications of as much as 7000 lb per hour, due to fuel crossflow through the fuel heat sink system.

With crossfeed closed, should fuel flows to the engines be mismatched, c.g. may move out of the desired range. If the right fuel flow is higher, c.g. will not move aft as fast as the normal schedule and if the left fuel flow is higher, the converse is true. If fuel flows are mismatched, c.g. should be closely monitored as the full capabilities of the transfer system may be needed to obtain optimum c.g.

Manual aft transfer is provided to augment the automatic aft transfer system. Manual aft transfer is only effective if tank 5 is less than full.

The following three methods are designed to move the c.g. aft. Early manual fuel management is especially important with a less-than-full fuel load during supersonic acceleration. Manual aft transfer has a lower rate of c.g. movement than Method a only, being equal to 71% of Method a rate.

Method a

Press tank 1 on to override the RHSO until the fuel level in tank 1 reaches 3000 pounds.

CAUTION

Do not manually select tank 1 with less than 3000 pounds remaining in that tank. The forward pumps would be operating without fuel cooling and lubrication.

Method b

Pressing tank 2 on and opening the crossfeed valve is effective until tank 3 is empty. The rate of c.g. movement of this method will vary depending on the amounts of fuel remaining in tanks 1 and 2.

NOTE

Using this method eliminates automatic aft transfer. Increased aft c.g. movement can be obtained by depressing the manual aft transfer switch once tank 5 starts to operate.

Method c

Depressing the manual aft transfer switch is the primary means for moving the c.g. aft when tank 2 and 5 are supplying fuel during normal fuel sequencing.

NOTE

With the crossfeed valve open and only one forward tank supplying fuel, the c.g. may move forward.

Engine Operation

Exhaust gas temperature and engine speed limits vary with compressor inlet temperature. Refer to Engine Operating Limits, Section V. When encountered, slow fluctuations in EGT, fuel flow and rpm should be reported for maintenance evaluation; however, fluctuations of ±1% are expected and are not detrimental to engine operation. Random fluctuations of ±4% are expected in nozzle position indication and are not detrimental to engine operation.

EFFECT OF ASYMMETRIC ENGINE FUEL FLOW ON AUTOMATIC CG SCHEDULING

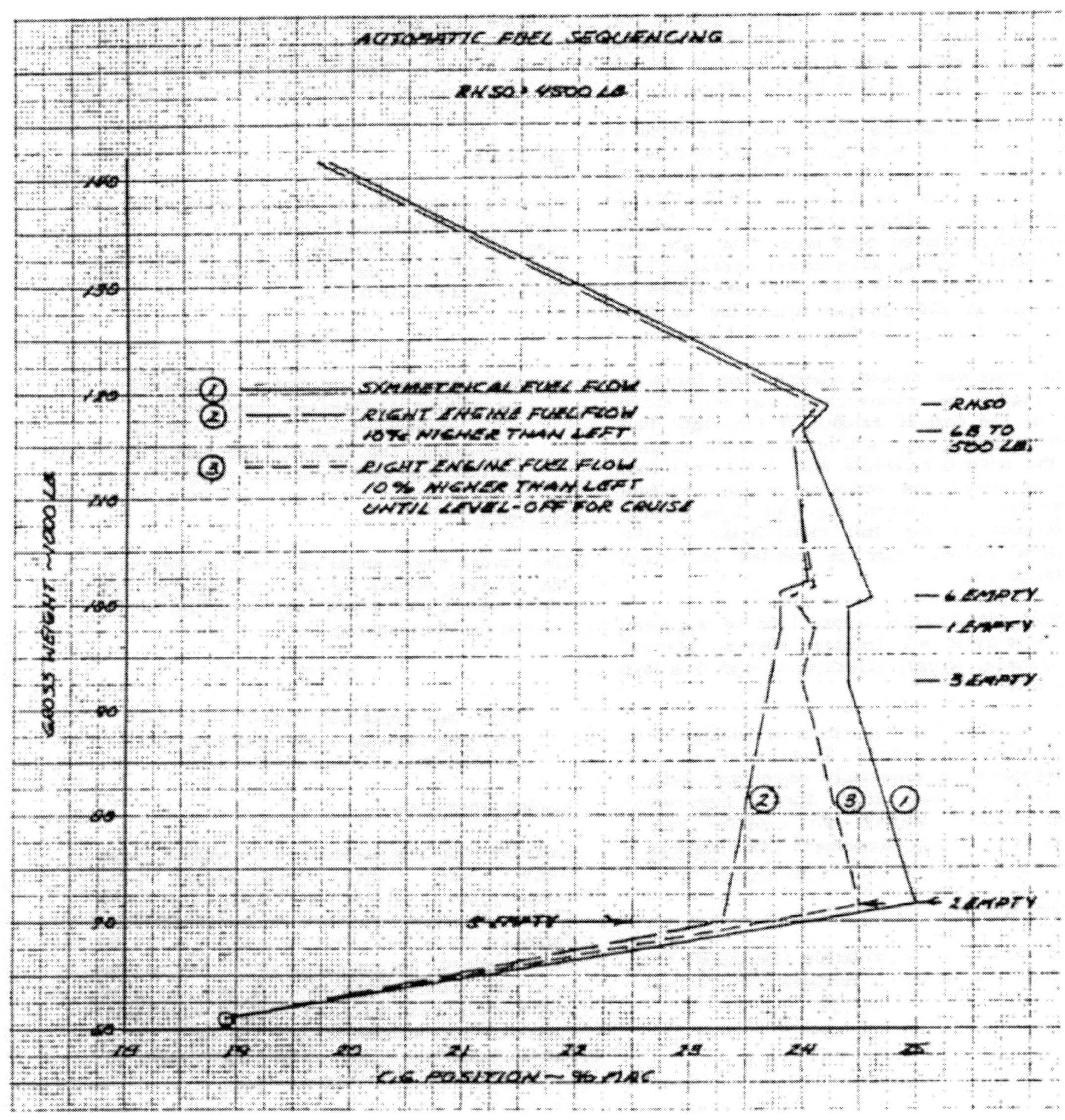

Figure 2-8

EFFECT OF AUTOMATIC AND MANUAL FUEL MANAGEMENT ON C.G. SHIFT RATE*

TANKS ON	PROCEDURE	Percent of Base C.G. Aft Shift Rate
1-3-6 to RHSO	HANDS OFF (NO ALTERNATE AVAILABLE)	+100%** (Base Rate)
HANDS OFF 1-3-6 after RHSO, >3000 lb in tank 1	 Press on tank 1 Press on tank 2, open crossfeed Open crossfeed Tank 5 on after tank 6 is empty	-19% +100% +67% +41% +35%
1-3-6 after RHSO <3000 lb in tank 1, before tank 5 on	HANDS OFF Press on tank 1 (See Note) Press on tank 2, open crossfeed Open crossfeed	+7% +100% +67% +41%
tank 5 on, before tank 2 on	HANDS OFF Press on tank 2, open crossfeed Open crossfeed Press on tank 2 Press on tank 2, manual aft transfer held on Press on tank 2, open crossfeed, manual aft transfer held on	+0.8% +11.3% -28.8% -28.8% +34.6% +49.7%
tank 2 on, before tank 4 on	HANDS OFF Open crossfeed Manual aft transfer held on Open crossfeed, manual aft transfer held on	+3.3% -14.1% +92.0% +29.7%

* Based on nominal, matched fuel flow consumption to each engine.

** This rate, which is 1% of C.G. movement for every 5000 lbs of fuel consumption, is used for comparison throughout all stages of tank sequencing. A (+) or (-) sign indicates that the C.G. is moving aft or fwd.

Note: Do not manually select tank 1 with less than 3000 lbs remaining in that tank. The forward pumps would be operating without fuel cooling and lubrication.

Figure 2-9

Effect of Engine Thrust Variation with EGT

For a given level of thrust, higher throttle settings and increased fuel flow are required as EGT is decreased, because combined burning efficiency of the engine and AB decreases with lowered EGT. Full throttle ceilings are therefore reduced. The degradation in thrust for all throttle settings, at Mach 3.2 and 80,000 feet, is approximately 1.3 percent per 10°C of EGT decrease. The trend is the same for other flight conditions.

Effect of RPM Suppression on Maximum Thrust

As EGT decreases, the engine nozzle opens to maintain scheduled rpm. At high Mach and maximum power, the nozzle may open fully and any EGT decrease will result in rpm suppression below schedule. When this occurs, engine speed will suppress approximately 50 rpm for each 10°C of EGT decrease. The airflow through the engine decreases due to the suppressed rpm, leading to a higher inlet bypass requirement and opening of the forward bypass doors. At Mach 3.2 this results in a thrust degradation and drag increase of approximately 3.5 percent per 10°C of EGT decrease for each engine. If Mach number decreases as a result of the change in thrust and drag, the spikes schedule more forward and the forward bypass doors open further. Performance will deteriorate rapidly under these cumulative effects. Cruise EGT should be maintained between 775°C and 805°C to avoid this situation.

Crew Comfort

Pressure suit ventilation air temperature tends to increase and flow decreases while approaching the end of long cruise periods at maximum speed. The increase in temperature is associated with increasing fuel manifold temperature as tank 3 empties and the quantity remaining is exposed to high skin temperatures. This results in less cooling of the engine bleed air as it passes through the fuel-air heat exhangers in the environmental control system. Comfortable suit vent temperatures are restored as soon as tank 2 is scheduled on.

If uncomfortably warm suit vent temperatures are encountered, insure that the suit heat rheostat in the forward cockpit is OFF. Each crewmember can open his air controller valve to increase flow through the suits. If this is not sufficient, the pilot can manually select the tank 2 boost pumps on if the condition can be associated with depletion of tank 3. The c.g. indication must be monitored and kept within limits. Premature use of tank 4 is not recommended as an alternate to using tank 2 fuel.

PRIOR TO DESCENT

▲1. Pilot's ANS distance display mode - DP/TURN.

a. Display Mode Select Switch - ANS.

b. Bearing Select switch - NORMAL.

(Tc) ANS DATA switch - TEST.

Press DISPLAY push-button switch to display data.

(Td) DP/TURN push-button switch - As desired.

RSO will coordinate the pilot's desired ANS distance display mode and will read the backset from DP or distance to turn, as applicable, over interphone.

2. IGV switches - LOCKOUT checked.

3. LN_2 quantity - Checked.

Check total liquid nitrogen quantity. If not sufficient for normal descent, refer to Fuel Tank Pressurization emergency procedure, Section III.

4. Inlet Controls - AUTO & CLOSE.

Inlet spike and forward bypass controls will be placed in AUTO and the aft bypass controls set at CLOSE unless manual inlet procedures are used. Refer to Section III, Figure 3-5 for manual inlet schedule.

(TS) INS altitude - Update as required.

Update to the after descent condition (air refueling, penetration or field elevation).

DESCENT

1. For inlet(s) in manual control, restart(s) - ON.

2. Throttles - 720°C EGT to Military.

Slowly retard both throttles to 720°C EGT to Military. 720°C EGT results in approximately 10 nm less descent distance than Military.

Expect a fuel vapor trail to occur while the afterburner fuel lines clear.

NOTE

- Pause at minimum A/B approximately 5 seconds if retarding from a high power setting.
- If 720°C EGT is selected, monitor EGT, RPM, and nozzle position. Expect EGT to decrease as Mach decreases. 700°C EGT should hold rpm at the Military schedule and maintain nozzle governing.

With IGV lockout inoperative:

a. Set Military.

For inlet(s) in restart:

b. Set 720°C EGT to Military.

For inlet(s) in restart with IGV lockout inoperative:

c. Set Military.

Refer to Schedule for Manual Inlet Control, Section III.

3. Airspeed - 365 KEAS (350 minimum).

Maintain cruise altitude while decelerating, or maintain cruise Mach number while descending, until approximately 365 KEAS. If KEAS hold is desired, engage the pitch autopilot and stabilize the aircraft at the desired KEAS for a few seconds before engaging KEAS hold. Maintain 350 KEAS minimum while decelerating to reduce the probability of unstart and minimize the possibility of engine stall or flameout.

4. Fuel tank pressure - Monitor.

Descend so that minimum tank pressure (-0.5) is not exceeded.

At Mach 2.5:

5. Throttles - Set 6900 rpm.

Retard throttles simultaneously. Some throttle misalignment may be required. An rpm decrease of 400-500 rpm can be expected from Mach 2.5 to 1.3. Maintain at least 6500 rpm while above Mach 2.0.

With IGV lockout inoperative:

a. Set 720°C EGT and maintain 700°C EGT minimum.

This procedure will increase descent distance approximately 25 - 30 nm. Maintain at least 700°C EGT while above Mach 1.3 to hold the military rpm schedule and maintain nozzle governing. ENP greater than 70% open will result in less than military rpm; in this event, advance the throttle as necessary to maintain military rpm. Maximum rpm occurs near 100°C CIT, as during climb. Mild compressor stalls may be encountered when near 85°C CIT (approximately Mach 1.8) as the IGV and internal bleeds shift.

For inlet(s) in restart:

DESCENT PROFILE

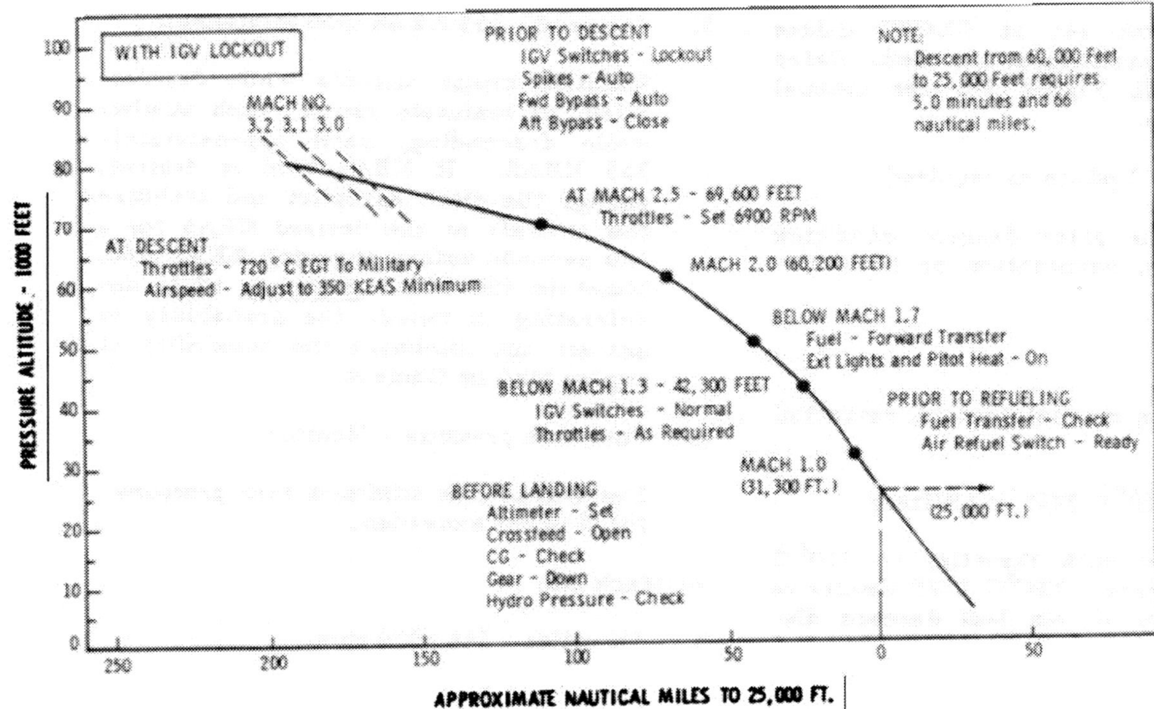

350 KEAS DESCENT

ALT	DISTANCE (NM) – CRUISE MACH 3.2	3.15	3.1	3.0	2.9	2.8	2.6	2.4	PROFILE CHECK – 350 KEAS DIST	IAS	ETE
84K	238	232	226								
82K	236	229	223	211							
80K	234	227	220	208	198	187			234	422	:15.5
78K	231	224	218	206	195	184					
76K	228	221	215	204	193	182	165		196	440	:14.0
74K	225	218	212	202	191	180	162				
72K			209	200	189	177	159	141			
70K				198	187	175	157	139	148	437	:12.0
68K					185	173	154	136			
66K						171	151	134			
64K						169	149	132	120	432	:11.0
62K							147	130			
60K								128	106	428	:10.0
55K									92	422	:09.5
50K									80	416	:08.5
45K	WITH LOCKOUT			W/O LOCKOUT					71	409	:08.0
40K	720° EGT ABOVE M2.5			MIL PWR ABOVE M2.5					63	401	:07.5
35K	6900 RPM AT M2.5			720° EGT BELOW M2.5 (700° MIN)					56	391	:06.5
30K									50	376	:06.0
25K									40	370	:05.0

NOTE

1. Distance and time includes 40 NM L/O prior to ARCP.
2. Add 35 NM for each IGV not locked out.
3. Subtract 25 NM for each restart on.
4. Subtract 30 NM for each inlet in restart w/o lockout.
5. Subtract 51 NM and :06 to FL 310 (IAF).
6. Subtract 40 NM and :05 to FL 250 (IAF).
7. Subtract 25 NM and :03 to FL 160.
8. Subtract 1 NM for each 10° of turn in descent.
9. 5500-6100 RPM below M 1.0.

F203-315(c)

Figure 2-10

b. Set 6500 rpm.

If the forward bypass doors are nearly full open (for any reason) and the engine internal bleeds shift before rpm is reduced, engine stall and flameout are possible.

Refer to Schedule for Manual Inlet Control, Section III.

For inlet(s) in restart with IGV lockout inoperative:

c. Set 6500 rpm. At Mach 2.0, set idle.

Refer to Schedule for Manual Inlet Control, Section III.

CAUTION

If protracted or non self-clearing stalls are encountered, follow procedures in Section III, Compressor Stall In Descent. Use restart and retard throttle to Idle on the affected engine only. Restarts on between Mach 2.5 and 1.3 near Military rpm may result in compressor stall. Check aft bypass doors closed.

At FL 600:

(T6.) IFF Mode C - Set as briefed.

(7.) DEF systems - As required.

The DEF systems may be turned off if descending to land, or placed in standby. If the systems are turned off, check that all System H automatic and manual mode legends and the W and S legends are off. Avoid operating System H below FL 500. The other systems should be in standby before reaching FL 500.

NOTE

System H can be maintained in the standby mode below FL 500 if immediate availability of the system is desirable. In this event, perform an automatic self test each half hour with transmit modes selected. Shut down the system when immediate availability is not required.

Below Mach 1.7:

8. Fuel forward transfer switch -On.

Transfer fuel to obtain c.g. within subsonic limits. Check tank 1 fuel quantity increasing.

9. Pitot heat switch - ON.

10. Exterior lights - On.

a. Anticollision/fuselage lights switch - ANTICOLLISION.

b. TAIL LT switch - STEADY.

Below Mach 1.3:

11. Inlet controls - Checked.

Check spike and forward bypass controls AUTO and aft bypass controls at CLOSE unless manual inlet control procedures are required.

12. IGV switches - NORMAL.

De-energizing the IGV Lockout System restores the engine to maximum thrust capability. The IGV should shift to axial and IGV lights illuminate if RPM is above 5500-6000 rpm. (See Figure 1-11.)

13. Throttles - As required.

Adjust descent profile as required. Reduce rate of descent, if necessary, to avoid low fuel tank pressure below FL 400.

When the desired c.g. is obtained:

14. **Fuel forward transfer switch -OFF.**

15. **Crossfeed switch - Set.**

    Crossfeed OPEN should be selected if immediate penetration for landing is to be accomplished, if tanks 5 and 6 are empty, or if the FUEL QTY LOW light illuminates.

AIR REFUELING DATA CARD			
TKR TCS			
AREA			
ARCP			
DP #			
TRACK			
ARCT			
ALT BLOCK			
A/R ALT			
A/A TAC			
HF FREQ			
UHF Prim.			
UHF Backup			
EXT. CODE			
ON LOAD			
FULL TANK CG			
MISSED A/R ALT	DP CH	DP CH	DP CH
BINGO FUEL			
PLANNED RES			

Figure 2-11

AIR REFUELING

The pitot-static flight instruments will be used for tanker rendezvous and in-flight refueling procedures. Check that the altimeter is set at 29.92 in. Hg.

Air Refueling Data Card

Air refueling data is recorded on checklist cards similar to Figure 2-11.

Pilot Director Lights (On Tanker)

For the KC-135, refer to Figures 2-12 and 2-13. Receiver director lights are located on the bottom of the tanker fuselage between the nose gear and the main gear. They consist of two rows of lights, the left row for elevation and the right row for boom telescoping.

The elevation lights consist of five colored panels with green strips, green triangles, and red triangles to indicate relative position. Two illuminated letters, D and U for down and up movement, respectively, indicate elevation corrections. Background lights are located behind the panels. The elevation lights are controlled by boom elevation during contact.

KC-10 pilot director lights, Figure 2-14, are similar to the KC-135 director lights. The KC-10 elevation director lights consist of a green square, amber triangles, red triangles, and amber D and U letters. The lights are controlled by boom elevation plus boom elevation rate during contact.

The colored panels which indicate KC-135 boom telescoping are not illuminated by background lights. An illuminated white panel between each colored panel serves as a reference. The letters A for aft and F for forward are visible at the ends of the boom telescoping panel. Figure 2-13 shows the panel illumination at various boom nozzle positions within the boom envelope. When contact is made, the lights are controlled by boom extension. There are no lights to indicate azimuth; however, a yellow line on the tanker indicates the centerline.

KC-135 AIR REFUELING BOOM LIMITS

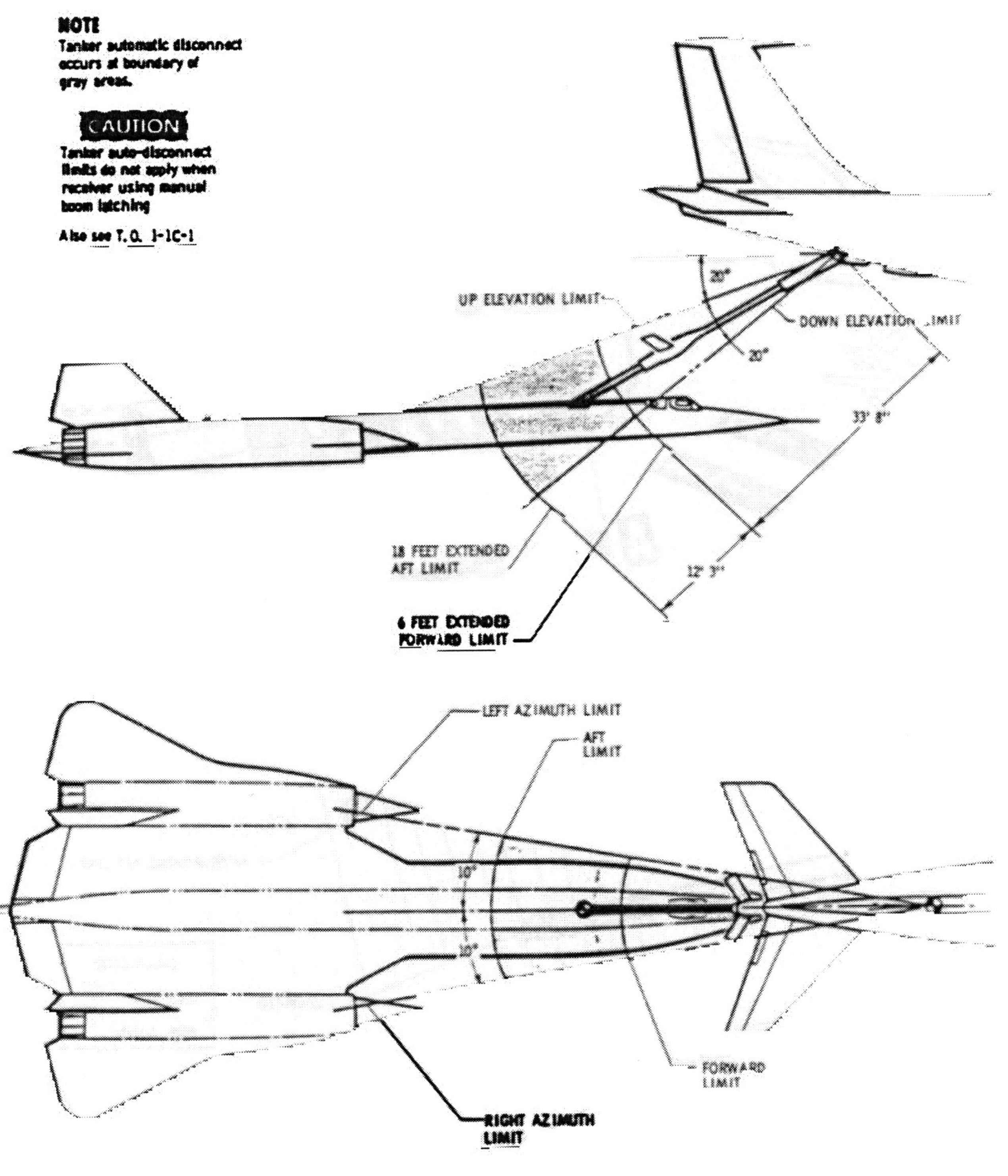

Figure 2-12

RECEIVER DIRECTOR LIGHTS AND ILLUMINATION PROFILE (KC-135)
(Also see T.O. 1-1C-1)

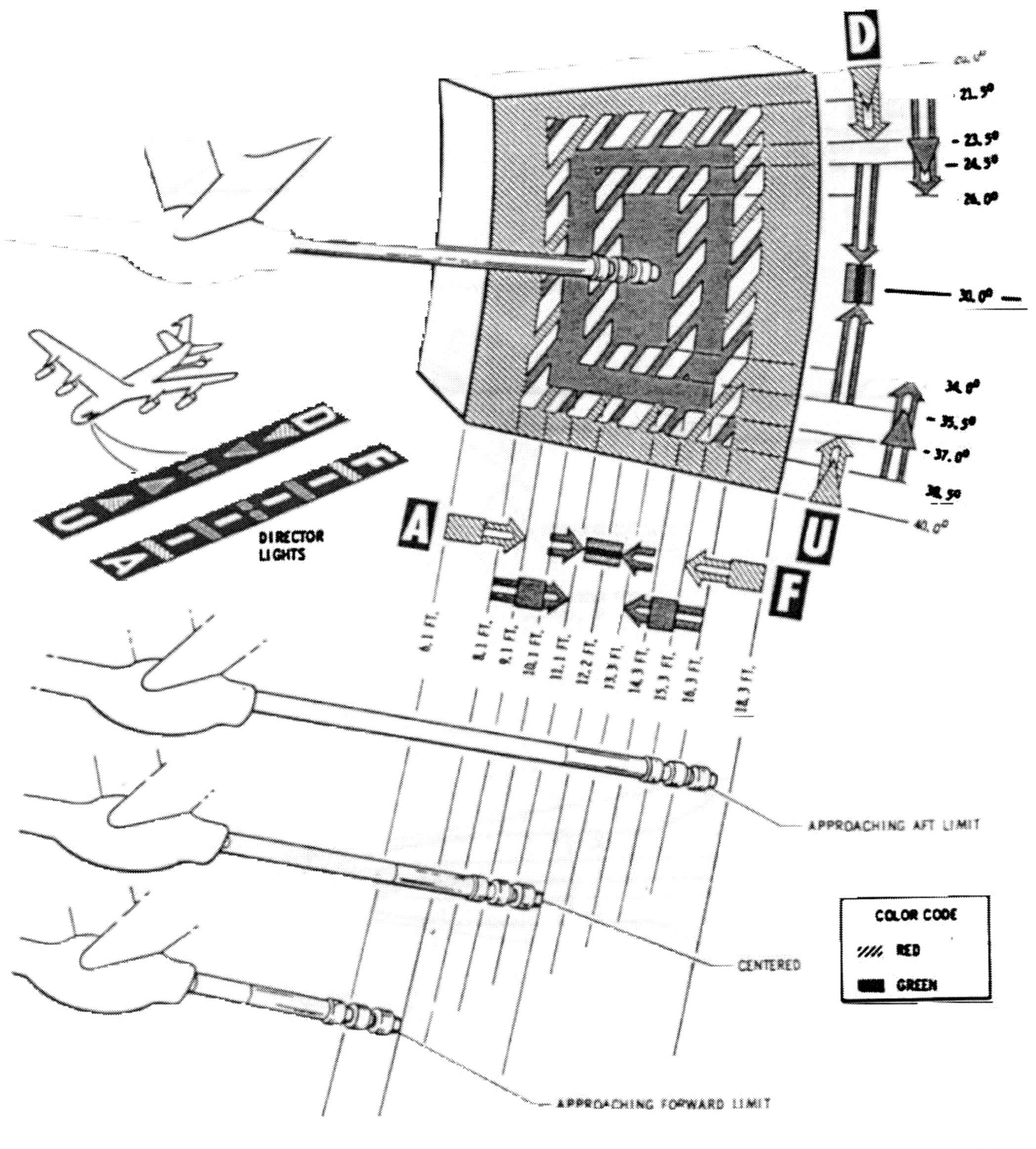

Figure 2-13

RECEIVER DIRECTOR LIGHTS AND ILLUMINATION PROFILE (KC-10)

(Also See T.O. 1-1C-1)

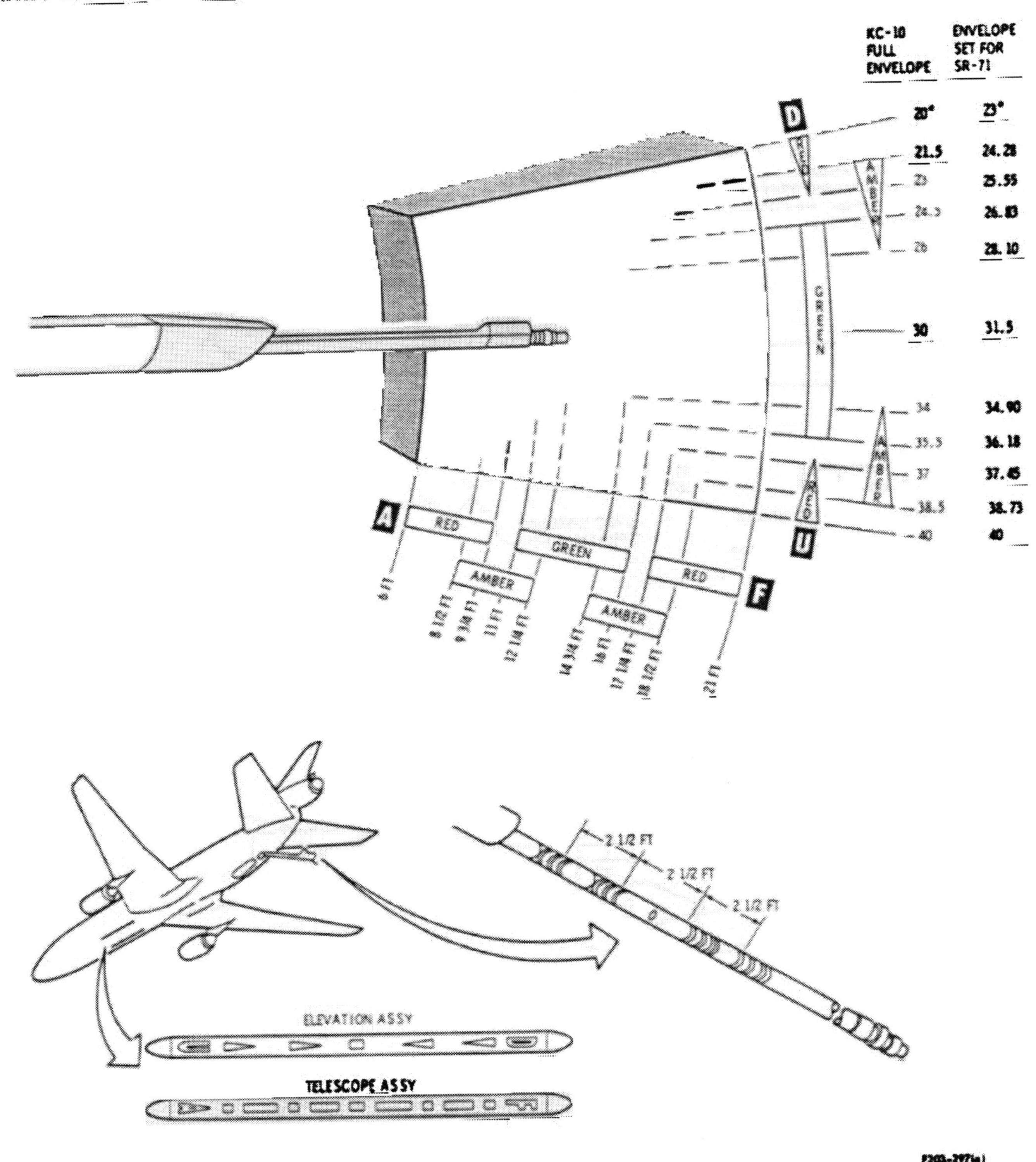

F303-297(a)

Figure 2-14

KC-10 AIR REFUELING BOOM LIMITS

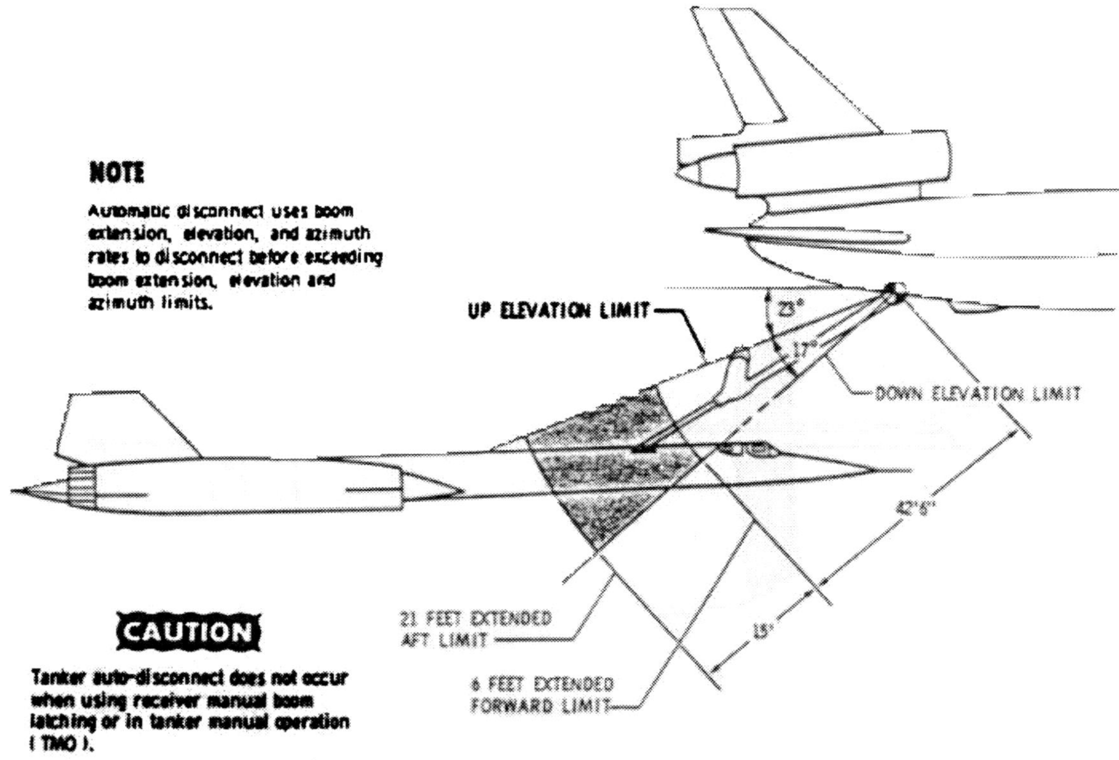

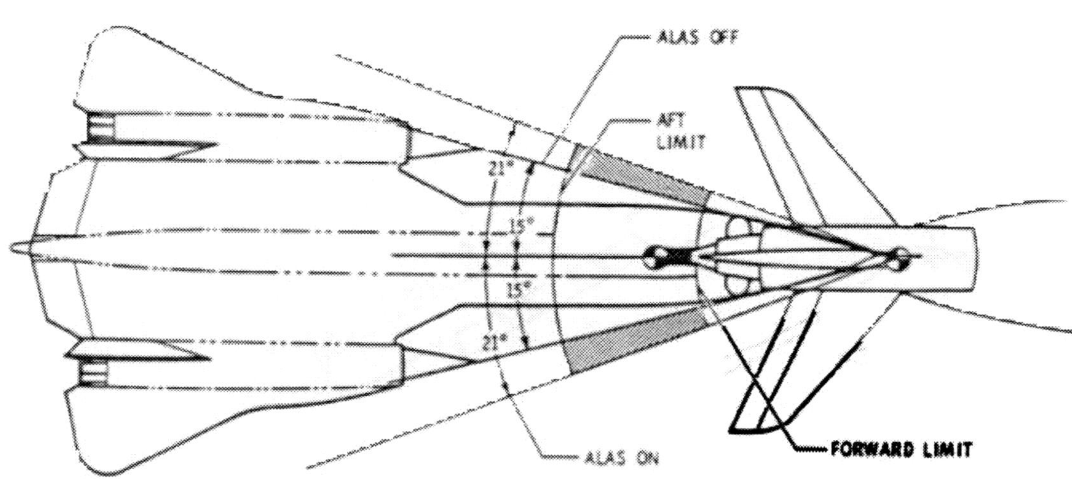

NOTE

Automatic azimuth disconnect set at ±21 degrees with the boom ALAS operating or ±15 degrees with ALAS off.

F203-298

Figure 2-15

The KC-10 lights which indicate boom telescoping are shown in Figure 2-14. When contact is made, the panels are controlled by boom extension plus extension rate.

During radio silence, when in the ready condition, the boom operator actuates the red panels of the receiver director lights to direct the receiver pilot. Illumination of the triangular panels nearest the U and D letters directs the pilot to move up or down respectively. The rectangular panels nearest the F and A letters direct the pilot to move forward or aft respectively. A steady red light indicates that a relatively large correction is desired in the indicated direction. A flashing red light indicates that relatively small correction is necessary.

Fuel Management Prior to Refueling

Transfer fuel as necessary to maintain c.g. within the subsonic limits and manually energize the tank 4 fuel pumps. Reducing the quantity in tank 4 provides more space for cool fuel from the tanker to reduce the temperature of tank 4 fuel and prevents the ullage pumps from cycling due to fuel level fluctuations during refueling. Forward transfer provides a more desirable center of gravity for unaugmented pitch stability and also replenishes tank 1 so that its boost pumps will supply the left fuel manifold during refueling. Fuel is also supplied to the right manifold when tank 4 is below 3600 ± 1200 pounds, or tank 1 is pressed on manually, or the tank 1 fuel level is above the right hand shutoff level. Forward transfer should be repeated if there is any appreciable delay before initial refueling contact or between subsequent refueling contacts. Do not transfer excessive amounts to tank 1, however, as cg forward of 17% reduces load factor limits. Forward transfer should be shut off before making contact.

NOTE

Crossfeed should be used if the FUEL QTY LOW light illuminates or tanks 5 and 6 are empty.

Approach For Refueling

Make the approach from behind and below. Deicing is only provided on the left windshield panel. The seat should be lowered prior to reaching the observation position. Upon completion of the Before Refueling procedure, maneuver to the pre-contact position about 50 feet to the rear and 10 feet below the refueling position. Stabilize and trim the aircraft. Maneuver as indicated by director light signals or by verbal instructions of the boom operator.

NOTE

- COMNAV 50 range and bearing information may be lost when below the tanker if the tanker has not switched to lower antenna.
- The air-to-air mode of the TACAN provides both range and bearing with all KC-10 and some KC-135Q aircraft, but range only with other KC-135Qs. TACAN information may be lost within one mile due to antenna blanking. TACAN signals are not resistant to meaconing, interference, and jamming.

Refueling Hookup

A successful hookup is confirmed by steady illumination of the director light panel and extinguishing of the ready light. Hookup can be maintained between the aircraft and tanker during a turn or in a descent. No adverse flight characteristics result from tanker downwash. Clear away aft and down along the axis of the refueling boom after disconnect.

Boom Elevation Limits

The KC-135 boom upper elevation limit is 20 degrees and the lower elevation limit is 40 degrees. For the SR-71 receiver, refueling above 25 degrees (green triangle) is not recommended. Any illumination of the red "down" triangle is outside the recommended SR-71 receiver envelope.

The upper disconnect limit of the KC-10 boom can be varied. Although the KC-10 can refuel between 20 degrees and 40 degrees, the upper limit for the SR-71 has been set at 23 degrees. With the 23 degree upper limit set, the director lights will show the center of the envelope at 31.5 degrees (Figure 2-14). During refueling, the KC-10 director lights are higher in the pilot's field of view than the KC-135 director lights. Below 36 degrees elevation, the SR-71 pilot may not be able to see the KC-10 director lights even with the seat full down.

Boom Extension Limits

	KC-135	KC-10
Range of Travel	12.2 ft.	15.0 ft.
Inner Disconnect	6.1 ft.	6.0 ft.
Outer Disconnect	18.3 ft.	21.0 ft.
Mid Boom	12.2 ft.	13.5 ft.

Boom Azimuth Limits

The KC-135 azimuth limit is ± 10 degrees. No automatic disconnect is provided in azimuth.

The KC-10 ALAS (Automatic Load Alleviation System) positions the boom elevator and rudders to reduce boom loading after contact. For the SR-71, the automatic disconnect azimuth limit is set to ±21 degrees with ALAS operating and ±15 degrees with ALAS off.

EGT Trimming

Manual EGT trimming may be used in lieu of the automatic trim system, if desired. This will avoid the possibility of "ratcheting." Manual trimming to temperatures above the nominal deadband schedule depicted in Figure 2-4 is permitted. If the engine should stall, or if EGT exceeds 830°C, <u>downtrim immediately</u>.

Disconnect

A disconnect may be accomplished:

1. Automatically.

   a. If boom envelope limits are exceeded.

   b. When fuel pressure exceeds 70 psi.

Pressure disconnects do not usually occur with the KC-10 since the refueling system reduces flow to maintain normal refueling pressure.

2. Manually.

   a. By the boom operator.

   b. By depressing the A/R DISC trigger on the control stick grip.

   c. By the KC-10 boom operator activating the Independent Disconnect System (IDS). The IDS retracts the KC-10 boom latch surface inward and frees the boom even with the SR-71 boom latches still extended.

   d. When pull on the boom becomes excessive ("tension" or "brute force" disconnect).

NOTE

The latch toggle hydraulic system incorporates a pressure relief valve which will permit the boom to be pulled out when a pullout force of approximately 5400 pounds is applied. Therefore, if a malfunction occurs which prevents disconnecting the boom, place the air refuel switch in the MAN O'RIDE position, depress the A/R DISC trigger switch, and proceed with brute-force pullout.

CAUTION

Disconnect in an aft and downward direction, and avoid forward relative motion during disconnect. Disconnect maneuvers which tend to pry the boom out of the receptacle will damage the receptacle or airframe.

Boom Control Failure (KC-10)

The KC-10 boom control system is a dual channel fly-by-wire system. If the two control channels disagree, the boom fails passive with controls in their last position.

CAUTION

If KC-10 boom control failure occurs, the receiver pilot should follow the boom operator's instructions. The boom operator will direct the receiver to a position that will allow a safe disconnect with the boom loaded slightly away from the receiver.

NORMAL AIR REFUELING PROCEDURE

Refueling can be accomplished from either cockpit of the SR-71B.

1. **Center of gravity - Checked.**

   Transfer fuel to maintain subsonic c.g.; the c.g. will tend to travel aft slightly during the initial portion of the refueling. After c.g. adjustment, the pitch trim should indicate at least 0° to 1° nose up. CG forward of 17% reduces load factor limits.

2. **Windshield Deice switch - Set.**

   If icing conditions are anticipated, set the Windshield Deice switch ON (down) to start hot air flow across the outside of the left windshield panel. Check that the WINDSHIELD DEICE ON caution light illuminates if deicing is selected.

3. **Air refuel switch AIR REFUEL, READY light on.**
4. **Tank 4 boost pump switch - Press on.**
5. (T5) IFF - As required.
6. Forward transfer - OFF.
7. ▲ Interphone - Set.
   a. Pull and set HOT MIC.
   b. Pull and set IFR COMM.

NOTE

If the tanker is not equipped with boom interphone, or if the tanker interphone is off, actuation of either cockpit INPH switch will result in disconnect.

8. ▲ **TACAN Mode selector switch - T/R.**
9. ▲ **Radios - Set.**

   Check UHF/VHF frequencies. Set UHF Mode switch to INT and power as required. Set interphone selector to radio with primary refueling frequency.

10. **Anticollision lights switch - FUS.**
11. **(Night only) FUS & TAIL switch - DIM.**

    For night refuelings, boom operators may request lights off.

CAUTION

While refueling, do not transmit on HF radio as the tanker HF transceiver may be damaged.

NOTE

After tanker hook-up the air refuel READY light should extinguish.

Total fuel quantity, pitch trim, and c.g. should be monitored.

CAUTION

Under normal conditions, if trimming is required, disconnect and retrim in the precontact position. If trimming on the boom, be alert for runaway trim. To avoid runaway trim due to a sticking trim switch, assure positive switch movement to neutral after each actuation.

Disconnect is indicated by illumination of the DISC light.

When refueling is complete:

12. Air refuel switch - OFF.

13. Speeds - Crosscheck and monitor.

    Use the ANS ground speed to confirm speed is increasing and as a systems crosscheck.

WARNING

Monitor airspeed and angle of attack after refueling and adjust attitude as necessary to remain within limits. A minimum of 300 KEAS is required.

(14.) Precomputed c.g. - Check.

    Compare c.g. precomputed for the end of refueling with actual c.g. Crosscheck tank quantity indications if c.g. is not as expected.

15. Pump release switch - Press and confirm normal tank lights on.

    After releasing the tank 4 pumps, only tanks 1,3, and 6 fuel tank boost pump switches remain illuminated.

NOTE

The tank 4 boost pumps must be released after refueling or premature depletion of that tank will occur.

16. Crossfeed - Closed.

(T17.) IFF - Set.

18. EGT trim switches - As required.

19. Windshield Deice switch - OFF.

    Check the WINDSHIELD DEICE ON caution light extinguished.

20. Anticollision lights - ANTI COLLISION.

21. (Night Only) FUS & TAIL switch -Bright.

▲22. Interphone control panel - Set.

▲23. Pilot's ANS distance display mode - DP/TURN.

(Ta) ANS DATA switch - TEST.

    Press DISPLAY push-button to display data.

(Tb) DP/TURN push-button switch - As desired.

    RSO will coordinate the desired pilot's ANS distance display mode.

POWER LIMITED REFUELING

Engine EGT may be manually uptrimmed. Uptrimming should be done prior to establishing tanker contact, but is permissible during contact. If CIT is below 5°C, the nominal EGT schedule in Figure 2-4 decreases rapidly and compressor surge (stall) is possible if the schedule is exceeded. Downtrim immediately if EGT exceeds 830°C or if the engine stalls.

If military power limit is reached prior to completion of fuel transfer, the receiver may request the tanker to initiate a "toboggan" maneuver, or may elect to use one afterburner. If afterburner is used, set one throttle in Minimum A/B. Left afterburner is recommended. A slightly downtrimmed EGT on the afterburning engine helps minimize power asymmetry. Use the opposite throttle to vary thrust. For actual and simulated single engine air refueling procedures, see Section III, Single Engine Air Refueling.

BREAKAWAY PROCEDURE

Any tanker or receiver crewmember, if an emergency arises, will transmit on the air refueling frequency the tanker aircraft call sign followed by "breakaway, breakaway, breakaway." The boom operator may signal breakaway by rapidly flashing the receiver director lights. The receiver pilot will actuate the A/R DISC trigger. Retard throttles and drop aft and down until the entire tanker is in sight.

WARNING

The pilot should use care not to overrun the tanker. If overrunning does occur, under no conditions should a turn be made until breakaway has been completed.

Breakaway Signal

To visually signal emergency breakaway, the tanker turns on the rotating beacon and rapidly flashes the director lights.

AIR REFUELING ALTERNATE PROCEDURE

If L hydraulic pressure is lost, R pressure may be utilized for refueling by moving the brake switch to ALT STEER & BRAKE.

Normal and manual boom latching air refueling procedures apply with ALT STEER & BRAKE selected.

CAUTION

Do not leave the brake switch in ALT STEER & BRAKE after refueling. R hydraulic pressure may be lost also if L system fluid loss is due to a malfunction of the steering or refueling system.

MANUAL BOOM LATCHING

A manual refueling procedure may be used if the signal amplifier fails. This procedure requires manual control of the refueling boom latches. The refueling boom latches are open with the air refuel switch in MAN O'RIDE and the A/R DISC trigger on the control stick grip depressed. The receiver pilot can usually feel the boom nozzle bottom in the receptacle. The READY light will extinguish when the boom nozzle is properly seated; then the pilot should latch the boom in the receptacle by releasing the trigger switch. When latching the boom manually:

1. Air refuel switch - MAN O'RIDE, READY light on.

2. Trigger switch - Press and hold.

After nozzle is seated in receptacle and READY light is off:

3. Trigger switch - Release.

When refueling is completed:

4. Trigger switch - Press and hold until boom is clear.

   READY light illuminates when boom is not seated. DISC light will not illuminate.

Subsequent procedures are the same as after normal refueling.

CAUTION

- If the A/R DISC trigger switch is released when the nozzle is not in the bottom of the receptacle (READY light is off), it is possible for the nozzle to damage or break the extended nozzle latches, preventing any further refueling.
- The boom limit switches are deactivated when using manual boom latching. The receiver pilot must initiate disconnect before exceeding the boom limits since the boom operator will be unable to release the nozzle latches. KC-10 boom operators can still use the Independent Disconnect System.

NOTE

There will not be a pressure disconnect when using manual boom latching, but the refuel manifold accepts tanker pressure with ample margin after tanks shut off automatically. The fuel vent manifold releases excess tank pressure if the tank shutoff valves malfunction.

RADIO SILENCE REFUELING PROCEDURE

Radio silence air refueling can be conducted if the following procedures are observed and both crews are experienced in normal air refueling procedures. The method, time, and place of rendezvous, and amount of fuel to be transferred, must be coordinated.

a. Air refueling checks will be completed before moving to the precontact position.

b. Before contact, maneuver as directed by tanker director lights. A steady red light indicates a large correction and a flashing red light indicates a small correction in the direction indicated by the red director lights. When contact is made, boom interphone may be used.

AIR REFUELING VISUAL SIGNALS

The following visual signals will be used for radio communication failure or radio silence.

Signals From Tanker

With boom in trail:

a. Ten foot extension of the probe means that the tanker is ready for contact.

When the ready signal is received, move from the observation position and stabilize in the precontact position, then move to the contact position.

b. Full extension of the probe means that the tanker is ready for contact but that he is in manual control, without disconnect control capability.

Close and open the Air Refueling door to acknowledge this signal.

NOTE

Boom interphone is inoperative when the tanker is in manual operation.

c. Full retraction of the boom indicates that offload has been completed.

Boom stowed:

a. Full retraction of the probe means that the tanker air refueling system is inoperative.

b. Five foot extension of the probe indicates that there is an air refueling system malfunction.

Check the air refueling system.

Director lights off:

A request to disconnect is signalled by turning the director lights off. Return to the precontact position after disconnecting.

Director lights flashing:

The BREAKAWAY command is signalled by turning the lower rotating beacon on and rapid flashing of the director lights.

Signal From Receiver

Cycle the Air Refueling Door:

Cycling the A/R door while in the precontact position indicates that the Manual Boom Latching procedure will be used. The tanker should signal acknowledgement by full extension of the refueling probe with the boom in trail, and then retract the probe to the ready position.

CAUTION

During manual boom latching, the receiver pilot must initiate all disconnects. KC-10 boom operators can still use the Independent Disconnect System.

Rock wings (daytime) and turn rotating beacon on/off at ten second intervals:

This signal indicates that the receiver must refuel. If not at a scheduled time or place for refueling (when the boom operator might not be in place) take a position where the signals are visible from the tanker cockpit.

FUEL DUMPING

Fuel dumping provides a means of reducing gross weight rapidly. The nominal dump rate is 2500 pounds per minute for both FUEL DUMP and EMER switch positions, but the rate varies with the amount of fuel remaining and the number of boost pumps operating.

Normally, fuel is dumped in the automatic fuel sequence. An additional tank can be selected in each tank group to increase the dump rate.

To dump fuel:

1. Fuel dump switch - FUEL DUMP.

All tanks containing fuel will empty in the normal usage sequence except tank 1. Tank 1 will dump automatically with the other tanks until its fuel level reaches approximately 4700 pounds, depending on aircraft attitude (see Fig. 1-36), then it will stop. The other tanks will continue dumping until the fuel level in tank 4 reaches 3700 pounds (again depending on aircraft attitude), then normal dumping is terminated automatically regardless of the fuel quantities remaining in the other tanks.

2. C.G. - Monitor.

Transfer fuel as necessary to maintain c.g. within limits. At heavy weight, with more than 40,000 pounds of fuel remaining, consider pressing tank 2 on manually to avoid any abnormal forward c.g. condition during dumping. Crossfeed can also be used and will tend to shift c.g. aft.

3. Fuel quantity - Monitor total and tank 4.

a. Tank 1 dumping should terminate automatically when 4700 pounds remain in that tank.

b. All dumping should terminate automatically when 3700 pounds remain in tank 4.

When the desired fuel quantity remains:

4. Fuel dump switch - OFF.

To dump fuel when tank 4 contains less than 4000 pounds, or if the FUEL DUMP switch position is inoperative:

5. Fuel dump switch - EMER.

Selection of the fuel dump switch emergency position overrides the stop-dump feature of the normal dump system.

6. Fuel quantities - Monitor tanks 1 and 4.

Emergency fuel dumping must be terminated by positioning the dump switch to OFF (or FUEL DUMP), or all tanks will empty.

When tank 1 is below 5000 pounds:

7. Fuel crossfeed switch - Press to OPEN.

When the desired fuel quantity remains:

8. Fuel dump switch - OFF.

BEFORE PENETRATION

The pitot-static flight instruments will be used when subsonic.

1. Display mode selector switch - Set.

T 2. Defog switch - Set.

T 3. Altimeter - Set.

(4) DEF systems power - Off.

(5) Sensor operate switches -STP.

(6) Sensor power switches -Off.

(7) V/H power switch - Off.

(8) Exposure power switch - Off.

(9) G-band Beacon switch - OFF.

NOTE

Do not shut down the MRS.

(10) INS altitude - Update.

Update to the field elevation.

PENETRATION

1. Crossfeed switch - OPEN.

CAUTION

Leave crossfeed open to assure fuel supply to both engines during landing and possible go-around operations.

2. Brake switches - DRY or WET, and ANTI SKID ON.

Use the DRY position for a RCR of 21 or more. Wet runway conditions shall be assumed to exist and the WET position used if RCR is less than 21. If RCR is not available, assume a wet runway condition if moisture is visible on the runway, particularly as evidenced by glare or reflections.

(T) a. Brake switch - OFF.

Below Mach 0.5:

3. Surface limiter control handle - Pulled, SURFACE LIMITER light out.

Pull and rotate the surface limiter handle 90 degrees to disengage the surface limiters, lock the handle, and extinguish the SURFACE LIMITER caution light.

▲4. UHF power selector - Set.

Set power 4 or lower, if making an ILS approach.

WARNING

ILS reception can be affected by UHF transmission at high power settings.

T 5. Defog switch - Set.

To dissipate fog in the windshield area, hold the defog switch OPEN for several seconds to provide hot air to the windshield, then select HOLD.

NOTE

Fog usually occurs in the rear cockpit first.

6. Landing light switch - On

BEFORE LANDING

Figure 2-16 depicts a typical landing pattern. At heavy weights, increase airspeed if necessary to maintain angle of attack less than 8 degrees for turns to base leg and 9 degrees for turns to final approach.

The design landing weight is 68,000 pounds with 10 fps sink rate. When landing at higher weights is required, the following speed and sink rate schedule applies.

LANDING PATTERN – Typical

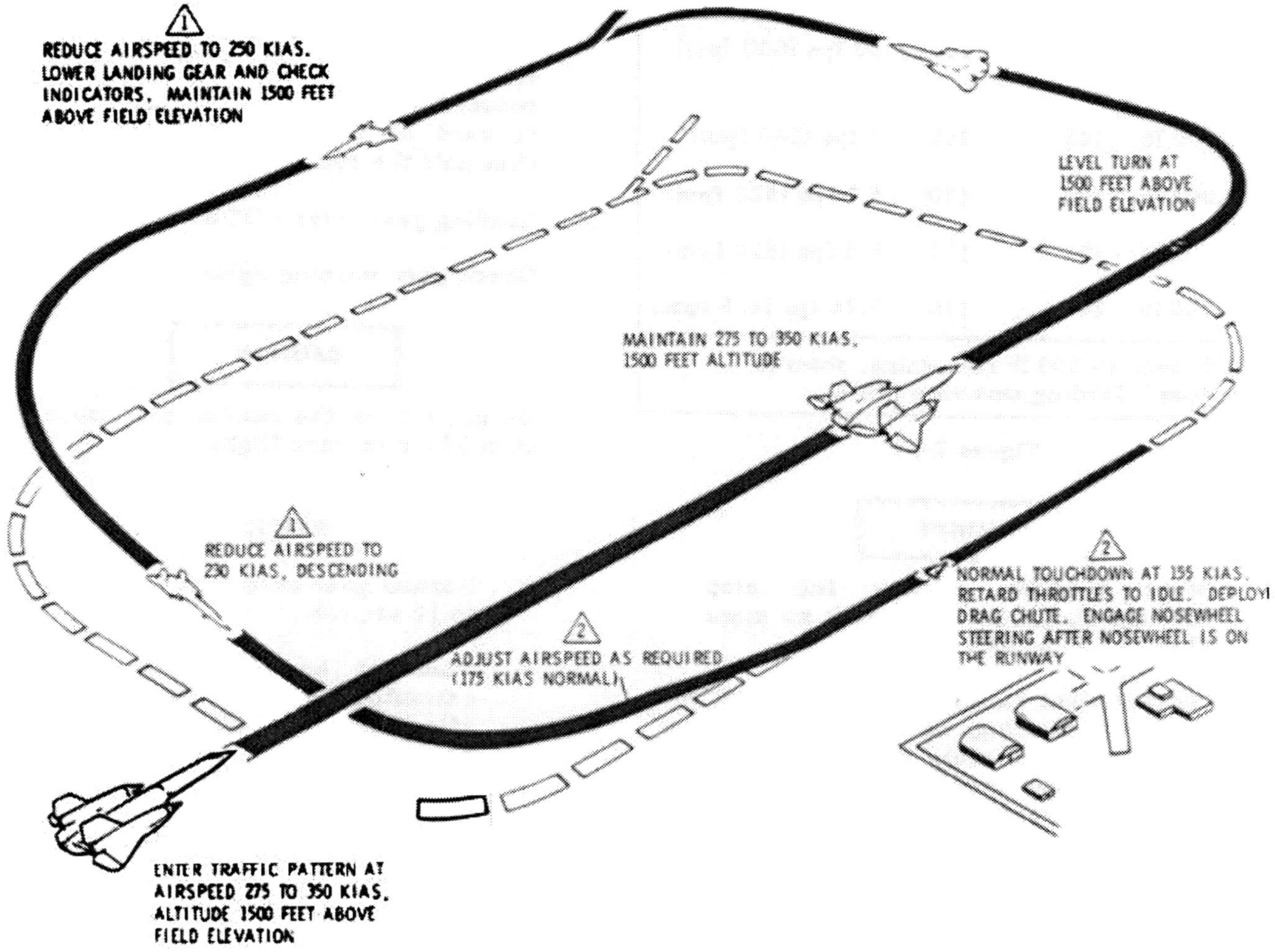

REDUCE AIRSPEED TO 250 KIAS. LOWER LANDING GEAR AND CHECK INDICATORS. MAINTAIN 1500 FEET ABOVE FIELD ELEVATION

LEVEL TURN AT 1500 FEET ABOVE FIELD ELEVATION

MAINTAIN 275 TO 350 KIAS, 1500 FEET ALTITUDE

REDUCE AIRSPEED TO 230 KIAS, DESCENDING

NORMAL TOUCHDOWN AT 155 KIAS. RETARD THROTTLES TO IDLE. DEPLOY DRAG CHUTE. ENGAGE NOSEWHEEL STEERING AFTER NOSEWHEEL IS ON THE RUNWAY

ADJUST AIRSPEED AS REQUIRED (175 KIAS NORMAL)

ENTER TRAFFIC PATTERN AT AIRSPEED 275 TO 350 KIAS, ALTITUDE 1500 FEET ABOVE FIELD ELEVATION

NOTE

For aircraft over 100,000 lbs. (more than 40,000 lb. fuel remaining), maintain 275 KIAS on downwind leg and 250 KIAS on base leg; and use an angle of attack of approximately 10.5° for final approach and landing.

Increase normal speed for final approach (175 KIAS) and landing (155 KIAS) by 1 knot per 1000 lb of fuel over 10,000 lb remaining. For maximum performance, the minimum landing speed is 10 KIAS less than the speed determined by this rule. See Appendix figure A2-15. The minimum final approach speed is 20 KIAS above the intended landing speed.

Figure 2-16

NORMAL LANDING SPEED SCHEDULES

Approx Fuel Remaining	Final Approach KIAS	Landing Speed KIAS	Max Sink Rate Allowable
10,000 lb or less	175	155	10 fps (600 fpm)
20,000 lb	185	165	9 fps (540 fpm)
25,000 lb	190	170	8.7 fps (522 fpm)
30,000 lb	195	175	8.5 fps (510 fpm)
40,000 lb	205	185	7.75 fps (465 fpm)
With over 40,000 lb remaining, observe Section V landing sink rate limits.			

Figure 2-17

CAUTION

When feasible, routine full stop landings should be made with no more than 10,000 pounds of fuel.

For heavyweight landings: With over 40,000 lb of fuel remaining, use the normal final approach speed schedule and maximum performance landing speed (10 KIAS less than normal landing speed). See Figure 2-17. Use the maximum performance landing technique for stopping.

NOTE

Maximum performance landing speeds result in touchdown angles of attack 1/2 to 1 degree greater than for normal landing speeds.

▲1. Approach and landing speeds -Computed.

Final approach and landing speeds are based on weight. Angle of attack will be approximately 10 degrees for a normal final approach.

Use the maximum performance landing speed schedule when conditions such as wet runway or short field length require minimum roll after touch down.

▲2. Center of gravity - Checked.

Transfer fuel as necessary to maintain subsonic c.g. limits. CG forward of 17% reduces load factor limits. If cg is forward of 17%, insure that no more than half the fuel remaining is in tank 1.

3. Landing gear lever - DOWN and checked.

Check gear warning lights.

CAUTION

Do not extend the landing gear more than 10 times each flight.

NOTE

- Normal gear extension time is 12 to 16 seconds.
- When at heavy weights, gear extension may be delayed until after turn to final approach course, if desired.

4. Hydraulic pressure - Checked.

5. Right refrigeration switch - OFF.

The pilot's R AIR SYS OUT caution light illuminates. Monitor E and R Bay temperatures and suit vent flow for adequate flow from the operating refrigeration unit. Turn the right refrigeration system on and the left refrigeration system off if flow is inadequate.

(T6.) Cockpit air handle - OFF

Place the cockpit air handle in the forward (valve closed) position to prevent cockpit fogging. The pilot's CKPT AIR OFF caution light illuminates.

NOTE

Refrigeration system shutoff and cockpit air shutoff are not normally required for low approaches.

7. Annunciator panel - Checked.

NOTE

Lowering the vision splitter during night landings reduces glare from reflections off the inside of the windshield.

NORMAL LANDING

Touchdown is made with the throttles in IDLE, and at approximately 9.5 degrees angle of attack. (Due to ground effect, angle of attack is nearly the same as for final approach.) Pitch angle is approximately 10.5 degrees, with the nose almost on the horizon. A high rate of sink will develop if airspeed becomes excessively low on final approach, and result in a hard landing.

NOTE

- o Throttle movement should follow the quadrant curvature so that the hidden ledge at the IDLE position can prevent inadvertent engine cutoff.
- o With cockpit air on, sudden fogging can occur when the throttles are retarded during the landing flare. Use the Cockpit Fog emergency procedure in this event.

NOTE

- o Angle of attack at touchdown must not exceed 14 degrees to avoid scraping the tail.

Use the maximum performance landing touchdown speed when wet or slippery runway conditions exist which degrade braking capability.

AFTER TOUCHDOWN

1. Drag chute - Deploy.

Deploy the chute when the main gear is on the runway and angle of attack is 10 degrees or less.

CAUTION

Deploying the drag chute at greater than 10 degrees angle of attack may result in the chute canopy contacting the runway and receiving scuff damage.

Pull the drag chute handle straight aft to the limit of its travel (approximately one inch).

WARNING

Avoid resting the hand on or near the drag chute handle after pulling it aft for normal deployment. Otherwise, the chute will be jettisoned if the handle is pushed forward inadvertently when the chute opens.

The initial forces caused by chute opening normally approximate one-half "g" deceleration. See Section VI, Figure 6-11.

If the chute does not deploy normally in approximately five seconds, turn the chute handle 90 degrees counterclockwise and pull aft, approximately six inches, to the limit of its travel.

NOTE

The drag chute switch in the aft cockpit of the SR-71B must be OFF to deploy from the forward cockpit; otherwise, the handle in the forward cockpit is inoperative.

Start the nose down at touchdown. Excessive nose gear loads may result on contact if a high angle of attack is maintained until airspeed is too low for positive control of attitude.

2. Nosewheel steering - Engage.

Engage nosewheel steering when the nosewheel is on the runway. Steering will not engage until the rudder pedals and nosewheel are aligned and aircraft weight is on any one gear.

Illumination of the nosewheel steering STEER ON light is a positive indication that steering has engaged.

It may be necessary to move the rudder pedals through a small range on each side of neutral to assure alignment with the nosewheel castering angle. In a crosswind, engagement will probably require momentarily moving the pedals in a direction opposite to that desired for steering.

Although the steering system includes a holding relay in the engagement circuit, the recommended method for positive engagement is to hold the nosewheel steering (CSC/NWS) button on the control stick depressed until steering is engaged. The button may be released after engagement.

Nosewheel steering is released by pressing and releasing the button a second time (whether actually engaged or not). If steering is inadvertently released, depress the button again and reengage steering as before.

3. Brakes - Checked.

Check for normal brake operation by light application prior to jettisoning the drag chute.

Anti-skid braking is available when aircraft weight is on at least one main gear; however, delay braking until the nosewheel is on the runway.

The normal performance procedure should be used on a dry runway or on a grooved runway with braking equivalent to a dry runway. (Refer to Wet/Slippery Runway Landings, this section, if landing on a runway where braking may be degraded). Apply brakes as required. Light braking is sufficient if the drag chute deploys normally. If the drag chute does not deploy, moderate braking force is necessary at normal landing weight.

4. Drag chute - Jettison.

Push the drag chute handle fully forward from the deploy position to jettison the chute.

The drag chute is normally jettisoned at or above 55 KIAS when on a dry runway or with equivalent braking action available; however, do not jettison the chute if the crosswind component exceeds 12 knots or if braking action is unsatisfactory.

(T) a. Drag chute switch - OFF.

WARNING

In the SR-71B, if the aft cockpit deploys the drag chute and the forward cockpit handle remains stowed, returning the aft cockpit drag chute switch to OFF will jettison the drag chute.

CAUTION

If the drag chute is not jettisoned, the elevons should not be moved during taxiing as the shroud lines may jam between the inboard elevons and fuselage and cause structural damage.

NOTE

The drag chute can not be jettisoned after using the emergency deployment system.

CROSSWIND LANDING

Refer to Landing Gear System, Crosswind Limits in Section V. Also refer to the Crosswind Component chart in the Appendix, Figure A2-1.

Runway alignment on final approach can be maintained by crabbing and/or dropping one wing. Remove any crab before touchdown, and use the wing-low technique to align the aircraft with the runway and prevent side drift.

Reduce sink rate to a minimum to accomplish a smooth touchdown. As crosswind components increase, sink rate must be minimized due to the increased side loads imposed on the landing gear.

CAUTION

It is essential to remove all crab before touchdown to minimize scuffing damage to the tires.

Crosswind Condition With Dry or Grooved Runway

Touchdown and try to remain on the upwind side of the runway. This provides more runway space on the downwind side, and puts the crosswind and runway "crown" effects in opposition. Deploy the drag chute early in the landing roll, as for a normal landing, but lower the nosewheel first and engage steering if the crosswind component is over 15 knots.

The chute's tendency to pull the aircraft off the runway in a crosswind decreases as speed is reduced, and the effect is easily controllable with nosewheel steering. Keep the stick forward to improve nosewheel steering effectiveness. Increasing rudder deflection and/or increasing elevon differential is required as speed decreases.

Do not shut down either engine when on a dry runway, or on a grooved runway which provides equivalent braking.

Crosswind Condition With Slippery Runway

For landing on a slippery runway with a crosswind, start the nose down immediately on landing and engage nosewheel steering before deploying the drag chute. After the nose is lowered, use lateral stick deflection and/or rudders to increase directional control. Use roll inputs in the same direction as rudder/nosewheel steering. This also increases braking on that side when combined with neutral or aft stick.

With a slippery runway, shutdown of one engine to assist in stopping is permissible if required due to drag chute failure. Shutdown the upwind engine when under 100 KIAS, and select ALT STEER & BRAKE if continuing on the right engine alone. Shutdown is not recommended if barrier engagement is available.

The nosewheel steering system provides adequate control in allowable crosswinds on slippery runways, even with damaged main gear tires. However, be careful not to overcontrol the aircraft and start a lateral skid. The nosewheel steering force can be very large and this force, combined with the reduced side reaction force capability of the main gear tires, may cause the main gear tires to "break away" and slide. The nosewheel steering force reaches a maximum at a 13-1/2 degree angle between the tires and the ground track. This corresponds to 6 degrees rudder deflection with the aircraft heading along the ground track.

WET/SLIPPERY RUNWAY LANDINGS

When landing on a runway where degraded braking is expected (i.e., on a wet runway without grooves), select the WET anti-skid braking mode. When crosswind is not a factor, use the maximum performance touchdown speed schedule and lower the nose while deploying the drag chute. Apply maximum braking as soon as the nosewheel is on the runway and engage nosewheel steering. Frequent anti-skid cycling may be felt. Retain the drag chute if stopping distance is critical. The chute can be jettisoned if a control problem or a lateral skid develops.

If the drag chute fails to deploy, use moderate up-elevon to increase drag and the load on the main gear. The WET anti-skid mode provides the best braking capability with or without the drag chute unless tire failure has occurred; in this event, it may be necessary to complete the stop with anti-skid OFF and the wheels locked. Refer to Flat Tire Landing emergency procedure, Section III. If the chute fails, shutdown of one engine is permissible when under 100 KIAS if required to assist in stopping, but shutdown is not recommended if barrier engagement is available.

Icy Runway Procedure

Use the same techniques as for the Wet/Slippery Runway landing.

MAXIMUM PERFORMANCE LANDING

Use the maximum performance schedule whenever minimum landing distance is desirable. Maximum performance touchdown speed is 10 KIAS less than normal touchdown speed. Start drag chute deployment as soon as the main gear is on the runway and angle of attack is 10 degrees or less.

WARNING

Do not deploy the chute in flight.

CAUTION

Deploying the drag chute at greater than 10 degrees angle of attack may result in the canopy contacting the runway and receiving scuff damage.

Lower the nose and apply maximum braking as soon as the nosewheel is on the runway. Engage nosewheel steering. Retain the drag chute. One engine may be shutdown after touchdown to reduce thrust and shorten the landing roll.

WARNING

Do not shutdown both engines, as it may result in the loss of brakes. Nosewheel steering will be lost when engine speed decays.

MINIMUM ROLL

Reduce fuel weight to 5000 pounds, if possible, and use the maximum performance landing procedure.

HEAVYWEIGHT LANDING

Landings with more than 40,000 pounds of fuel remaining should be avoided, but can be accomplished if necessary. Use normal final approach speeds, but do not exceed 11 degrees angle of attack. Use the maximum performance touchdown speed schedule and observe touchdown rate of sink limits from Section V. When touchdown speed is less than the chute deploy limit speed (210 KIAS), lower the nose and deploy the drag chute as soon as the main gear is on the runway. If touchdown speed is higher than 210 KIAS, hold the nose off until 210 KIAS is reached, then lower the nose and deploy the drag chute. To minimize the possibility of tire

GO AROUND - Typical

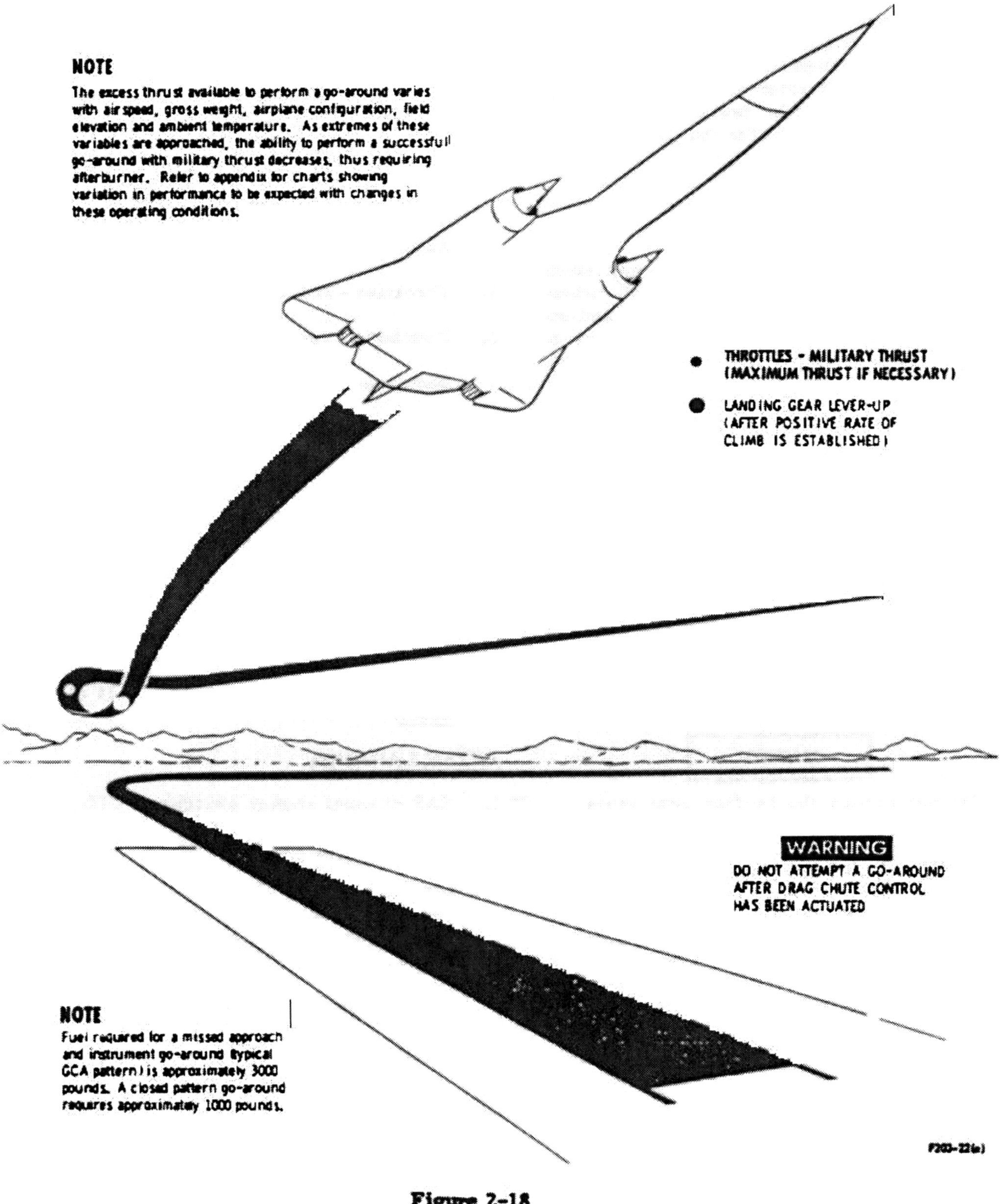

Figure 2-18

failure at heavy weight, the brakes should be applied early in the landing roll. This reduces the distance travelled at high speed. Retain the drag chute. Barrier engagement should be anticipated, since the brake energy rating may be exceeded. Refer to Figure 5-9, Section V.

GO-AROUND

A go-around may be initiated at any time during the approach or landing roll when sufficient runway remains for takeoff and no attempt to deploy the drag chute has been made. (See Figure 2-18.) For go-around after touchdown, reduce pitch attitude to approximately 5 degrees pitch angle (5 degrees above what the attitude would be with the nose on the ground) then adjust attitude to takeoff at 210 KIAS.

TOUCH-AND-GO LANDING

Normal Before Landing and After Takeoff procedures apply to touch-and-go operations. The maximum fuel load recommended is 25,000 pounds remaining and cg aft of 17%. The limit sink rate is 8.7 fps (522 fpm) with 25,000 pounds of fuel.

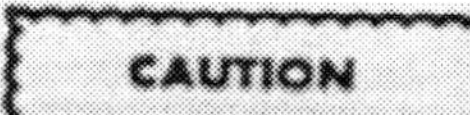

Do not extend the landing gear more than 10 times each flight.

At least ten complete retract and extend cycles of the gear can be made with normal hydraulic quantity and reservoir nitrogen pressure servicing. If reservoir nitrogen pressure is depleted, cycling the landing gear may cavitate the L hydraulic system pump and cause complete loss of L hydraulic system pressure.

NOTE

Manual EGT trim may be used when making successive low approaches or touch and go landings. This prevents auto EGT up-trim "ratcheting" which could occur because of power manipulation during approach, and which might result in EGT overtemperature while at or above Military power during go-around.

1. Throttles - IDLE.
2. Touchdown speed - As required.

   Base touchdown speed on fuel remaining.

After touchdown:

3. Pitch angle - Approximately +5°

   Reduce pitch attitude to approximately 5 degrees pitch angle (5 degrees above what the attitude would be with the nose on the ground). The nosewheel should not touch the runway.

4. Throttles - Military
5. Pitch attitude - Adjust to fly off at 210 KIAS.

AFTER LANDING

T 1. SAS channel engage switches - Off.

2. Landing light - OFF.

   Select TAXI LT if lighting is required; otherwise, select OFF. The landing light should not be operated without airstream cooling or it will burn out prematurely.

3. Right refrigeration switch - ON.

(T4) Cockpit air shutoff handle - ON.

(T5) HF radio - OFF.

(T6) IFF - OFF.

When clear of the runway:

7. Pitot heat switch - OFF.

8. Crossfeed - Closed.

T 9. EGT trim switches - HOLD.

T 10. Periscope - Stowed.

(11) Viewsight power - Off.

ENGINE SHUTDOWN

1. Wheel chocks - Installed.

2. Nosewheel steering - Disengaged.

(3) OBC Power Control switch - OFF.

4. C.G. - Forward of 17%.

Transfer fuel forward of 17% for easier downloading of sensor equipment.

NOTE

If Tank 4 is not on, press Tank 4 boost pumps on before transferring fuel forward. Otherwise, with crossfeed off, a reduction in fuel flow to approximately 3600 lb per hour will occur on the right side. This is less than the desired value for normal operation of the fuel heat sink system. Release the tank after completing fuel transfer.

In the SR-71B, use the DAFICS BIT check after landing to check aft cockpit switch inputs to DAFICS.

(T) 5. APW switch - PUSHER/SHAKER.
APW switch position in the cockpit that does not have APW switch control does not affect the DAFICS BIT.

(T) 6. SPIKE DOOR control transfer - Take control.
Check SPIKE DOOR transfer light illuminated on aft cockpit control transfer panel. Inlet switches in the cockpit that does not have SPIKE DOOR control are not functional and do not affect the DAFICS BIT.

(T) 7. AFCS control transfer - Take control.

T 8. DAFICS Preflight BIT - Check.

T a. SAS channel engage switches - ON.

T b. SENSOR/SERVO lights - Check off.

T c. Forward cockpit (aft cockpit in SR-71B) switch positions for DAFICS PREFLIGHT BIT - Set.
- Autopilot pitch and roll engage switches - ON.
- ATT REF SELECT switch - ANS
- KEAS HOLD switch - ON
- HEADING HOLD switch - ON

T d. DAFICS PREFLIGHT BIT switch - ON.

The BIT TEST light illuminates steady green while the test is running.

Pressure from A hydraulic system is required to engage the DAFICS PREFLIGHT BIT. Low pressure or flow from A, B, L or R hydraulic system will cause the DAFICS preflight BIT to fail.

If the DAFICS PREFLIGHT BIT switch will not engage, recheck:

1) CSC/NWS switch - Released.
2) ATT REF SELECT switch - ANS
3) APW switch - PUSHER/SHAKER
4) SPIKES & FWD BYPASS doors - AUTO
5) RESTART switches - Off
6) Throttle Restart switch - Off
7) SAS channel engage switches - ON.
8) AUTOPILOT PITCH & ROLL engage switches - ON
9) KEAS HOLD switch - ON
10) HEADING HOLD switch - ON

NOTE

If at BIT completion the FAIL light, any SENSOR light, any SERVO light, or any CMPTR OUT light illuminates, notify maintenance.

After one minute:

e. Check BIT TEST light flashing green, sensor and servo lights extinguished, BIT FAIL light extinguished, and OFF flags in both TDI's. The CIP barber pole reads zero.

f. Check autopilot pitch and roll engage switches, KEAS HOLD switch, and HEADING HOLD switch -Off. AUTOPILOT OFF and SAS OUT lights illuminated.

g. Check DAFICS PREFLIGHT BIT switch - OFF (guard down).

h. SENSOR/SERVO recycle switches - Press one of the six.

Pressing one of the six SENSOR/SERVO recycle switches resets the DAFICS system to the flight mode. Check SENSOR/SERVO lights, BIT TEST light, and SAS OUT lights are out. Check both spikes have returned to the full forward position and the CIP barber pole has returned to normal. Both TDI's will initiate resynchronization and run up to 55,000 ft., Mach 2.0, and 300 KEAS. AOA will indicate 10°. AOA will return to 0° in approximately 1 min 15 sec and TDI indications will return to normal in approximately 2 min 15 sec after the DAFICS system has been reset to the flight mode. The A, B, and M CMPTR OUT annunciator panel lights flash momentarily when the DAFICS system is reset.

9. Exterior lights - OFF.

▲10. TACAN and ILS - OFF

11. PVD - OFF.

▲12. Loose items - Secured.

▲13. Canopy seal pressure lever - OFF.

▲14. Canopy - Open

CAUTION

The pilot should notify the RSO when he opens the canopy. If either canopy is open, the aft canopy latch handle must be in the aft position or the cockpit air handle must be in the forward (off) position for adequate equipment cooling. Otherwise, most of the cooling air would exit through the cockpit openings instead of the bays.

15. SENSOR/SERVO lights - Check off.

All pitch, yaw, and roll SAS channels should be engaged before checking the effects of generator switching.

(T16.) ANS MODE switch - OFF.

Prior to ANS shutdown, place system in DEAD RECKON MODE and record LAT/LONG from Present Position Display. Coordinates will be used for system evaluation.

17. Right generator switch - OFF.

NOTE

With transfer of electrical power while on the ground, the DAFICS may undergo ground reinitialization indicated by momentary illumination of the A, B, and M CMPTR OUT caution lights, OFF flags in both TDIs, and TDI resynchronization to 55,000 ft., Mach 2.0, and 300 KEAS.

18. **Bus tie pushbutton - Press.**

Split the buses to obtain satisfactory MRS records.

NOTE

The tank 4 boost pump indicator light may extinguish.

The tank 4 boost pump indicator light is associated with the power supply to pumps 4-3 and 4-4 only. If pumps 4-2 and 4-4 are on and pumps 4-1 and 4-3 are off (tank 5 empty, tank 2 feeding, tank 4 not manually selected), when the right generator is turned off and the buses are split, the tank 4 boost pump indicator light will extinguish as pump 4-4 is de-energized. In this condition, pumps 2-1 and 4-2 continue to supply the engines. The tank 4 boost pump indicator light should illuminate again when the right generator is returned to service.

19. **Right generator switch - NORM.**

Check the GEN BUS TIE OPEN caution light on and the R GEN OUT caution light extinguished.

20. **APW System switch - OFF.**

Ⓣa. **Aft cockpit - CONT FWD.**

21. **Fuel derich system - Both checked, rearmed, and OFF.**

a. **Set both engines 400 rpm above idle speed.**

b. **Actuate the derich test switch until 860°C EGT is exceeded with LEFT and then RIGHT selected.**

When the EGT indications exceed 860°C:

c. **Verify that the EGT gage warning lights are on and that the Fuel Derich lights are on.**

d. **Note that engine speeds decrease between 50 and 400 rpm.**

e. **Cycle the fuel derich switch to REARM, and then OFF.**

Verify that each engine returns to 400 rpm above idle, and EGT indications are normal.

f. **Reset the throttles to IDLE.**

22. **Tanks No. 1 & 4 boost pump switches - Press on.**

23. **Left generator switch - OFF.**

After fifteen seconds for MRS recording:

24. **Left generator switch - NORM.**

25. **Right generator switch - OFF.**

After fifteen seconds:

26. **Left generator switch - OFF.**

Check that both GEN OUT caution lights are on. DAFICS computers should automatically reset after power transfer, and no SAS SENSOR or SERVO lights should illuminate.

Check that the following lights are off:

SAS OUT
MRS power switch (RSO)

Check that the following lights are on:

INSTR INVERTER ON
GEN BUS TIE OPEN
L and R GEN OUT
L and R XFMR RECT OUT

CAUTION

This step should be completed and ac power restored to the fuel system boost pumps without delay. An engine fuel-hydraulic pump can be damaged by cavitation if operation is continued for any significant period with a low fuel pressure condition.

27. Left and right generator switches EMER.

The following annunciator panel lights should not be illuminated:

L and R GEN OUT
L and R XFMR RECT OUT
EMER BAT ON

Fuel panel lights for empty tanks should illuminate EMPTY.

INS - Check normal operation.

INS continues to operate on aircraft battery and instrument inverter power with both generators OFF or in EMER.

a. INS REF annunciator light - Check not illuminated.

b. ADI - Check attitude indication is unchanged and no OFF flags in view.

(c.) RSO attitude indicator:

With S/B R-2595 check attitude is unchanged and no OFF flag.

Without S/B R-2595 attitude not valid and OFF flag is in view.

(Td.) INS FUNCTION switch - Set ATT and check heading slew.

Set the INS Function switch to ATT, INS REF annunciator light illuminates, flag at top of ADI comes in view (heading not valid), ADI attitude remains valid. Push and turn the heading slew knob and confirm HSI and BDHI compass card rotation.

(T29.) INS FUNCTION switch - OFF.

(T30.) INS PWR switch - Press (off).

31. One generator switch - NORM.

Resume normal operation with the generator corresponding to the engine which is to be shut down last.

32. Remaining generator switch - OFF.

Turn off generator corresponding to the engine which is to be shut down first (usually the engine that was started first). Check that the L and R FUEL PRESS warning lights are off to assure that normal pressure exists in the fuel supply manifolds.

T 33. SAS channel engage switches - Off.

34. Brake switch - Set.

Set ALT STEER & BRAKES if the left engine is to be shut down first.

35. First engine throttle - OFF.

Confirm with ground personnel that area under engine is clear before shutting down the engine.

36. Flight control system - Checked.

Check nosewheel disengaged. After flight control (A or B) hydraulic steady-state pressure from the first engine is below 1500 psi, individually check each axis for full deflection and freedom of travel in both directions. Confirm correct ground crew observation, using the following sequence: nose up, nose down, left roll, right roll, nose left and nose right.

NOTE

Rapid control surface deflection while near idle rpm may result in temporary illumination of an A or B HYD warning light. The light should extinguish when flow demands on the system diminish and normal pressure is restored.

37. Brakes and steering - Checked.

Check brakes and nosewheel steering operate with only one hydraulic system (L or R) operating. Pump brakes and check normal pressure while crew chief visually confirms brake actuation on both trucks. Nosewheel STEER ON light illuminates when nosewheel steering engaged. Nose should swing as rudder pedals are moved slightly.

38. Second generator switch - OFF.

Confirm with ground personnel that area under engine is clear.

CAUTION

Do not delay engine shutdown after generator power to the boost pumps is removed.

39. Second engine throttle - OFF.

40. Instrument inverter - Checked and OFF.

Check that the following lights are on:

INSTR INVERTER ON
EMER BAT ON

NOTE

A relay delays EMER BAT ON light illumination for 10 seconds after loss of T-Rs (Step 38).

Press the Indicator and Lights Test switch to check AØ and BØ (bright illumination of the left and right FIRE lights, respectively) of the instrument inverter. TDI off flag (Pilot and RSO) remaining out of view with normal TDI indications (or TDI values increasing or decreasing in response to DAFICS resynchronization) is a check of CØ instrument inverter power.

▲41. Seat and canopy safety pins -Installed.

▲42. UHF and VHF radios - OFF

43. Battery switch - OFF.

NORMAL GROUND EGRESS – Pressure Suit

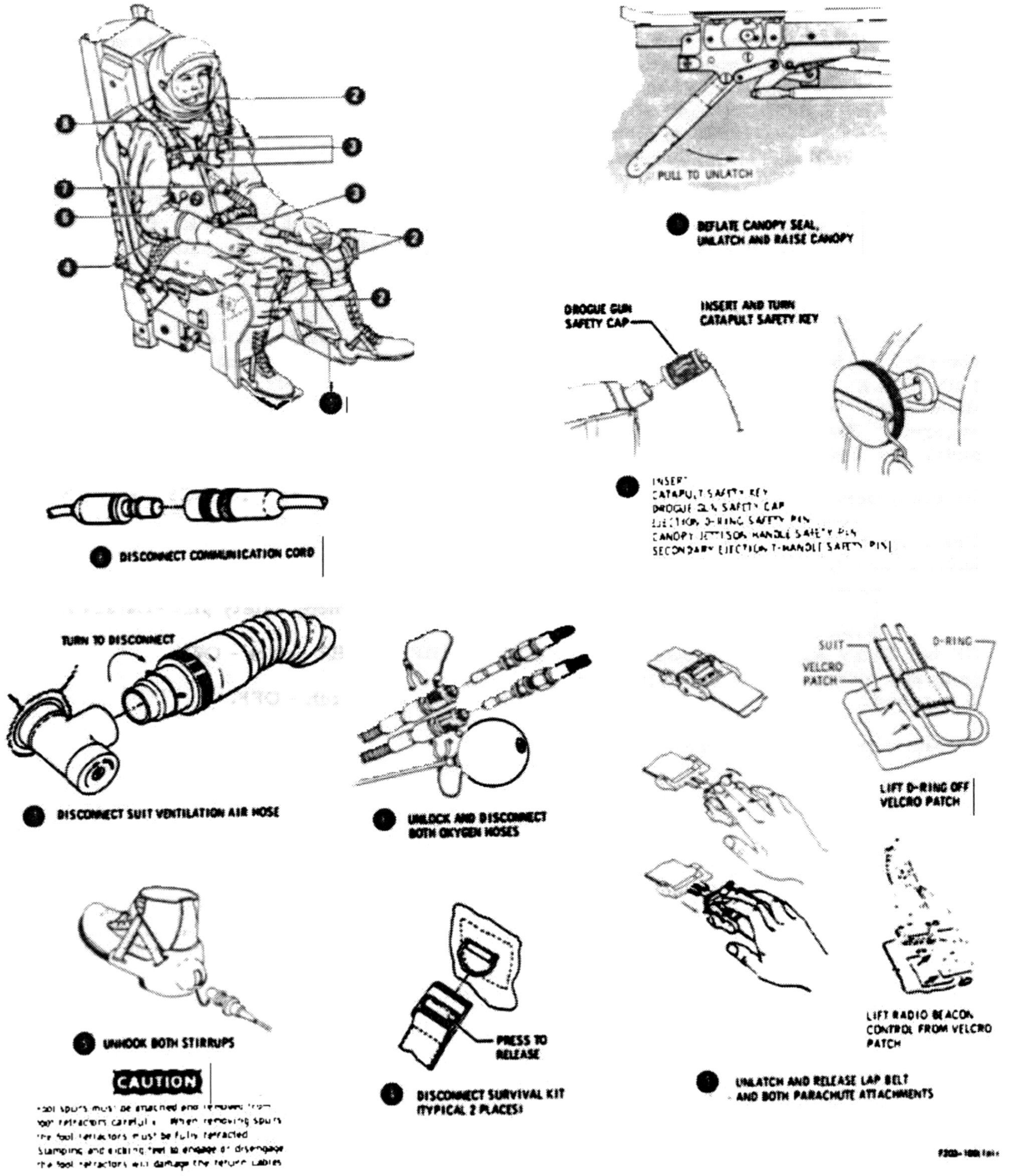

PULL TO UNLATCH

DEFLATE CANOPY SEAL, UNLATCH AND RAISE CANOPY

DROGUE GUN SAFETY CAP

INSERT AND TURN CATAPULT SAFETY KEY

INSERT
CATAPULT SAFETY KEY
DROGUE GUN SAFETY CAP
EJECTION D-RING SAFETY PIN
CANOPY JETTISON HANDLE SAFETY PIN
SECONDARY EJECTION T-HANDLE SAFETY PIN

DISCONNECT COMMUNICATION CORD

TURN TO DISCONNECT

DISCONNECT SUIT VENTILATION AIR HOSE

UNLOCK AND DISCONNECT BOTH OXYGEN HOSES

SUIT

D-RING

VELCRO PATCH

LIFT D-RING OFF VELCRO PATCH

UNHOOK BOTH STIRRUPS

CAUTION

Foot spurs must be attached and removed from foot retractors carefully. When removing spurs the foot retractors must be fully retracted. Stamping and kicking feet to engage or disengage the foot retractors will damage the return cables.

PRESS TO RELEASE

DISCONNECT SURVIVAL KIT (TYPICAL 2 PLACES)

LIFT RADIO BEACON CONTROL FROM VELCRO PATCH

UNLATCH AND RELEASE LAP BELT AND BOTH PARACHUTE ATTACHMENTS

Figure 2-19

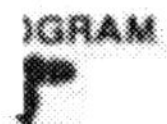

SURVIVAL QUICK LAUNCH

WARNING

Quick Launch procedures are not intended for normal operations. Quick Launch procedures will only be used when directed by the commander.

Takeoff using Survival Quick Launch procedures should be used only to avoid destruction of the aircraft.

QUICK LAUNCH SETUP

The Quick Launch Setup procedures require that all normal procedures through Before Taxiing (or Before Takeoff) have been completed before the Quick Launch Setup checklist is initiated.

QUICK LAUNCH SETUP PROCEDURE

After Before Taxiing (or Before Takeoff) checks complete:

(T1.) HF Radio - OFF.

(T2.) IFF mode 4 code select switch - HOLD.

Place the switch in the momentary HOLD position for 15 seconds, then wait another 15 seconds before turning equipment OFF.

After 15 seconds:

(T3.) **IFF - OFF.**

(4.) **Sensor and OBC power - Off.**

5. **Pitot heat switch - OFF.**

6. **EGT trim switches - Downtrim, if desired, then AUTO.**

If engine run to check automatic EGT trim has not been completed, consider downtrimming EGT slightly. Return EGT trim switch to AUTO so that automatic EGT trimming will trim EGT into the nominal band during takeoff.

7. **C. G. - 18%.**

Transfer fuel to 18% so that c.g. will be at 18% to 20% for takeoff.

8. **PVD - OFF.**

▲9. **Loose items - Secured.**

(T10.) **Cockpit air - Off (forward).**

▲11. **Canopy seal switch - OFF.**

▲12. **Canopy - Open.**

(T13.) **ANS MODE switch - OFF.**

(T14.) **INS FUNCTION switch - OFF.**

(T15.) **INS PWR switch - Press (Off).**

16. **Right generator switch - OFF.**

17. **Right throttle - OFF.**

Confirm with ground personnel that area under engine is clear before shutting down the engine.

18. **Left generator switch - OFF.**

Confirm with ground personnel that area under engine is clear.

CAUTION

Do not delay engine shutdown after generator power to the boost pumps is removed.

19. **Left throttle - OFF.**

20. **Instrument inverter - Checked and NORM.**

21. **Cabin pressure switch - 10,000 FT, if desired.**

▲22. **Seat and canopy pins - Installed.**

23. **Pilot's A, B, and M CMPTR circuit breakers (3 total) - Pull.**

T24. RSO's A, B, and M COMPUTER circuit breakers (9 total) - Pull.

T25. INS FUNCTION switch - STOR HDG.

▲26. Oxygen - OFF.

27. Battery switch - OFF.

QUICK LAUNCH START

While subject to Quick Launch, ensure that nobody has access to the aircraft unless authorized by the aircrew.

If Quick Launch Setup procedures were not completed prior to start or if the aircraft is removed from Quick Launch status, use normal procedures for launch.

When routine aircraft servicing is required, cockpit access requires crew authorization and the crew should accompany maintenance personnel (to remain aware of aircraft status and confirm that cockpit setup is not changed).

Quick Launch Setup and Quick Launch Start procedures require ANS Ground Hot Start and INS Stored Heading procedures. If the aircraft is moved after the ANS and INS are shutdown, these alignments are invalidated and normal procedures for ANS and INS alignment should be used.

Survival Quick Launch procedures assume external power is available for start. If external power fails the engines can be started but engine instrument indications, except for rpm, will not be available until a generator is turned on. If the crew chief does not use a headset during start, the aircrew must coordinate the hand signals to be used prior to assuming Quick Launch status.

QUICK LAUNCH START PROCEDURES

After external power applied:

T1. INS PWR switch - Press (On).

2. Battery switch - BAT.

3. Right engine - Start.

T4. RSO's A, B, and M COMPUTER circuit breakers (9 total) - Push in.

These circuit breakers are pulled to keep the PTAs from being powered until cooling air is available. The circuit breakers may be reset as soon as the start procedures are in progress. Since the left and right refrigeration switches are still on (from Before Taxiing checks), cooling air to the PTAs will be available as soon as the right engine starts.

The DAFICS circuit breakers in the front cockpit are reset after the DAFICS circuit breakers in the aft cockpit to prevent the DAFICS computers from operating (sensing and storing power faults) until DAFICS has proper power. If the forward cockpit circuit breakers are reset first, DAFICS memory will indicate transient power faults, however DAFICS operation and reliability is not degraded.

After right engine is started:

5. Right generator - On (NORM), light off.

Check R GEN OUT light extinguishes

T6. ANS Mode switch - INERTIAL ONLY.

The ANS is not turned on until after the right engine is started so that the ANS has cooling and the LIMIT light will not flash.

The MAL light will flash until the HOT switch is pressed.

T7. ANS HOT switch - Press.

8. External power - Disconnected.

9. Left engine - Start.

10. Left generator - On (NORM), light off.

Check L GEN OUT light extinguishes.

11. Pilot's A, B, and M CMPTR circuit breakers (3) - Push In.

Setting the 3 dc CMPTR circuit breakers in the forward cockpit starts the DAFICS computers. Check the A, B, and M CMPTR OUT annunciator lights extinguish.

▲12. Ejection seat and canopy pins Removed.

▲13. Canopy - Closed and locked.

▲14. Canopy seal switch - ON.

(T15.) Cockpit air - On (aft).

16. Nosewheel steering - Engaged.

With NAV RDY light flashing:

(T17.) INS FUNCTION switch - NAV.

With F/A in mode window:

(T18.) ANS MODE START switch - Press.

The chronometer may not be charged for Quick Launch procedures; if not, the ANS will not star track if Astro-Inertial mode is selected.

(T19.) MRS - ON.

QUICK LAUNCH TAXI

1. Brakes - DRY or WET and ANTI SKID ON.

▲2. Circuit breakers - Checked.

3. Flight controls and trim setting - Check.

4. Fuel - Check tanks 1, 3, and 5 (or 6) on.

▲5. CG - Checked.

▲6. Oxygen - ON and pressure checked.

7. Pitot heat switch - ON.

(T8.) IFF - NORMAL.

(T9.) HF radio - On.

QUICK LAUNCH TAKEOFF

T 1. SAS - Engaged, lights off.

▲2. Warning and caution lights - Checked.

3. Tank 4 - Press on.

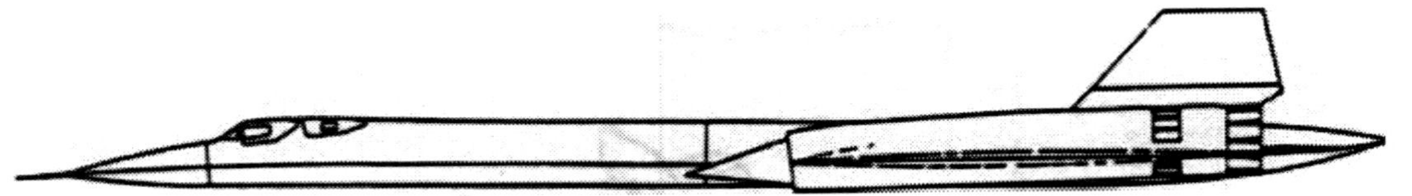

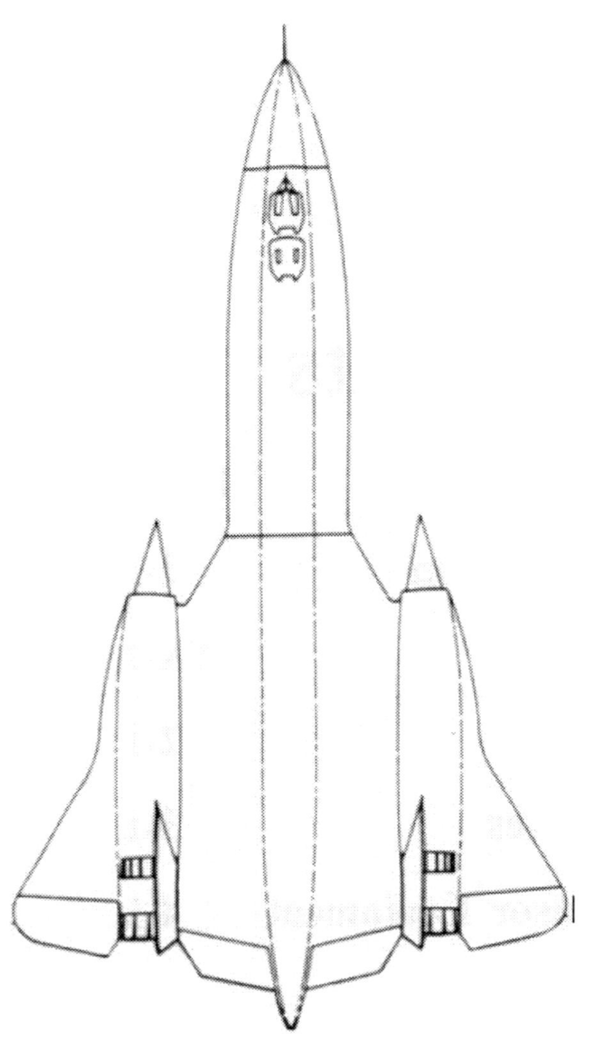

THIS PAGE INTENTIONALLY LEFT BLANK OR STILL CLASSIFIED.

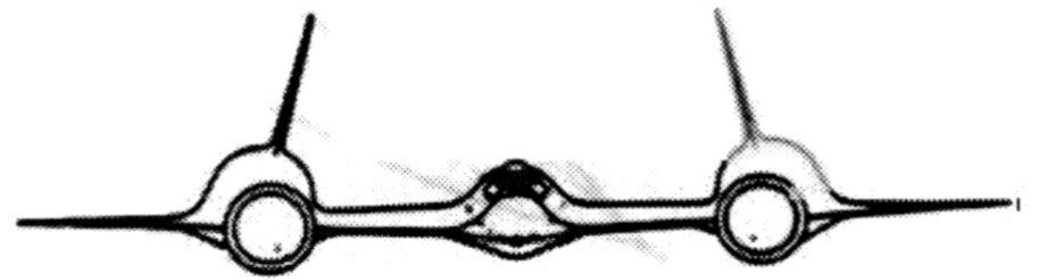

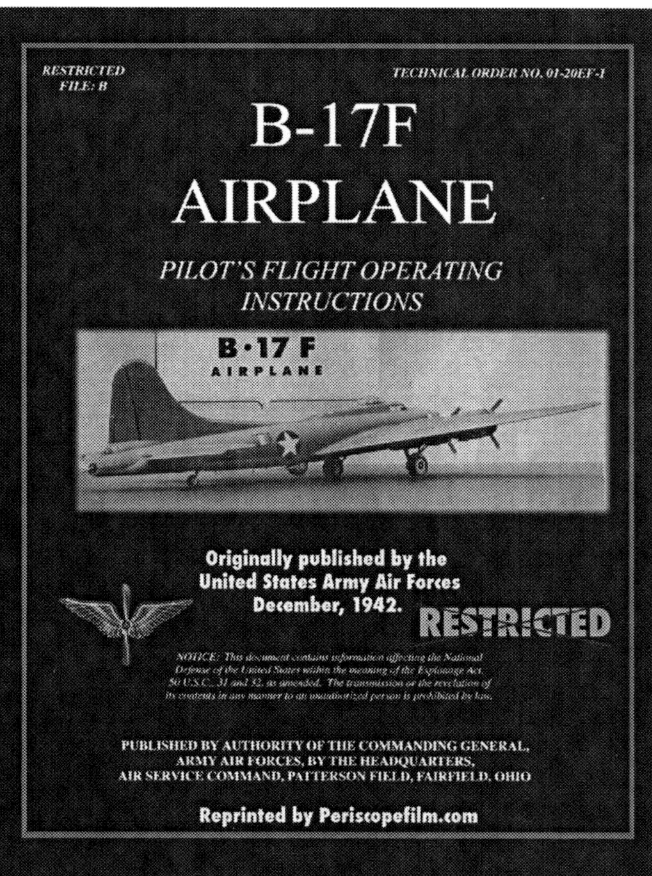
RESTRICTED
FILE: B
TECHNICAL ORDER NO. 01-20EF-1
B-17F
AIRPLANE
PILOT'S FLIGHT OPERATING
INSTRUCTIONS
B-17 F
AIRPLANE
Originally published by the
United States Army Air Forces
December, 1942.
RESTRICTED
PUBLISHED BY AUTHORITY OF THE COMMANDING GENERAL,
ARMY AIR FORCES, BY THE HEADQUARTERS,
AIR SERVICE COMMAND, PATTERSON FIELD, FAIRFIELD, OHIO
Reprinted by Periscopefilm.com

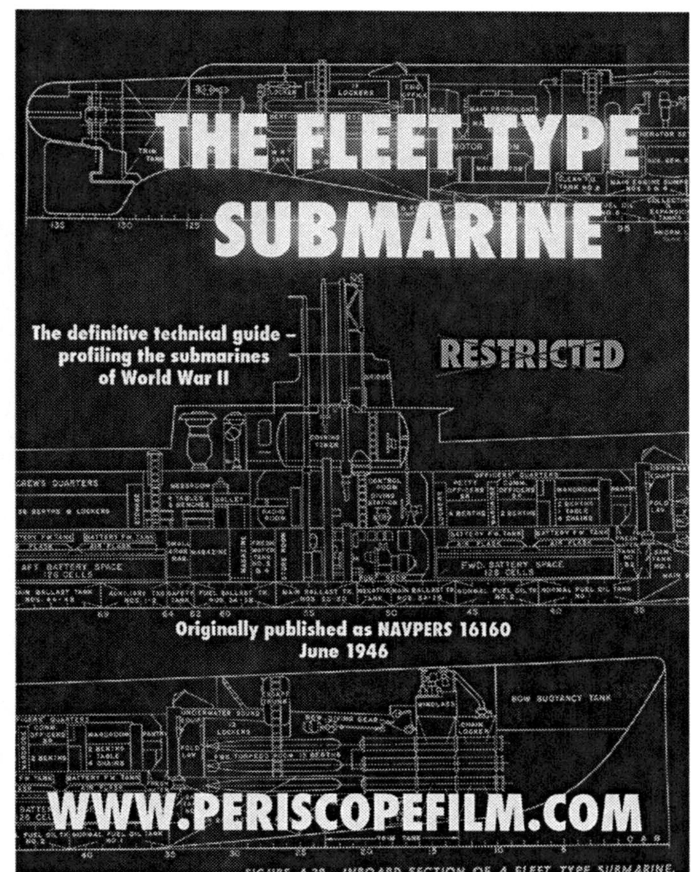
THE FLEET TYPE
SUBMARINE
The definitive technical guide –
profiling the submarines
of World War II
RESTRICTED
Originally published as NAVPERS 16160
June 1946
WWW.PERISCOPEFILM.COM

RESTRICTED
FILE: B
TECHNICAL ORDER NO. 01-20EF-1
NORTHROP YB-49 FLYING WING
PILOT'S FLIGHT OPERATING
INSTRUCTIONS
FLYING WING
RESTRICTED

REVISED 1 FEBRUARY, 1945
REPRINTED 1 MARCH, 1945 WITH ADDITIONS AS INDICATED ON THE BACK COVER
PILOT'S FLIGHT OPERATING INSTRUCTIONS
RESTRICTED
THE
B-29
Originally published by the
United States Army Air Forces
January, 1944.
RESTRICTED
PLANE COMMANDER TRAINING MANUAL FOR THE SUPERFORTRESS
THIS REVISED EDITION SUPERSEDES THE ORIGINAL (BLUE COVERED) AIRPLANE COMMANDER TRAINING MANUAL FOR THE SUPERFORTRESS. ALL COPIES OF THE LATTER ARE RESCINDED.